Introduction to ENGINEERING

An Assessment and Problem Solving Approach

Introduction to ENGINEERING

An Assessment and Problem Solving Approach

Quamrul H. Mazumder

CRC Press
Taylor & Francis Group
Boca Raton London New York

CRC Press is an imprint of the
Taylor & Francis Group, an **informa** business

CRC Press
Taylor & Francis Group
6000 Broken Sound Parkway NW, Suite 300
Boca Raton, FL 33487-2742

© 2016 by Taylor & Francis Group, LLC
CRC Press is an imprint of Taylor & Francis Group, an Informa business

No claim to original U.S. Government works

Printed on acid-free paper
Version Date: 20160113

International Standard Book Number-13: 978-1-4987-4748-6 (Hardback)

This book contains information obtained from authentic and highly regarded sources. Reasonable efforts have been made to publish reliable data and information, but the author and publisher cannot assume responsibility for the validity of all materials or the consequences of their use. The authors and publishers have attempted to trace the copyright holders of all material reproduced in this publication and apologize to copyright holders if permission to publish in this form has not been obtained. If any copyright material has not been acknowledged please write and let us know so we may rectify in any future reprint.

Except as permitted under U.S. Copyright Law, no part of this book may be reprinted, reproduced, transmitted, or utilized in any form by any electronic, mechanical, or other means, now known or hereafter invented, including photocopying, microfilming, and recording, or in any information storage or retrieval system, without written permission from the publishers.

For permission to photocopy or use material electronically from this work, please access www.copyright.com (http://www.copyright.com/) or contact the Copyright Clearance Center, Inc. (CCC), 222 Rosewood Drive, Danvers, MA 01923, 978-750-8400. CCC is a not-for-profit organization that provides licenses and registration for a variety of users. For organizations that have been granted a photocopy license by the CCC, a separate system of payment has been arranged.

Trademark Notice: Product or corporate names may be trademarks or registered trademarks, and are used only for identification and explanation without intent to infringe.

Visit the Taylor & Francis Web site at
http://www.taylorandfrancis.com

and the CRC Press Web site at
http://www.crcpress.com

Contents

Preface xvii
Acknowledgments xxi
Author xxiii

Section I Engineering Education

1. Engineering and Technology Professions 3
   1.1 Engineering and Engineering Technology 3
        1.1.1 Scientists 4
        1.1.2 Engineers 4
        1.1.3 Engineering Technologists 5
   1.2 Employment Outlook 6
   1.3 Skills and Competencies (What Employers Are Looking For) 6
        1.3.1 Engineering Grand Challenges for the Twenty-First Century 7
   1.4 Engineering Disciplines 9
        1.4.1 Aerospace and Aeronautical Engineering 10
        1.4.2 Civil Engineering 11
        1.4.3 Computer Engineering 12
        1.4.4 Electrical and Electronic Engineering 13
        1.4.5 Environmental Engineering 14
        1.4.6 Industrial and Operations Engineering 15
        1.4.7 Mechanical Engineering 15
   1.5 Professional Engineering License 17
        1.5.1 Steps to Become a PE 17
   1.6 Professional Engineering Organizations 19
   1.7 Summary 22
   Bibliography 23

2. Engineering Education 25
   2.1 What Is Engineering? 25
   2.2 Engineering Education 26
   2.3 ABET Accreditation 26
        2.3.1 Criterion 1: Students 27
        2.3.2 Criterion 2: Program Educational Objectives 28
        2.3.3 Criterion 3: Student Outcomes 28
        2.3.4 Criterion 4: Continuous Improvement 29
        2.3.5 Criterion 5: Curriculum 29
   2.4 Curriculum Mapping 29
   2.5 Assessment and Continuous Improvement 31
   2.6 High-Impact Educational Practices 33
        2.6.1 First-Year Seminars and Experiences 34
        2.6.2 Common Intellectual Experiences 34
        2.6.3 Learning Communities 34

		2.6.4	Writing-Intensive Courses	34
		2.6.5	Collaborative Assignments and Projects	34
		2.6.6	Undergraduate Research	34
		2.6.7	Diversity/Global Learning	35
		2.6.8	Service Learning and Community-Based Learning	35
		2.6.9	Internships	35
		2.6.10	Capstone Courses and Projects	35
	2.7	Plan of Study		35
		2.7.1	Find Your Catalog Year	36
		2.7.2	Understand Your Program Requirements	36
		2.7.3	Determine Prerequisite Requirements for Each Required Course	37
		2.7.4	Organize All Materials to Develop Your Plan of Study	37
		2.7.5	Meet with Your Advisor	37
		2.7.6	Balance Course Load with Work and Life	38
		2.7.7	Take Courses from Different Areas of the Curriculum	38
	2.8	Summary		39
	Bibliography			40

3. **How to Learn** 41
   3.1 Introduction 41
   3.2 Metacognition 42
        3.2.1 Motivation 43
              3.2.1.1 Self-Efficacy 43
              3.2.1.2 Self-Reflection 44
              3.2.1.3 Self-Assessment 44
   3.3 Types of Knowledge 45
   3.4 Study Skills/Metacognitive Learning Strategies 45
   3.5 Learning Styles 46
   3.6 Teaching Styles 47
   3.7 Time Management 48
   3.8 Bloom's Taxonomy 50
   3.9 Learning Pyramid 51
   3.10 Skill-Based Learning 53
   3.11 Summary 53
   References 54

Section II Engineering Fundamentals

4. **Computer-Aided Design** 57
   4.1 Introduction 57
   4.2 Advantages of CAD 57
   4.3 CAD Software 59
        4.3.1 Installing Google SketchUp Pro 59
        4.3.2 Starting SketchUp Pro 59
        4.3.3 Toolbar 60
        4.3.4 Dynamic Viewing Function 62
        4.3.5 Abbreviations 62

		4.3.6	Sketch and Extrusion... 65
		4.3.7	Modeling 3D Object (Bracket Tutorial) 65
		4.3.8	Adding Dimensions to the Model.. 68
	4.4	Drawings in SketchUp Pro 70
		4.4.1	Part Tutorial... 71
			4.4.1.1 Adding Rounds... 72
	4.5	Orthographic Drawings in CAD 74
	4.6	Preparing Model for LayOut... 78
		4.6.1	Opening Model in LayOut 79
	4.7	Dimensioning Orthogonal View 84
	4.8	L-Bracket Design... 88
	ABET Program Outcomes 92
	Review Questions 93

5. Statics 99
5.1 Introduction 99
5.2 Scalars and Vectors 99
5.3 Equilibrium of Forces 103
5.4 Moment, Couple, Torque 106
5.5 Free-Body Diagram.. 107
5.6 Centroids and Center of Gravity... 110
5.7 Moment of Inertia 114

6. Materials Engineering... 123
6.1 Introduction 123
6.2 Metals... 123
    6.2.1 How Do We Obtain Metals?.. 126
    6.2.2 Properties of Metals.. 128
    6.2.3 Processing of Metals... 133
    6.2.4 How Does Processing Affect the Properties of Metals?............. 135
           6.2.4.1 Annealing... 136
           6.2.4.2 Quenching.. 136
           6.2.4.3 Precipitation Hardening.. 137
    6.2.5 What Technologies Are Made Possible by Metals?.................... 137
6.3 Ceramic Materials 138
    6.3.1 How Do We Obtain Ceramic Materials? 139
    6.3.2 What Are the Properties of Ceramic Materials? 139
    6.3.3 How Do We Process Ceramic Materials? 141
    6.3.4 How Does Processing Affect the Properties
           of Ceramic Materials?... 143
    6.3.5 What Technologies Are Made Possible by Ceramic Materials? 144
6.4 Polymers 145
    6.4.1 What Is a Polymer? 145
    6.4.2 What Are the Properties of Polymers?.. 147
    6.4.3 How Do We Obtain Polymers? 148
    6.4.4 How Do We Process Polymers into Usable Forms? 149
    6.4.5 How Does Processing Affect Polymer Properties?..................... 150
    6.4.6 What Technologies Are Made Possible by Polymers? 151

6.5	Composite Materials		152
	6.5.1	What Are Composite Polymer Materials?	152
	6.5.2	How Do We Obtain/Manufacture Composite Materials?	153
	6.5.3	What Are the Properties of Composite Materials?	153
	6.5.4	How Does Manufacturing Affect the Properties of Composites?	154
	6.5.5	What Technologies Are Made Possible by Composites?	154
6.6	Conclusion		154

7. Design and Analysis ... 157

7.1	Introduction		157
7.2	Stress and Strain		158
	7.2.1	Stress	158
	7.2.2	Strain	159
	7.2.3	Hooke's Law	159
	7.2.4	Poisson's Ratio (v)	159
7.3	Uniaxial Loading and Deformation		160
7.4	Maximum Normal Stress Theory		160
7.5	Pressure Vessel		162
	7.5.1	Hoop Stress	162
	7.5.2	Axial Stress	163
7.6	Shear Stress		164
7.7	Stress in Beams		165
7.8	Buckling in Column		166
7.9	Thermodynamics		167
	7.9.1	First Law of Thermodynamics	167
	7.9.2	Second Law of Thermodynamics	168
	7.9.3	Third Law of Thermodynamics	168
7.10	Heat Transfer		168
	7.10.1	Conduction	169
	7.10.2	Convection	170
	7.10.3	Radiation	170
7.11	Fluid Mechanics		170
	7.11.1	Properties of Fluids	170
	7.11.2	Barometers	172
	7.11.3	Forces on Submerged Plane Surfaces	173
		7.11.3.1 Buoyancy	173
	7.11.4	Conservation of Mass	174
	7.11.5	Conservation of Energy	175
	7.11.6	Reynolds Number	175
		7.11.6.1 Laminar Flow	175
		7.11.6.2 Turbulent Flow	175
	7.11.7	Flow Distribution	176
	7.11.8	Drag	177
	7.11.9	Lift	177

Contents

8. Electric Circuits and Components 181
- 8.1 Introduction 181
- 8.2 Understanding Electricity 182
  - 8.2.1 Electric Charge 182
  - 8.2.2 Conduction and Insulation 182
  - 8.2.3 Coulomb's Law 182
  - 8.2.4 Electric Field 183
  - 8.2.5 Charge Distribution 183
  - 8.2.6 Charged Particles 183
  - 8.2.7 Potential Difference 183
  - 8.2.8 Current 185
- 8.3 Circuit Components 185
  - 8.3.1 Resistors 185
  - 8.3.2 Capacitors 186
  - 8.3.3 Diodes 186
  - 8.3.4 Transistors 186
  - 8.3.5 Bipolar Junction Transistor 186
  - 8.3.6 n-Type Semiconductor 188
  - 8.3.7 p-Type Semiconductor 188
  - 8.3.8 BJT Operating Regions 188
  - 8.3.9 Field Effect Transistors 188
  - 8.3.10 Junction FET 188
  - 8.3.11 Metal Oxide Semiconductor FET 189
  - 8.3.12 Integrated Circuits 189
- 8.4 Circuit Symbols and Schematic Design 189
  - 8.4.1 Getting Started with Schematics 190
  - 8.4.2 Reference Designators in a Circuit 191
  - 8.4.3 Circuit Design 191
  - 8.4.4 Simplified Ground Connections 192
  - 8.4.5 Simplified Power Connections 192
- 8.5 Understanding Circuits 193
  - 8.5.1 Series Circuits 193
  - 8.5.2 Parallel Circuits 195
- 8.6 Resistors 195
  - 8.6.1 Ohm's Law 196
  - 8.6.2 Resistor Color Codes 197
  - 8.6.3 Reading Color Code of a Four-Band Resistor 197
  - 8.6.4 Reading Color Code of a Five-Band Resistor 198
  - 8.6.5 Resistor Tolerance 198
  - 8.6.6 Power Dissipation Ratings 199
  - 8.6.7 Resistors in Series 199
  - 8.6.8 Resistors in Parallel 200
  - 8.6.9 Series and Parallel Combination 200
  - 8.6.10 Current Limiting with Resistors 200
  - 8.6.11 Voltage Divider Circuit with Resistors 201

8.7	Capacitors		201
	8.7.1	Common Types of Capacitors	203
	8.7.2	Capacitors in Series	203
	8.7.3	Capacitors in Parallel	204
	8.7.4	Series and Parallel Combination	204
	8.7.5	RC Time Constant	205
	8.7.6	Energy Stored in a Capacitor	206
	8.7.7	Designing a Low-Pass RC Filter	207
8.8	Inductors		208
	8.8.1	Series and Parallel Connection	209
	8.8.2	Energy Stored in an Inductor	210
	8.8.3	RL Time Constants	211
	8.8.4	Designing an RL Filter	211
8.9	Diodes		212
	8.9.1	Doping of Silicon Crystal	213
	8.9.2	Depletion Zone	213
	8.9.3	Half-Wave Rectification	214
	8.9.4	Full-Wave Rectification	214
	8.9.5	Zener Diodes	214
8.10	Transistors		215
	8.10.1	Important Datasheet Parameters	215
	8.10.2	Operating Regions	217
	8.10.3	Forward Active	217
	8.10.4	Reverse Active	217
	8.10.5	Saturation	217
	8.10.6	Cutoff	218
	8.10.7	Avalanche Breakdown	218
	8.10.8	Using a Transistor as a Switch	218
	8.10.9	Using an n–p–n Transistor as a NOT Gate	218
	8.10.10	Using a p–n–p Transistor as a NOT Gate	218
	8.10.11	Inverter Circuit with a MOSFET	219
8.11	Integrated Circuits		219
	8.11.1	IC Manufacturing Process	220
	8.11.2	Wafer Production	220
	8.11.3	Fabrication of Silicon Wafer	220
	8.11.4	Deposition	220
	8.11.5	Masking	220
	8.11.6	Etching	221
	8.11.7	Doping	221
	8.11.8	Passivation	221
	8.11.9	Electrical Testing	221
	8.11.10	Assembly	221
	8.11.11	IC Datasheets	221
	8.11.12	Supply Voltage	222
	8.11.13	Supply Current	222
	8.11.14	Functional Description	222
	8.11.15	Characteristic Graphs	222
	8.11.16	Absolute Maximum Ratings	223

		8.11.17	555 Timer	223
		8.11.18	Monostable (One-Shot) Multivibrator	223
		8.11.19	Astable Multivibrator	225
		8.11.20	Operational Amplifiers	225
		8.11.21	Noninverting Amplifier	226
		8.11.22	Unity Gain Buffer	227
		8.11.23	Inverting Amplifier	227
	8.12	Summary		228

9. Engineering Economics ... 231

	9.1	Introduction		231
	9.2	Nominal and Effective Interest Rate		232
		9.2.1	Solving Engineering Economics Problems	233
		9.2.2	Effective Interest Rates for Annual and Any Time Period	237
	9.3	Effect of Time and Interest on Money		238
		9.3.1	Cash Flow Diagram	238
		9.3.2	Single Amount Factors (F/P and P/F)	239
		9.3.3	Uniform Series PW Factor and Capital Recovery Factor (P/A and A/P)	241
		9.3.4	Sinking Fund Factor and Uniform Series Compound Amount Factor (A/F and F/A)	242
		9.3.5	Arithmetic Gradient Factors (P/G and A/G)	244
		9.3.6	Geometric Gradient Series Factors	245
	9.4	PW, AW, and FW Analysis		246
		9.4.1	PW Analysis	246
		9.4.2	FW Analysis	246
		9.4.3	AW Analysis	246
	9.5	Rate of Return Analysis		247
	9.6	Decision Making		247

10. Probability and Statistics ... 251

	10.1	Introduction		251
	10.2	Probability		252
		10.2.1	Probability Theorems	252
		10.2.2	Permutations and Combinations	258
	10.3	Statistics		259
		10.3.1	Statistical Difficulties	260
	10.4	Statistics in Organizations		260
		10.4.1	Descriptive and Inferential Statistics	260
		10.4.2	Sample versus Population	261
	10.5	Data		262
		10.5.1	Measures of Central Tendency	262
		10.5.2	Spread: Measures of Dispersion	263
		10.5.3	Central Limit Theorem	265
		10.5.4	Normal Frequency Distribution	266
		10.5.5	Standard Normal Probability Distribution: Z Tables	266
	10.6	Summary		268

11. Computer Programming 273
- 11.1 Importance of Programming 273
- 11.2 Languages and Applications 274
  - 11.2.1 High-Level Languages 274
  - 11.2.2 Machine Code and Assembly Language 275
  - 11.2.3 Compiling and Executing a Program 275
- 11.3 Algorithm Development 276
  - 11.3.1 General Guidelines for Creating Simple Algorithms 276
  - 11.3.2 Flowcharts 277
  - 11.3.3 Pseudocode 278
- 11.4 Comments and Documentation 280
- 11.5 Reserved or Keywords 280
- 11.6 Variables, Constants, and Their Data Types 280
  - 11.6.1 Initialization 282
  - 11.6.2 Naming Conventions 283
- 11.7 Input and Output 284
  - 11.7.1 PRINT and WRITE 284
  - 11.7.2 GET and READ 284
- 11.8 Assignment and Arithmetic Operators 285
  - 11.8.1 Order of Execution 286
- 11.9 Conditional Expressions 286
  - 11.9.1 Relational Operators 286
  - 11.9.2 Logical Operators 287
- 11.10 Selection Statements 288
  - 11.10.1 IF-THEN 288
  - 11.10.2 SELECT…CASE 289
- 11.11 Loop Structures 290
  - 11.11.1 FOR Loops 290
  - 11.11.2 WHILE…DO Loops 290
  - 11.11.3 DO…UNTIL Loops 291
- 11.12 Functions and Procedures 291
- 11.13 Testing and Debugging 292
  - 11.13.1 Types of Errors 292
- 11.14 Summary 292

Section III Product Design and Development

12. Product Design and Development 297
- 12.1 Introduction 297
- 12.2 Overview of the Design Process 298
  - 12.2.1 Product Discovery 298
  - 12.2.2 Project Planning 299
  - 12.2.3 Product Definition 299
  - 12.2.4 Conceptual Design 299
  - 12.2.5 Product Development 299
  - 12.2.6 Product Support 299
- 12.3 Definition of the Problem 301

12.4	Product Development Process			301
12.5	Product Design			303
	12.5.1	Steps in Product Design		303
	12.5.2	Factors to Consider in Product Design		305
		12.5.2.1	Design for Manufacture	305
		12.5.2.2	Product Life Cycle	305
		12.5.2.3	Concurrent Engineering	306
		12.5.2.4	Remanufacturing	307
12.6	Objectives of Design and Development Techniques			307
12.7	Techniques That Can Be Used in Product Design and Development			308
	12.7.1	Techniques and Tools for Design Improvement		308
		12.7.1.1	Concurrent Engineering	308
		12.7.1.2	Quality Function Deployment	308
		12.7.1.3	Design for X	309
		12.7.1.4	Failure Mode and Effects Analysis	310
	12.7.2	Computational Techniques and Tools		311
		12.7.2.1	Computer-Aided (CAx) Systems	311
		12.7.2.2	Engineering/Product-Based Data Management	311
		12.7.2.3	Knowledge-Based Engineering	312
		12.7.2.4	Finite Element Analysis	312
		12.7.2.5	Rapid Prototyping	313
		12.7.2.6	3D Printing	313
12.8	Summary			314
Bibliography				314

13. Manufacturing Processes 317

13.1	Introduction to Machines			317
13.2	Metal Forming Processes and Metal Cutting Processes			317
	13.2.1	Hot Forging		318
	13.2.2	Cold Forging		318
	13.2.3	Rolling		318
13.3	Machines and Tools			319
	13.3.1	Engine Lathe and Tools		319
	13.3.2	Head Stock		320
	13.3.3	Tool Post		320
	13.3.4	Tail Stock		320
	13.3.5	Carriage Mechanism		321
	13.3.6	Lathe Machine Specification		321
	13.3.7	Lathe Machine Tools		322
	13.3.8	Lathe Operations		322
	13.3.9	Drilling Machine		324
		13.3.9.1	Construction	324
		13.3.9.2	Types of Drills	325
		13.3.9.3	Operations Performed on a Drill Machine	326
	13.3.10	Milling Machine		328
		13.3.10.1	Construction of Milling Machine	328
		13.3.10.2	Milling Cutters	328
		13.3.10.3	Operations Performed on Milling Machine	330
13.4	Machine Process Flow			333

13.5 Welding and Cutting 335
    13.5.1 Arc Welding 335
    13.5.2 Welding Process Fundamentals 336
        13.5.2.1 Shielding and Fluxing 336
        13.5.2.2 Shielded Metal Arc Welding 337
        13.5.2.3 Gas Metal Arc Welding 337
    13.5.3 Use of Laser in Manufacturing 338
        13.5.3.1 Welding 338
        13.5.3.2 Cutting 338
        13.5.3.3 Surface Treatment/Modification 339
        13.5.3.4 Hardening 340
        13.5.3.5 Remelting 340
        13.5.3.6 Coating 340
        13.5.3.7 Drilling/Piercing 341
Review Questions 341

Section IV Engineering Profession

14. Engineering and Society 345
    14.1 Social Influence in a Globalized World 345
    14.2 Evolution of Social Norms and Technology 346
    14.3 Emerging Social Norms in Social Networking 346
    14.4 Political Influence 348
    14.5 Learning from Tragedy: The Fukushima Disaster 350
    14.6 Lessons from Fukushima 353
    14.7 Pursuit of Social Understanding 353
    References 355

15. Engineering Ethics 357
    15.1 Introduction 357
    15.2 Ethics in Engineering Education 358
    15.3 Ethics in the Engineering Profession 360
        15.3.1 NSPE Code of Ethics for Engineers 360
            15.3.1.1 Preamble 360
        15.3.2 The Path to Resolution of an Ethical Dilemma in Employment 365
        15.3.3 Seeking Advice and Reporting Concerns and Violations 366
        15.3.4 Other Questions to Consider before the Employee Takes Action 367
    References 368

16. Communication and Teamwork 369
    16.1 Essential *Soft* Skills in a Globalized Workforce 369
    16.2 Communication 370
        16.2.1 Oral Communication 372
            16.2.1.1 Know Your Audience 372
            16.2.1.2 Know Your Content 373
            16.2.1.3 Presentation Software 374
            16.2.1.4 Nonverbal Messages 375
            16.2.1.5 Rehearsing Presentations 376

Contents xv

		16.2.1.6	Presenting with Groups	376
		16.2.1.7	Concluding Remarks on Oral Communication	377
	16.2.2	Written Communication		377
		16.2.2.1	Letters and Memos	377
		16.2.2.2	Email and Electronic Communications	379
		16.2.2.3	Reports	380
		16.2.2.4	Citations and Bibliographies	381
		16.2.2.5	Source Reliability	382
		16.2.2.6	Figures, Images, and Tables	386
	16.2.3	Other Technical Communication Mediums		386
	16.2.4	Communicating by Listening		387
16.3	Teamwork Essentials			387
	16.3.1	Fundamental Characteristics of Effective Teams and Team Members		387
	16.3.2	Leadership in Teams		388
	16.3.3	Embracing Diversity		389
References				389

Appendix A: Unit Conversion 391

Appendix B: Rubrics and Key Performance Indicators 393

Appendix C: ABET Outcomes for Each Chapter 407

Appendix D: Equation and Graph 409

Appendix E: Z-Tables 421

Index 425

Preface

This book was motivated by the desire to further the evolution of the core course in engineering and engineering technology. This text is intended to provide students a comprehensive introduction to the field of engineering. To provide the best coverage of all the components, contributions from several instructors from different institutions were compiled to create the content of this book.

This book is also written for engineering faculty and department chairs as a practical guide to improve the assessment processes for undergraduate engineering education towards improved student learning. It is written by engineering faculty and assessment professionals who have many years of experience in the assessment of engineering education and of working with engineering faculty.

The book reflects the emphasis on ABET, Inc., which is based on the assessment of student outcomes. ABET, Inc., is the organization that accredits most US engineering, computer science, and technology programs, as well as providing substantial equivalency evaluations to international engineering programs. Additionally, the high-impact education practices by Association of American Colleges and Universities (AAC&U) are also integrated in this book.

The book begins with a brief overview of assessment theory and introduces readers to key assessment resources. It illustrates—through practical examples that reflect a wide range of engineering disciplines and practices at both large and small institutions, and along the continuum of students' experiences, from first year to capstone engineering courses through to the dissertation—how to go about applying formative and summative assessment practices to improve student learning at the course and program levels. This book concludes with a vision for the future of assessment for engineering education.

The author covers six basic themes:

1. Use of assessment to improve student learning and educational programs at both undergraduate and graduate levels
2. Understanding and applying ABET criteria to accomplish differing program and institutional missions
3. Illustration of evaluation/assessment activities that can assist faculty in improving undergraduate and graduate courses and programs
4. Description of tools and methods that have been demonstrated to improve the quality of degree programs and maintain accreditation
5. Using high-impact educational practices to maximize student learning
6. Identification of methods for overcoming institutional barriers and challenges to implementing assessment initiative

Approach

As the title of this book implies, "Introduction to Engineering: An Assessment and Problem Solving Approach" uses a comprehensive approach toward student learning in engineering and engineering technology majors. Over the years, engineering educators have been overwhelmed with challenges associated with motivation, engagement, retention, and graduation rates of students in engineering majors. This may be attributed to inadequate academic preparation prior to undergraduate education, among several other factors. However, engineering educators and students must collaborate and cooperate in the learning process to improve the current situation. This book uses experiential and problem- and activity-based learning to engage students and empower them in their own learning.

The learning objectives at the beginning of each chapter are aligned with ABET learning outcomes and AAC&U high-impact educational practices. The topics and examples in the chapters are carefully developed toward achieving the learning outcomes. Real-world examples of theory and applications are embedded to spark student interest in the topics. The goal is to convince students about the perceived values of the topics in their personal and professional lives so that they become interested in learning. The end-of-chapter problems are designed to assess learning objectives with close correlation to ABET learning outcomes. The chapters and topics are selected to assess all 11 student learning criteria as outlined by ABET.

The book will also include several engineering problems for students to work in teams and develop skills in successful team performance. The instructor of this course can use the materials presented in the book to assess a number of ABET outcomes using the sample rubrics provided in the solutions manual for the book. This book will help both instructors and students to develop competencies and skills required by ABET, AAC&U, and future employers. Sample rubrics and performance indicators for assessment of each outcome are provided in Appendix B. The learning outcomes listed at the beginning of the chapter also include corresponding ABET outcomes. Appendix C tabulates ABET outcomes that can be assessed in each chapter. Most of the ABET outcomes can be assessed except outcome (b), which requires experimental investigation or laboratory experience.

Organization

The book is divided into four major sections with three to five chapters in each section. To provide a foundation in engineering, Section I (Chapters 1 through 3) describes engineering and engineering technology careers, skills, and competencies desired to succeed; engineering education; ABET learning outcomes; metacognition; study skills; and learning strategies to develop skills necessary for success. This section will not only describe what competencies are desired but also describe how to develop them effectively. Students will be required to develop individual success plans and strategies that are most appropriate based on their goals. Chapters 4 through 11 comprise Section II and address fundamental engineering topics such as computer-aided design, statics, materials, design and analysis, electric circuits, engineering economics, probability and statistics, and computer programming. To ensure availability to everyone, any of the book's examples and exercises that require outside software are based on publicly available software, such as Google SketchUp instead of proprietary software. Other general engineering topics useful to all engineering and engineering technology disciplines will be presented in a manner so that both students with and without a background in physics or calculus will benefit.

Section III of the book includes Chapters 12 and 13 and delves into the arena of product design and development. Topics such as invention, discovery, quality engineering, and rapid prototyping using 3D printers will be included. Design optimization, reducing product development cycle, product life cycle management, design for environment, sustainability, and manufacturing processes will also be included.

Section IV of the book (Chapters 14 through 16) provides an overview of the relationship between engineering and society, social responsibilities of engineers, and the goods and evils of technology. Engineering ethics, the professional code of conduct, and the engineer's creed will be covered in the engineer's ethics chapter of this section. The last chapter of the book will focus on teamwork and communication, an important skill needed by an engineering workplace as well as one of the ABET learning outcomes. Topics such as how to become a productive and contributing team member, making effective oral presentations, and written communication will be included.

Instructors' Resources

To assist in teaching with this text, a *Solutions Manual* is available. This manual includes the teaching philosophy that was a foundation for this book, syllabus examples for the introductory engineering course, chapter-by-chapter teaching resources and exercises, a chapter-by-chapter test bank with both objective and essay questions, suggestions for term-long group projects, and methods of obtaining feedback from students about several aspects of the course. The end-of-chapter questions can be used to assess ABET learning outcomes relevant to the chapter topics.

The textbook website has additional information for both instructors and students. Instructors adopting the book for their classes can receive PowerPoint presentations, the Instructor's Manual, solutions to chapter questions, and other resources for teaching the course; these can be requested by contacting the publisher by email at for crctextbook@taylorandfrancis.com. The website is regularly updated with new features and content.

Quamrul H. Mazumder
University of Michigan–Flint

Acknowledgments

The author would like to express his sincere thanks to many who dedicated their time in the development of this book. First, he would like to thank Dr. James Lookadoo of Pittsburgh State University for his encouragement to write this book to help first-year students and engineering and engineering technology faculty better understand the assessment process. Dr. Ulan Dakeev contributed in the development of Chapter 4: "Computer-Aided Design" and Chapter 10: "Probability and Statistics." Laura Sutton spent countless hours in the development of Chapter 11: "Computer Programming," "Chapter 14: "Engineering and Society," and Chapter 16: "Communication and Teamwork." Marsha Nottigham, a senior metallurgist at Delphi Automotive, contributed in the development of Chapter 6: "Materials Engineering" and Ishtiaque Amin, an electrical engineer of Nexteer Automotive, contributed in the development of Chapter 8: "Electric Circuits and Components." Gaurav Verma of Indian Institute of Skill Development (in Gurgaon, India) contributed in the development of Chapter 13: "Manufacturing Processes."

The author would also like to acknowledge the contribution of Kawshik Ahmed and Siwen Zhao, two engineering graduates, for their careful review of the materials for appropriateness to first-year engineering students. They also prepared some figures, tables, and charts for the book. The author thanks Ashley O'Brien, another student, for reviewing and editing the chapters without any background in engineering to ensure that the materials can be understood by students of nonengineering majors.

Author

Dr. Quamrul H. Mazumder received his Bachelor's degree in engineering from Bangladesh Agricultural University. After graduation, he worked for three years and then returned to graduate school to pursue his Master's degree in mechanical engineering in the United States. He worked at different engineering positions for 18 years with increasing responsibilities. During his employment as an engineering manager, he completed his Master's degree in Business Administration and his PhD in mechanical engineering.

Dr. Mazumder is passionate about learning, and he joined a full-time teaching position in the engineering technology department at Pittsburgh State University. Later he joined the University of Michigan–Flint where he currently serves as the associate chairman for the department of computer science, engineering, and physics.

Dr. Mazumder has 15 years of teaching experience in engineering and engineering technology. His research includes student learning, motivation, metacognition, teaching and learning styles, quality of higher education, globalization of education, and other related areas. He is recognized as a global leader and expert in quality and assessment of higher education. He has been providing his services and expertise in higher education quality improvement initiatives of Bangladesh where he conducted numerous workshops, seminars, and training to faculty, administrators, and students. He served as a Fulbright Scholar in Bangladesh during 2013 where he developed curriculum and assessment processes for a graduate engineering program.

Section I

Engineering Education

1

Engineering and Technology Professions

CHAPTER OBJECTIVE AND STUDENT LEARNING OUTCOMES

After completing the chapter, students will be able to

1. Describe the role of engineers in changing the world and human life.
2. Develop a better understanding of employment outlooks of different engineering disciplines in local, regional, national, and global marketplaces.
3. Describe the knowledge, skills, and competencies required for a meaningful engineering career.
4. Describe the roles, responsibilities, challenges, and opportunities associated with major engineering disciplines. [ABET outcome h, see Appendix C]
5. Develop understanding of the process to become a registered professional engineer and need of professional engineering licensure. [ABET outcome f, see Appendix C]

1.1 Engineering and Engineering Technology

To design, develop, and support products and services in today's technologically advanced industry, professionals must work together as an *engineering team*.

Primary team members include scientists, engineers, and technologists among others (Figure 1.1).

The role and importance of engineering technologist is not clearly understood by many, and therefore, clarification of this role is needed. There are similarities and differences between engineering technologists and engineers, although they work toward a common goal of engineering problem solving. Engineering technology (ET) education emphasizes on problem solving using hands-on experience, laboratories, and technical skills, preparing students for application-oriented careers in industry such as manufacturing, after-market service, marketing, and technical sales. Engineering programs typically require additional, higher level mathematics, including multiple semesters of calculus and calculus-based theoretical science courses. ET programs typically focus on algebra, trigonometry, applied calculus, and other courses that are more practical than theoretical in nature.

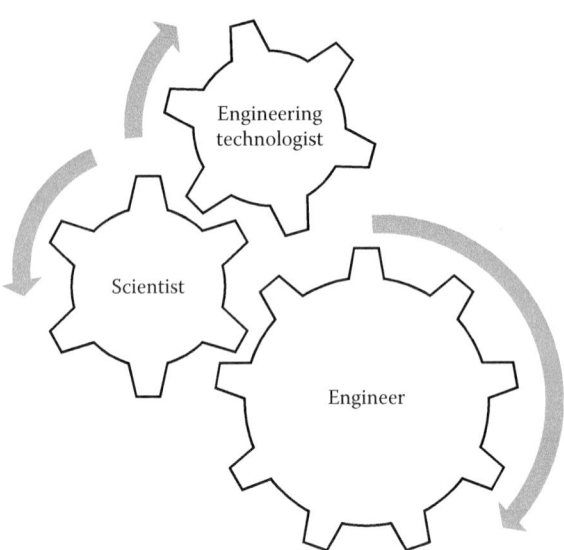

FIGURE 1.1
Interrelationship among scientists, engineers, and engineering technologists.

According to a national accrediting agency (TAC/ABET), graduates of baccalaureate-level engineering technology (BET) programs are called *engineering technologists*, and graduates of associate degree (AS) programs in ET are called *engineering technicians*. The upper-division coursework of BET programs is designed to provide additional analytical and problem solving beyond those learned at the two-year level. Most BET programs are accredited by TAC/ABET and are designed to accept appropriate coursework in math, science, and a technical specialization completed at approved AS programs. For ABET accreditation, engineering and ET programs are reviewed and accredited by two separate accreditation commissions, using two separate sets of accreditation criteria: the Engineering Accreditation Commission (EAC) and the Engineering Technology Accreditation Commission (TAC).

The following paragraph provides definitions of scientists, engineers, and engineering technologists in engineering careers to shed some light on the similarities and differences among these three professions.

1.1.1 Scientists

Engineers working as scientists are the most theoretical of the team members. They typically work in the research and development division of large corporations to develop and improve products and services through innovative technologies. They typically seek ways to apply new discoveries to advance technology for mankind. Most engineering scientists have earned a doctorate in engineering.

1.1.2 Engineers

Engineers use the knowledge of science and mathematics gained through engineering education, experience, and practice, applied with judgment, to develop ways to economically utilize materials and resources to the benefit of mankind. Engineering involves activities such as the conception, design, development, and formulation of new systems and products through the implementation, production, and operation of engineering systems. Engineers often work closely with scientists and technologists in developing new products and technologies. A minimum of four years of study is required to become an engineer. Most BET programs are accredited by EAC/ABET.

1.1.3 Engineering Technologists

Engineering technologists are graduates from a bachelor-level program in ET. They apply engineering and scientific knowledge combined with technical skills to support engineering activities, especially those application oriented. They typically concentrate their activities on the applied design, using current engineering practices. Technologists in the engineering team play key roles; they are typically involved in product development, manufacturing, product assurance, sales, and program management. TAC/ABET specifies that faculty who teach in these programs have a minimum of a master's degree in engineering or ET or equivalent, or a PE license and a Master's degree.

Engineering technicians work with equipment, primarily assembling and testing component parts of devices or systems that have been designed by others, usually under direct supervision of an engineer or engineering technologist. In order to assemble, repair, or improve technical equipment, engineering technicians work by learning its characteristics, rather than by studying the scientific or engineering basis for its original design. They may carry out standard calculations, serve as technical sales people, make estimates of cost, assist in preparing service manuals, or perform design-drafting activities. As a group, they are important problem-solving individuals whose interests are directed more to the practical than to the theoretical aspect of a project. They are frequently employed in laboratories and/or manufacturing facilities where they may set up experiments, accumulate scientific or engineering data, and/or service or repair engineering or production equipment.

Ishtiaque Amin is a circuit design engineer in the Electrical Systems Group at Nexteer Automotive in Saginaw, Michigan. He has been in this role for four years where his primary responsibilities include design and analysis of embedded electronic applications in electric power steering (EPS) systems. His expertise lies in the areas of magnetic sensors for torque and rotary position-sensing applications. He is also the lead electrical design engineer for multiple current and future EPS applications for automotive manufacturers throughout the world. He has won several awards in this position for his commitment to flawless design launches. He is an expert in robust engineering practices for steering applications by holding Design for Six Sigma (DfSS) Black Belt, Six Sigma Green Belt, and Fast X Journeyman certifications. Simultaneously, he serves as a supervisor for student co-ops and interns, assisting them with work-study programs and summer projects. His previous experiences include engineering positions at ArcelorMittal, Cooper Industries, and NewPage Inc.

In addition, he currently holds Lecturer I position at University of Michigan–Flint, in the Department of Computer Science, Engineering and Physics, and Adjunct Faculty position at Saginaw Valley State University, in the Department of Electrical and Computer Engineering. He teaches courses related to electric circuits, digital logic, CAD, and digital signal processing. He received BS and MS degrees in Electrical Engineering from Michigan Technological University in 2009 and 2011, respectively. His graduate thesis was on the detection of complex point targets with distributed assets and partially correlated signals, in the area of signal processing. During undergraduate studies his concentration was on Electronics. He is an active member of the honor societies Tau Beta Pi, Omicron Delta Kappa, and Eta Kappa Nu.

Engineer Amin believes that in order to prepare for a successful engineering career, students must be diligent and thorough in their efforts of pursuing college education. They should look out for internship/co-op opportunities that will add valuable industry experiences to their resumes, be actively involved in university research programs with a commitment to generating new ideas, ask questions to be well informed about subject matters, and keep an open mind when searching for solutions to problems they face.

1.2 Employment Outlook

According to Bureau of Labor Statistics of the US Department of Labor, "Professional and related occupations and service occupations are expected to create more new jobs than all other occupational groups from 2008 to 2018 in addition, growth will be faster among occupations for which postsecondary education is the most significant form of education or training, and, across all occupations, replacement needs will create many more job openings than the job growth."

Architecture and engineering occupations are expected to add roughly 270,600 jobs, representing a growth rate of 10.3% over the 2008–2018 period. About 178,300 of these jobs, more than 6 of 10, are expected to be for engineers, and the growth of civil engineers is anticipated to be especially robust. As a greater emphasis is placed on improving the nation's infrastructure, civil engineers will be needed to design, implement, and upgrade transportation, water supply, and pollution control systems. In addition, it is estimated that the occupation of drafters, engineers, and mapping technicians will increase by roughly 52,200 jobs and that architects, surveyors, and cartographers will increase by 40,100. Table 1.1 shows the projected job openings in STEM clusters between 2012 and 2022. Table 1.2 shows the number of engineers in thousands in 2008 and expected growth between 2008 and 2018.

1.3 Skills and Competencies (What Employers Are Looking For)

Regardless of the role in which you will be working, there are a common set of intangible skills that employers look for across all engineering disciplines:

TABLE 1.1

Occupations Assigned to the Science, Technology, Engineering, and Mathematics (STEM) Cluster That Are Projected to Have the Most Job Openings, 2012–2022

Occupation	Job Openings, Projected 2012–2022 (thousands)
Environmental scientists and specialists, including health	39.7
Electrical engineers	44.1
Architectural and engineering managers	60.6
Industrial engineers	75.4
Mechanical engineers	99.7

Sources: National Association of State Directors of Career Technical Education Consortium (Career Cluster), Silver Spring, MD; Bureau of Labor Statistics, Employment Projections program (projected job openings), Washington, DC.

- *Effective communication skills*: With an increase in the documentation and instructions that engineers use in the workplace, clear and concise communication is required.
- *Interpersonal skills*: You need to know work effectively as part of a team and work with customers to identify needs and provide solutions.
- *Technical knowledge*: Whatever technical expertise is vital to your job, you need to understand how to that knowledge to solve practical problem.
- *Organizational skills*: Resource planning, being able to prioritize tasks, and managing time effectively are key skills for engineers.
- *Enthusiasm and commitment*: Learning new skills is part of every engineer's role, so you need to be adept at assimilating a lot of new information.

More importantly, employers are looking for evidence that you are well rounded, and take an active interest in and have an understanding of the engineering industry. Furthermore, they want to know that you have the motivation, drive, and ambition to make an impact within their company.

1.3.1 Engineering Grand Challenges for the Twenty-First Century

Engineering students need to understand and prepare to address the global engineering issues of the next century. The National Academy of Engineers (NAE) announced 14 grand engineering challenges for the twenty-first century that, if met, would greatly improve how we live. The final choices were presented into four themes that are essential for humanity to flourish—sustainability, health, reducing vulnerability, and joy of living. The goal was to identify what needs to be done to help people and the planet thrive. The results of that poll are ranked here by the number of votes each engineering challenge received during the period January 15 through March 28, 2008:

1. Make solar energy economical
2. Provide energy from fusion
3. Provide access to clean water
4. Reverse-engineer the brain

TABLE 1.2
Employment Trend of Different Engineering Disciplines

Matrix Code	2008 National Employment Matrix Title	Employment Number 2008	Employment Number 2018	Percent Distribution 2008	Percent Distribution 2018	Change 2008–2018 Numeric	Change 2008–2018 Percent	Total Job Openings due to Growth and Replacement Needs
17-2000	Engineers	1571.9	1750.3	1.0	1.1	178.3	11.3	531.3
17-2011	Aerospace engineers	71.6	79.1	0.0	0.0	7.4	10.4	22.3
17-2021	Agricultural engineers	2.7	3.0	0.0	0.0	0.3	12.1	0.9
17-2031	Biomedical engineers	16.0	27.6	0.0	0.0	11.6	72.0	14.9
17-2041	Chemical engineers	31.7	31.0	0.0	0.0	−0.6	−2.0	7.8
17-2051	Civil engineers	278.4	345.9	0.2	0.2	67.6	24.3	114.6
17-2061	Computer hardware engineers	74.4	77.5	0.0	0.0	2.8	3.8	23.5
17-2070	Electrical and electronics engineers	301.5	304.6	0.2	0.2	3.1	1.0	72.3
17-2071	Electrical engineers	157.8	160.5	0.1	0.1	2.7	1.7	38.9
17-2072	Electronics engineers, except computer	143.7	144.1	0.1	0.1	0.4	0.3	33.4
17-2081	Environmental engineers	54.3	70.9	0.0	0.0	16.6	30.6	27.9
17-2110	Industrial engineers, including health and safety	240.4	273.7	0.2	0.2	33.2	13.8	94.6
17-2111	Health and safety engineers, except mining Safety engineers and inspectors	25.7	28.3	0.0	0.0	2.6	10.3	9.2
17-2112	Industrial engineers	214.8	245.3	0.1	0.1	30.6	14.2	85.4
17-2121	Marine engineers and naval architects	8.5	9.0	0.0	0.0	0.5	5.8	2.3
17-2131	Material engineers	24.4	26.6	0.0	0.0	2.3	9.3	8.1
17-2141	Mechanical engineers	238.7	253.1	0.2	0.2	14.4	6.0	75.7
17-2151	Mining and geological engineers, including mining safety engineers	7.1	8.2	0.0	0.0	1.1	15.3	2.6
17-2161	Nuclear engineers	16.9	18.8	0.0	0.0	1.9	10.9	5.4
17-2171	Petroleum engineers	21.9	25.9	0.0	0.0	4.0	18.4	8.6
17-2199	All other engineers	183.2	195.4	0.1	0.1	12.2	6.7	50.2
17-3000	Drafters, engineering and mapping technicians	826.2	878.3	0.5	0.5	52.2	6.3	220.0

Source: Bureau of Labor Statistics, US Department of Labor, Washington, DC, 2015.

5. Advance personalized learning
6. Develop carbon sequestration methods
7. Restore and improve urban infrastructure
8. Engineer the tools of scientific discovery
9. Advance health informatics
10. Prevent nuclear terror
11. Engineer better medicines
12. Manage the nitrogen cycle
13. Enhance virtual reality
14. Secure cyberspace

Addressing these challenges must be prioritized, so it is necessary to redesign engineering education, integrating these challenges into the curriculum, particularly in capstone projects.

1.4 Engineering Disciplines

There are 17 engineering disciplines listed in the Bureau of Labor Statistics. Of these disciplines, 80% of engineers work in seven major disciplines. The percent of electrical and electronic engineers combined is highest (19%) followed by civil (18%) and mechanical (15%) (Figure 1.2). The engineering disciplines are listed below with the seven major disciplines in italics.

1. *Aerospace Engineering*
2. Agricultural Engineering
3. Biomedical Engineering

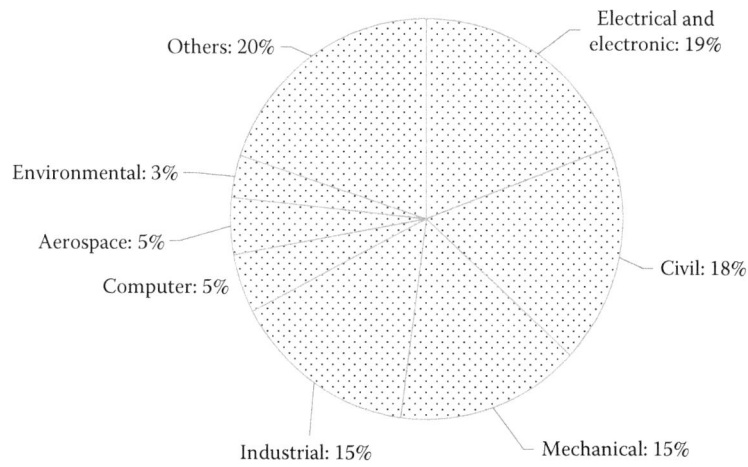

FIGURE 1.2
Distribution of engineering disciplines in percent.

4. Chemical Engineering
5. *Civil Engineering*
6. *Computer Engineering*
7. *Electrical and Electronic Engineering*
8. *Environmental Engineering*
9. *Industrial and Operations Engineering*
10. Marine Engineering and Naval Architecture
11. Materials Engineering
12. *Mechanical Engineering*
13. Mining Engineering
14. Nuclear Engineering
15. Safety Engineering
16. Petroleum Engineering

A brief description and overview of different engineering disciplines are provided in the following sections.

1.4.1 Aerospace and Aeronautical Engineering

Aerospace engineers design aircraft, spacecraft, satellites, and missiles. In addition, they test prototypes to make sure that they function according to design. The main difference between aerospace and aeronautical engineering can be summed up very simply, according to Bruce R. White, Dean of the College of Engineering at University of California, Davis. "Aeronautical engineering tends to focus on flight and activities within an atmosphere," White said, "while aerospace engineering includes the atmosphere, but also extends into applications in space, where there is no atmosphere." Both aeronautical and aerospace engineers are involved in areas such as aerodynamics, flight stability, and aircraft control, as well as traditional engineering not only issues related to aircraft systems. The following is a list of activities performed by an aerospace or aeronautical engineer:

- Design, develop, manufacture, and test aircraft and aerospace products and systems.
- Maintain, overhaul, and repair aircraft parts and systems for reliable operation of the aircraft.
- Develop in-depth understanding of aerospace materials and quality standards, such as aerospace material standards.
- Understand and comply with regulatory requirements and standards such as Federal Aviation Regulations, Canadian Aviation Regulations, and European Aviation Safety Agency.
- Develop acceptance standards of aerospace parts, systems, and components.
- Inspect parts for malfunction, damage, or repair ability for safe and reliable operation of aircraft.

Aircraft can be grouped into five major categories:

1. *Private business jets*.
2. *Small commercial aircraft*: Turbo propeller-type aircraft or small jets carrying 40–150 passengers. Examples include DE Havilland Dash-8, Embraer regional jet, Canadian regional jet, and so on.
3. *Large commercial aircraft*: These aircraft use two to three large jet engines and carry 150–500 passengers. Examples include Boeing 737, Airbus 340, and so on.
4. *Military aircraft*: These are used by defense systems and usually can reach supersonic speed (speed equal to or more than the speed of sound, i.e., greater than 750 miles per hour). Examples include F-16, F-18, and so on. Other military products include missiles and rockets.
5. *Space shuttle*: Space shuttle is used to explore different planets and stars in the galaxy by NASA and other space agencies. The space shuttle consists of two solid rocket boosters used for launch, an external fuel tank that carries fuel for the launch, and an orbiter that carries astronauts and payload. A typical space shuttle mission typically lasts for 7–10 days, but can extend for longer time depending on the mission. Aerospace engineers may become specialists in aerodynamics, thermodynamics, celestial mechanics, flight mechanics, propulsion, acoustics, and guidance and control systems.

1.4.2 Civil Engineering

Civil engineers are involved in the design, construction, supervision, operation, and maintenance of construction projects ranging from small buildings to large projects and systems, including roads, buildings, airports, tunnels, dams, bridges, piping for water supply, oil, gas, and sewage treatment. Civil engineers can also work in research, but that role usually requires a graduate degree. Typical activities and tasks performed by civil engineers include the following:

- Conduct and analyze survey reports, maps, and other data to plan projects and make design decisions.
- Prepare estimates of construction costs, and ensure compliance with government regulations, potential environmental hazards, and other factors in planning stages and risk analysis.
- Understand regulatory requirements and submit permit applications to local, state, and federal agencies verifying that projects comply with pertinent regulations.
- Evaluate cost of materials, equipment, or labor to determine a project's economic feasibility.
- Use modern engineering computational tools, such as computer-aided design software, to plan and design transportation systems, hydraulic systems, and structures in line with industry and government standards.
- Develop, plan, and oversee the repair, maintenance, and replacement of public and private infrastructure such as bridges and highways.

For the safety of citizens and public infrastructures, civil engineers are required to possess professional engineering license or registration. Civil engineers may also hold supervisory or administrative positions ranging from a supervisor of a construction site to a city engineer. The federal government employs civil engineers to inspect projects for regulatory and safety compliances. Civil engineers can perform jobs as construction, geotechnical, structural, and transportation engineers.

> *Construction engineers*: Construction engineers may also hold title as construction managers and are responsible for managing construction projects, meeting schedules, plans, and specifications. During construction, they are also responsible for safety of workers and temporary structure in the job site.
>
> *Geotechnical engineers*: The interaction of structures with earth (including soil and rock) is the focus of geotechnical engineers. Geotechnical engineers perform soil tests to ensure the strength present meets that required for foundations to carry their desired loads. Seismic loads must be assessed for building roads and tunnels.
>
> *Structural engineers*: Structural engineers design reinforcements to ensure strength and reliability of major projects, such as buildings, bridges, and dams.
>
> *Transportation engineers*: Transportation engineers plan, design, operate, and maintain transportation systems such as mass transit, airports, ports, harbors, and traffic systems.

1.4.3 Computer Engineering

Computer engineers perform a wide range of duties related to design and development of computer hardware and development of programs to operate the computers. Computer engineering jobs can be categorized in two major areas: hardware engineers and software engineers.

Computer hardware engineers: Computer hardware engineers design, develop, manufacture, and test computer systems and components, such as microprocessors, circuit boards, memory devices, networks, and routers. They create innovative technologies and systems to advance computer technology. Typical activities performed by hardware engineers include the following:

- Design, develop, analyze, and test new computer hardware and systems.
- Upgrade existing computer equipment to make it compatible with new software.
- Plan and manage the manufacturing process for computer hardware and peripheral equipment.

Computer hardware engineers also design devices and systems that are embedded in other products, such as automobiles, aircraft, and medical devices, which interact with a computer system or the Internet. Computer hardware engineers work closely with software engineers to ensure compatibility of hardware and software systems.

Software engineers or software developers: Computer software developers develop applications that allow users to perform specific tasks on a computer or other devices. They also develop programs for the underlying computer systems that run the devices (operating system) or control networks. Typical activities performed by software developers include the following:

- Understand, evaluate, and analyze users' needs, and then design, test, and develop software to meet those needs.

- Develop application software or system software and plan how the pieces will work together.
- Develop flowcharts, architecture, and procedures to write the program or code.
- Test software and hardware for reliable function of the code and hardware.
- Prepare a software user manual and other documentation for the programmer and user of the code.
- Manage software development projects.

Software engineers begin the development process by asking how the customer plans to use the software. Using that customer input, the programmer develops the program and writes the code. After the code is written, software engineers test the program for reliability, user-friendliness, and other requirements as specified by the customer. The software developers can be grouped into two major areas:

Applications software developers: Application software developers develop computer applications, such as word processors, spreadsheet, and games, for consumers. Application software may include custom software for a specific customer or commercial software to be sold to the general public.

Systems software developers: Systems software developers design and develop the systems, such as operating systems, that keep computers functioning properly. They also develop system interfaces that allow users to interact with the computer.

1.4.4 Electrical and Electronic Engineering

There are similarities and differences in roles and responsibilities of electrical and electronic engineers. Most of these engineers hold a position as an electrical engineer. Electrical engineers design, develop, analyze, test, and manufacture electrical equipment, such as electric circuits, transformers, electric motors, generators, and navigation and communication and electric power distribution systems. Electrical engineers also design the electrical systems or circuits of large buildings and structures, automobiles, and aircraft. Typical activities performed by electrical engineers include the following:

- Develop innovative solutions to problems using electrical equipment, systems, and devices.
- Perform detailed electrical design calculations to optimize design, manufacture, and application of electrical systems or subsystems.
- Develop electrical specifications, codes, standards, and design guidelines for safe and reliable operation of electrical equipment and systems.
- Manage projects that require installation, maintenance, manufacture, and testing of electrical equipment to ensure that products meet specifications, codes, and standards.

Electronic engineers design and develop electronic equipment used in computers, cell phones, smart TVs, broadcast and communications systems, portable music players, global positioning systems, and so on. Typical activities performed by electronic engineers include the following:

- Develop innovative solutions through design of electronic components, software, products, or systems for personal, commercial, industrial, medical, military, or scientific applications.
- Develop specifications, codes, and standards for safe and reliable operation of electronic equipment or systems.
- Analyze system requirements, estimate cost, and select components to meet desired needs and capacity.
- Develop maintenance and testing procedures for electronic components and equipment for safety and reliability.
- Perform design modifications and upgrades for system optimization, repair, and maintenance.
- Monitor performance and reliability of electronic equipment, instruments, and systems to comply with regulatory requirements.

1.4.5 Environmental Engineering

Environmental engineers develop solutions to problems related to air and water pollution, contamination, renewable energy, and so on. These solutions often require knowledge in the areas of soil science, biology, chemistry, and so on. They are involved in global issues such as climate change, sustainability, design for environment, and design of recycling. Typical activities performed by environmental engineers may include the following:

- Design and develop innovative solutions leading to environmental protection, such as water recycling facilities, air pollution control systems, and recycling waste products to useful energy.
- Develop codes, standards, and specifications related to environmental protection and sustainability.
- Enforce policies and procedures to protect the environment and advocate legal issues related to violations of such policies by individual, industry, and nation.
- Provide technical support for environmental remediation projects and for legal actions.
- Monitor industrial and municipal facilities to ensure compliance with environmental regulations.
- Prepare procedures for corporations and government agencies for cleaning up contaminated sites.

Environmental engineers conduct hazardous waste management studies in which they evaluate the significance of hazards and advise on treating and containing them. They also design systems for municipal and industrial water supplies and industrial wastewater treatment, and research the environmental impact of proposed construction projects. Environmental engineers in government develop regulations to prevent mishaps.

1.4.6 Industrial and Operations Engineering

Industrial engineers develop solutions to improve efficiencies in production by reducing waste. They use principles of Six Sigma and lean manufacturing to optimize production and manufacturing processes. Industrial engineers are also involved in monitoring the health and safety of workers in manufacturing and other industries. Typical activities performed by industrial engineers may include the following:

- Develop innovative methods to maximize efficiency in the production processes.
- Review engineering drawings and specifications to develop process flows, methods, and production schedules to reduce cycle time and production costs.
- Develop and implement quality management procedures to improve product quality and reliability.
- Address issues related to production systems and perform time studies to determine standard time required to perform tasks.
- Coordinate with design, manufacturing, and production personnel to eliminate waste.
- Review and improve equipment and systems, moving toward ergonomic designs.
- Perform simulation of production systems to improve efficiency and reduce cost.
- Coordinate with suppliers as well as internal and external customers to develop a robust supply chain management system.

1.4.7 Mechanical Engineering

Mechanical engineering is one of the broadest engineering disciplines. Mechanical engineers are involved in the design, development, building, and testing of mechanical and thermal devices, including tools, engines, and machines. Mechanical engineers are required to work with electromechanical systems and devices such as embedded systems, robotics, and automation. Mechanical engineers also work in research and development of new equipment, systems, and processes. A graduate degree is normally required for research and development positions. The following is a list of some activities performed by mechanical engineers:

- Identify, analyze, and solve problems using mechanical, electrical, and thermal systems.
- Design and develop mechanical and thermal equipment, systems, and processes using computational methods and tools such as finite element analysis and computational fluid dynamics analysis.
- Optimize design using simulation and validate prototype through testing and analysis of test results.
- Improve the design of existing product performance and reliability to meet desired needs.
- Design for manufacturing and develop efficient manufacturing processes.

Mechanical engineers design, develop, and manufacture products such as automobile, aircraft, medical devices, wind turbines, electrical motors, generators, internal combustion engines, refrigeration and air conditioning equipment, and medical devices.

Mechanical engineers also design products and systems inside buildings such as elevators and escalators, plumbing, heating, ventilation and air conditioning systems, and sprinkler systems.

Most employers require a bachelor's degree in mechanical engineering from an ABET accredited program with some work experience.

ELISIA GARCIA KOWALSKI—PRODUCT INTEGRATION ENGINEER, US DEPARTMENT OF DEFENSE

I support the Marine Corps Program Office of Light Armored Vehicles (LAV) as a Product Integrator for the Department of Defense. I use my technical background as an engineer to aid in development and execution of upgrade programs to the legacy LAV fleet. This includes overseeing cost, schedule, and performance of efforts, as well as monitoring activity required for long-term sustainability and readiness. My position requires interfacing across multifunctional areas such as engineering, logistics, contracting, and business/finance within various organizations within the government. Once a new project is identified as a requirement, a detailed Scope of Work (SOW) is written to capture the project requirements and include technical performance specifications and deliverables. Contracts are then awarded to complete the SOW and are closely monitored through various acquisition phases of the life cycle. After development, production, and fielding are completed, the project moves into the sustainment phase that will maintain components for the users for as long as the service life requires, which can be upward of 50 years.

The engineering degree provided me with the discipline and technical aptitude to problem solving and research information. I also completed a Master of Business Administration degree to complement my portfolio for versatility in the workforce. I continue to expand my knowledge base through training and certifications offered within the Department of Defense to ensure that I remain relevant and competitive. When researching the various degree options after high school, I initially did not know what I wanted to do. My father is an engineer for General Motors and he persuaded me into giving it a chance and I am so happy I took his advice. I also wanted to ensure that I would be marketable after graduation. Female minorities in the engineering field are highly sought after. After I graduated, I was working as a controls engineer for an automotive contractor.

I am also a member of the Industrial Advisory Board for the University of Michigan–Flint. There are always surprises! I continue to learn something new every day. I love that my career field continues to challenge me and provide me with opportunities for career growth. I enjoy working in Program Management because we work directly for the warfighter and are able to see development efforts come to fruition. I plan to continue gaining experience with the intention of an eventual promotion into a leadership role.

A successful career as an engineer requires flexibility, willingness to learn something new, and being unafraid to ask questions. Teamwork is the key in the real-world environment. You can always learn from others and people are generally more than willing to share their knowledge. Also, do not limit yourself and always explore your options. While you may have advisors to help guide you, you are the only one ultimately in charge of your career path.

1.5 Professional Engineering License

Professional engineer, or PE, is the title used by professional engineers throughout the United States. *State licensing boards* grant the PE title to engineers who meet certain requirements in education, experience, and exams.

State laws limit the practice of engineering to licensed professional engineers, which means that a PE is required for such things as

- Stamping and sealing designs
- Bidding for government contracts
- Owning a firm
- Consulting
- Offering expert witness testimony
- Advertising services to the public

Engineers who do not perform the above functions can also benefit from holding a professional license. As a PE, you are likely to reach managerial positions more quickly and earn a higher salary than your peers. The PE acts as a standard that shows you have met a series of stringent professional requirements and are a member of a select group of practicing engineers.

1.5.1 Steps to Become a PE

While each *state licensing board* has its own laws regarding engineering licensure, there is a general four-step process for licensure candidates:

- Earn a degree from an *ABET*-accredited engineering program.
- Pass the *FE exam*.

- Gain acceptable work experience (typically a minimum of four years). In most cases, this must be completed under the supervision of a PE.
- Pass the *PE exam* in the appropriate discipline. Check your state's requirements; some require experience to be earned before you can take the PE.

DR. NEIL G. MURRAY JR., ENGINEERING MANAGER TRW

As an engineering manager I am responsible for leading and mentoring the engineering team at TRW Automotive that develops electronic crash sensors for automotive airbag control systems. My team consists of men and women with degrees in mechanical, electronic, and industrial engineering. We develop both the product and manufacturing process in parallel from the very beginning of each program. As one element of my mentoring role, each week I devote two to three hours to training my team and other employees in DfSS tools such as Axiomatic Design, Quality Function Deployment, TRIZ, and Optimization. Teaching these courses is fun and the learning experience helps to reinforce the knowledge and team dynamic.

I have engaged in a lifetime of learning, finally completing a doctorate in Manufacturing Systems Engineering at Lawrence Technological University at age 61. Initially, my BS in Mechanical ET provided an entry point into a career as a cost estimator and mold designer in the rubber molding industry. Engineering as a career has substantially changed over the years, and had I not reinvented myself, I would not have been in a position to take advantage of opportunities as they came along. MS in Manufacturing Systems Engineering provided the knowledge and credentials to be considered for and transition into engineering management at a start-up electronics company in the automotive industry. Each step in my career has prepared me for the next step. I transitioned from Advanced Manufacturing Engineer to the Six Sigma continuous improvement group where I gained further knowledge and experience in problem solving and improvement. While in the Six Sigma role I also entered a doctorate of engineering program. Success in this role led to being considered for my current responsibilities as an engineering design manager position leading the crash sensors development team.

I grew up on a farm and developed an interest in complex mechanisms such as the knot tying fingers that tied the twine on hay bales. These devices are so complex and yet simple at the same time being entirely mechanical. This interest led to studies in mechanical engineering. I was in the Six Sigma group, and as a pilot project for Design for Six Sigma, we redesigned an automotive tire pressure sensor to reduce cost and increase reliability. One of the directors was pleased with this work and asked me to take on a role in engineering management developing crash sensors. I hold a number of patents for new developments in crash sensing, ASQ certification as a Six Sigma Black Belt, and membership in the Society of Plastic Engineers and the Institute of Electrical and Electronic Engineers. I believe that membership in these professional societies is important in keeping current with advances in technology. That I am a member of both an electronic and plastics molding organization reflects the broad nature of my professional growth over the years. This broad scope has kept my career challenging and interesting. I am also an active member of the Engineering Industrial Advisory Board for the engineering programs at the University of Michigan at Flint, Michigan.

The extent to which engineering has evolved has been a pleasant surprise. This is not just a matter of technology. Technology is just a tool that makes us more productive and more capable of solving interesting problems. Really, it is the amazing change from individual work to a culture of collaborative work where team members may be located almost anywhere on earth. This has been exciting and tremendous fun. And if we are doing it right, engineering should be fun, creative work. My intent is to begin new responsibilities as a part-time lecturer at a local university. My advice is to pursue a plan of continuous learning. The knowledge that one gains from that first degree is only a starting point in an exciting career. You must keep current as your industry and technology evolve. The other point of advice is that the enthusiasm that you bring to your career and a spirit of taking on new challenges and solving new problems is critical in being recognized as a valued member of an engineering organization.

1.6 Professional Engineering Organizations

Engineers are recommended to become members of professional societies and organizations as they provide important support to engineers. Professional organizations advocate for engineers, provide networking opportunity with other engineers in their community, offer courses and seminars for professional development, publish newsletters and journals, and organize conferences to update engineers on the latest developments in their field. Professional organizations also provide mentorships to new engineers by introducing them to experienced engineers in their profession. A list of major engineering professional organizations are listed as follows:

American Institute of Aeronautics and Astronautics (www.AIAA.org): AIAA's mission is to inspire and advance the future of aerospace for the benefit of humanity. AIAA's vision is to be the voice of the aerospace profession through innovation, technical excellence, and global leadership. Its tagline is *Shaping the Future of Aerospace*. It reflects our belief that AIAA members are continually shaping the future of aerospace through their creativity, ingenuity, and passion for aerospace engineering and science.

American Society for Engineering Education (www.asee.org): The American Society for Engineering Education is a nonprofit organization of individuals and institutions committed to furthering education in engineering and ET. It accomplishes this mission by

- Promoting excellence in instruction, research, public service, and practice
- Exercising worldwide leadership
- Fostering the technological education of society
- Providing quality products and services to members

American Society of Civil Engineers (www.ASCE.org): The oldest engineering organization in the United States supports civil engineers in their profession while also serving the public.

American Society of Mechanical Engineers (www.ASME.org): This professional organization for mechanical engineers sets the codes and standards for mechanical devices to ensure quality, reliability, and safety. The ASME stresses the importance of change while also embracing the long history of mechanical engineering. The organization also promotes education and innovation in this industry.

Institute of Electrical and Electronics Engineers (www.IEEE.org): This nonprofit organization advocates for innovation, both in theory and practice, in electrical engineering and other disciplines. A primary focus of this organization is providing professional development and continued learning for engineers.

Institute of Industrial Engineers (www.IIE.org): This association is the world's largest society dedicated to the development of industrial engineering. Focused on the design, improvement, and installation of systems of people and processes, its mission is to help advance the field of industrial engineering through networking and sharing knowledge.

National Society of Professional Engineers (www.nspe.org): The NSPE advocates for the needs of the professional engineering community and attracts professionals with a high degree of integrity. The NSPE promotes innovation, professional growth, teamwork, ethics, and professional licensure for its members. Engineers are recommended to comply with NSPE Code of Ethics as listed below:

As a Professional Engineer, I dedicate my professional knowledge and skill to the advancement and betterment of human welfare.
  I pledge:
- To give the utmost of performance;
- To participate in none but honest enterprise;

Engineering and Technology Professions 21

- To live and work according to the laws of man and the highest standards of professional conduct;
- To place service before profit, the honor and standing of the profession before personal advantage, and the public welfare above all other considerations.

In humility and with need for Divine Guidance, I make this pledge.

Adopted by National Society of Professional Engineers, June 1954
Source: See more at http://www.nspe.org/resources/ethics/code-ethics/engineers-creed#sthash.bYmLTU51.dpuf

Society of Automotive Engineers (www.sae.org): This organization advocates for engineers in automotive, aerospace, and commercial vehicle sectors. SAE provides seminars and certification programs for real-world learning to support career development through online and classroom modes. Most of the courses are offered through Internet so engineers from around the world can enroll.

Society of Women Engineers (www.swe.org): This nonprofit organization supports the education and advancement of women engineers and recognizes the significant achievements women make in this profession.

MARSHA NOTTINGHAM, MATERIALS ENGINEER

As a materials engineer, I solve three types of problems related to materials. The perennial problem is *what material should we make a part from?* This problem requires knowledge or material properties such as strength, corrosion resistance, electrical, thermal, and optical properties of metals, ceramics, and plastics. The second problem is often *why did this part fail?* This requires analytical work to determine why the part cracked, malfunctioned, corroded, and otherwise degraded. It is a lot like CSI forensics, except that we are dealing with automotive components instead of people. The third type of problem requires *development of a materials specification that guarantees good materials in the production process*. This involves working with materials' suppliers to understand the variations in their processes as well as defining the limits that will be acceptable for our parts.

The best part of my job is the opportunity to work with a diverse group of engineers and technicians. I also enjoy the problem-solving part of the job as it challenges me with new opportunities every day. It would be better if I could deal with less paper work. However, I understand that documentation is necessary for compliance and substantiation of my decision. To become a materials engineer, one must have at least a bachelor's degree in materials science or engineering with some job experience.

Although I was originally pursuing an aerospace engineering degree, I fell in love with material science during an introductory course in materials engineering.

I really liked the hands-on, concrete nature of materials engineering compared to computational or analytical methods used in aerospace engineering education. I enjoyed pulling a metal through tensile tester, breaking a part in the impact testing process.

During my first professional job, I realized that everyone in the organization is not as data driven as engineers are. Different groups of people view from different perspectives to make decisions. Engineering students should choose a career with a company or product that they enjoy working on. Some jobs require people skills and most of the jobs require good communication and team working skills that students must acquire while they are at school. Students need to get actively involved in students' organizations, undergraduate research, and extracurriculum activities to develop these professional skills that will help them succeed in their career and personal life. It is also important to earn a good salary. However, you should not be motivated by money alone; you must enjoy what you are doing.

1.7 Summary

Engineering is a learned and noble profession. Engineers develop products and services to make the world a better place to live. Engineers, engineering technologists, and technicians work together in a team to achieve common goals and objectives. The future of engineering lies among products and services that focus on sustainability, health, and well-being of individuals and society. The employment outlook for all engineering disciplines is positive with projected growths of 5%–20% in different areas. Engineers need to develop competencies beyond technical skills such as communication and team working skills to succeed in the competitive global environment.

Engineers and engineering technologists must join professional societies in their disciplines as these organizations provide professional development opportunities for career growth. Engineers are also required to abide by the code of ethics to perform their jobs.

Problems

1. The NAE announced 14 grand engineering challenges for the twenty-first century that are essential for sustainability, health, and humanity. Select any five of these grand challenges and describe how you will contribute as a future engineer in these areas.
2. Investigate the employment outlook for engineers from the US Department of Labor, Bureau of Labor Statistics (www.bls.gov).
    a. Describe the employment outlook of your discipline for next 5–10 years.
    b. What is the average salary and starting salary in your field of engineering?

3. List five desirable skills and competencies employers of engineering graduates look for.
   a. Self-assess your level of competencies in all of these five areas with scales of 0 to 10, where 0 is the lowest level of competency and 10 is the highest level of competency.
   b. Present a plan for how you will improve your competencies in two of the lowest levels of competencies as listed in part (a).
4. Select the engineering discipline that you are currently enrolled in and list the duties performed by engineers of that discipline. If your discipline is not listed above, investigate from online resources about the duties performed by engineers of your discipline.
5. Describe the steps required to become a registered professional engineer. What are the benefits of becoming a registered professional engineer?
6. List the NSPE code of ethics for engineers.
7. How will you change the world as a future engineer? What products and services will you develop to contribute toward advancement of technology and society?

Bibliography

Accreditation Board of Engineering and Technology, http://www.abet.org/accreditation/new-to-accreditation/engineering-vs-engineering-technology/.

Engineering Grand Challenges by National Academy of Engineers. (March 2008), http://www.engineeringchallenges.org/GrandChallengeScholarsProgram.aspx.

McCurdy, L. (March 1995). *What Is the Difference Between Engineering and Engineering Technology?* Engineering Liaison Committee, Pomona, CA.

National Council of Examinations for Engineers and Surveyors, www.ncees.org.

2

Engineering Education

CHAPTER OBJECTIVE AND STUDENT LEARNING OUTCOMES

After completing the chapter, students will be able to

1. Describe the importance of engineering education to prepare for a professional career.
2. Describe knowledge, skills, competencies, and abilities to be developed during the course of study at higher education institutions. [ABET outcome a, see Appendix C]
3. Develop a better understanding of accreditation, educational objectives, and student learning outcomes, as well as how to achieve them.
4. Develop a strategy to succeed using curriculum mapping, plan of study, and high-impact practices.

2.1 What Is Engineering?

The word *engineer* is derived from the Latin words ingenius, *ingeniare* (*to contrive, devise*), and *ingenium* (*cleverness*). Engineers apply scientific discoveries in developing products and services to improve human life and make the world a better place to live. An engineer can be described as a problem solver, innovator, designer, creator, inventor, researcher, and so on. Among all these characteristics, problem solving is the fundamental role of most engineers. For example, the problems related to global warming and fossil fuel depletion, geopolitical issues with control of energy sources, were addressed by engineers who developed renewable energy devices and systems such as solar panels, wind turbines, and geothermal energy. These renewable energy systems helped solve a multidimensional problem by harnessing energy from natural sources and reducing carbon emission.

An engineer may develop the next generation communication device such as a cellphone, or medical equipment to monitor patients or help doctors treat illnesses, or a spacecraft that will carry humans to a new planet that can bring clean water to an underdeveloped region, or a new power source that provides clean energy, or a device that can detect toxic agents and chemicals, or a new building that is earthquake safe.

By using mathematical and scientific principles, engineers design and develop new products, processes, and systems that make our lives better. Engineers are at the forefront of cutting-edge technology, who through innovation, creativity, and change provide for our safety, health, security, comfort, and recreation. Becoming an engineer is challenging and rewarding as it requires development of novel solutions to problems that were never solved before.

2.2 Engineering Education

Engineering education can be defined as the activities involved in teaching and learning the principles and practices related to engineering profession. Engineering education is required to become an engineer and offered by institutions around the world.

To become an engineer, students must develop both interest and ability in mathematics and science. This interest can be developed at early stages from elementary and middle school by taking exploratory courses in science, attending science fairs, science competitions, summer camps, and so on. The first step in developing an interest requires students to explore what engineers do and see if they are interested in performing similar activities in their professional lives. Students must also develop necessary knowledge and skills by taking mathematics and science courses in middle school and high school.

Some high schools offer courses such as pre-engineering, drafting, and material science that students should consider taking to prepare them for engineering education. High school students also should consider selecting electives in math and science such as physics, chemistry, computer science, trigonometry, geometry, algebra II, and calculus. These classes not only prepare students for university level courses, they also develop critical thinking skills necessary to succeed in an engineering profession.

Engineering education in the United States, Canada, and other countries follows a set of guidelines developed by stakeholders such as professional societies, engineering educators, and employers of engineering graduates. Engineering educators integrate the input of these stakeholders in the curriculum to provide the knowledge, skills, and abilities necessary for engineering students to succeed in their profession. Several organizations and agencies in different countries and regions develop the guidelines and standards to ensure quality and consistency among different engineering programs. The engineering programs meet the quality standards through an accreditation process. The leading engineering accreditation organization that is recognized around the world is the *Accreditation Board of Engineering and Technology (ABET)* of the United States.

2.3 ABET Accreditation

ABET is recognized as the leading organization that accredits engineering and technology programs in the United States and around the world. Accreditation assures students, parents, and other stakeholders of the program quality in higher education because accredited programs are required to follow a set of guidelines defined by accreditation agencies. Although there are different accreditation agencies in different regions of the world, most of them recognize ABET-accredited programs and accept them.

Earning a degree is a significant achievement and an important investment for every student, especially as the cost and time required for engineering education continually increase.

As students' professional success depends on their educational foundation, the quality of the education can make a significant difference. Earning a degree from an ABET-accredited program

> Verifies that the quality of the educational experience you've received meets the standards of the profession, increases and enhances employment opportunities, permits and eases entry to a technical profession through licensure, registration, and certification, establishes eligibility for many federal student loans, grants, and/or scholarship

In 2015, more than 3400 programs at 700 institutions in 28 countries are accredited by ABET. An ABET-accredited program assures prospective students that

1. The institution is committed to improving their educational experience, and the program is committed to using best practices and innovation in education.
2. The program is guided by its industry, government, and academic constituents through formal feedback.
3. The program considers the students' perspective as part of its continuous quality improvement process.

ABET accreditation can be of great value to students. It is often required for eligibility for federal student loans, grants, and scholarships. Many forms of professional licensure, registration, and certification also require graduation from ABET-accredited programs as a minimum qualification. In addition, many employers, including the federal government, require graduation from ABET-accredited programs to be eligible for employment in certain fields. Also, multinational corporations are increasingly listing graduation from an accredited program as a requirement for employment. ABET accredits programs reviewed by four different commissions:

1. Applied Science Accreditation Commission (ASAC)
2. Computing Accreditation Commission (CAC)
3. Engineering Accreditation Commission (EAC)
4. Engineering Technology Accreditation Commission (ETAC)

ABET accreditation requires meeting seven different criteria: students, program education objectives, student outcomes, continuous improvement, curriculum, faculty, and facilities. Five of the seven criteria are described briefly in the following sections.

2.3.1 Criterion 1: Students

Student performance must be evaluated. Student progress must be monitored to foster success in attaining student outcomes, thereby enabling graduates to attain program educational objectives. Students must be advised regarding curriculum and career matters.

The program must have and enforce policies for accepting both new and transferred students, awarding appropriate academic credit for courses taken at other institutions, and awarding appropriate academic credit for work in lieu of courses taken at the institution. The program must have and enforce procedures to ensure and document that students who graduate meet all graduation requirements.

2.3.2 Criterion 2: Program Educational Objectives

The program must have published program educational objectives that are consistent with the mission of the institution, the needs of the program's various constituencies, and these criteria. There must be a documented, systematically utilized, and effective process, involving program constituencies, for the periodic review of these program educational objectives that ensures that they remain consistent with the institutional mission, the program's constituents' needs, and these criteria.

An example of program educational objectives for a mechanical engineering program is listed as follows:

1. Graduates use engineering principles and standards to develop innovative solutions when they model, analyze, and design mechanical, thermal, and automotive components and systems.
2. Graduates are successfully employed in mechanical engineering and other related professions.
3. Graduates successfully demonstrate ethical, professional, leadership, teamwork, communication, and technical competencies.
4. Graduates engage in lifelong learning demonstrated by professional development, publication, continuing education, and other similar activities.

2.3.3 Criterion 3: Student Outcomes

The program must have documented student outcomes that prepare graduates to attain the program educational objectives. Student outcomes include (a) through (k) below plus any additional outcomes that may be articulated by the program:

a. An ability to apply knowledge of mathematics, science, and engineering
b. An ability to design and conduct experiments, as well as to analyze and interpret data
c. An ability to design a system, component, or process to meet desired needs within realistic constraints such as economic, environmental, social, political, ethical, health and safety, manufacturability, and sustainability
d. An ability to function on multidisciplinary teams
e. An ability to identify, formulate, and solve engineering problems
f. An understanding of professional and ethical responsibility
g. An ability to communicate effectively
h. The broad education necessary to understand the impact of engineering solutions in a global, economic, environmental, and societal context
i. A recognition of the need for and an ability to engage in lifelong learning
j. Knowledge of contemporary issues
k. An ability to use the techniques, skills, and modern engineering tools necessary for engineering practice

2.3.4 Criterion 4: Continuous Improvement

The program must regularly use appropriate, documented processes for assessing and evaluating the extent to which student outcomes are being attained. The results of these evaluations must be systematically utilized as input for the continuous improvement of the program. Other available information may also be used to assist in the continuous improvement of the program.

2.3.5 Criterion 5: Curriculum

The curriculum requirements specify subject areas appropriate to engineering but do not prescribe specific courses. The faculty must ensure that the program curriculum devotes adequate attention and time to each component, consistent with the outcomes and objectives of the program and institution. The professional component must include the following:

1. One year of a combination of college-level mathematics and basic sciences (some with experimental experience) appropriate to the discipline. Basic sciences are defined as biological, chemical, and physical sciences.
2. One and one-half years of engineering topics, consisting of engineering sciences and engineering design appropriate to the student's field of study. The engineering sciences have their roots in mathematics and basic sciences but carry knowledge further toward creative application. These studies provide a bridge between mathematics and basic sciences on the one hand and engineering practice on the other. Engineering design is the process of devising a system, component, or process to meet desired needs. It is a decision-making process (often iterative), in which the basic sciences, mathematics, and the engineering sciences are applied to convert resources optimally to meet these stated needs.
3. A general education component that complements the technical content of the curriculum and is consistent with the program and institution objectives.

Students must be prepared for engineering practice through a curriculum culminating in a major design experience based on the knowledge and skills acquired in earlier coursework and incorporating appropriate engineering standards and multiple realistic constraints. One year is the lesser of 32 semester hours (or equivalent) or one-fourth of the total credits required for graduation.

2.4 Curriculum Mapping

Curriculum mapping is the process of displaying a curriculum to identify and address academic gaps, redundancies, and misalignments for purposes of improving the overall coherence of a degree program to meet student learning outcomes. A curriculum typically includes all courses in a program and other requirements for completion of a degree or certificate program. In general, curriculum mapping refers to the alignment of student learning outcomes and courses required in the program for a student to graduate—that is, how well and to what extent the courses in the program meet the content that students are actually taught with the expected level of achievement for the learning outcomes. Curriculum mapping may also refer to the mapping and alignment of all elements that are entailed in educating students, including assessment of learning outcomes, textbooks, assignments, lessons, and instructional techniques. The best practice in curriculum design is to develop a coherent curriculum with the following characteristics:

1. Well organized and purposefully designed to facilitate learning
2. Contains all required courses and activities with little or no repetitions
3. Courses show good alignment with desired student learning outcomes

The goal of a curriculum map is to ensure that the topics covered in the courses match the learning outcomes as learning outcomes are assessed using the work performed by students in the course. Ideally, as the first step in the process of designing a degree or certificate program, program objectives and student learning outcomes should be finalized. For engineering and engineering technology programs, educators use ABET learning outcomes or develop learning outcomes that meet ABET outcomes. After finalizing the learning outcomes, the second step in the process is to develop courses with content aligned with these learning outcomes. A curriculum map will provide a clear picture with relationship between courses and student learning outcomes.

Several approaches can be used to map a curriculum with a common goal to maximize student learning with a number of specific, measurable outcomes to be assessed. Three commonly used approaches are vertical coherence, horizontal coherence, and subject-area coherence.

> *Vertical coherence*: When a curriculum is *vertically aligned* or *vertically coherent*, what students learn in one topic, subject, or course prepares them for the topics offered in higher level subjects or courses. In some cases, courses are numbered or the course description is used to communicate this logical sequence and structure. For example, a first-year calculus course may have course number as Math 111: Calculus I, and the second-year Math course as Math 211: Calculus II. It can be observed that students must learn the topics in Math 111 to prepare them for Math 211.
>
> *Horizontal coherence*: When a curriculum is *horizontally aligned* or *horizontally coherent*, what students learn in one computer-aided design course mirrors what they are learning in a different computer-aided design course. Curriculum mapping aims to ensure that the assessments, tests, and other methods used to evaluate student learning are based on what was expected from the students and taught in the course.
>
> *Subject-area coherence*: When a curriculum is coherent within a subject area—such as engineering economics, fluid mechanics, or electric circuits mathematics—it may be aligned both within and across different levels of students (i.e., first year, second year, third year, etc.). Curriculum mapping for subject-area coherence aims to ensure that instructors are working toward the same learning outcomes standards in similar courses (say, three different sections of introduction to engineering courses taught by different instructors). Students in all sections of the same course should learn the same amount of content and receive the same quality of instruction, across subject-area courses.

TABLE 2.1

Curriculum Mapping for a Mechanical Engineering Program

	Student Learning Outcomes										
Courses	a	b	c	d	e	f	g	h	i	j	k
Introduction to Engineering	x					x		x	x		
Computer-Aided Design							x				x
Statics	x			x		x					
Mechanics of Materials	x			x		x					
Engineering Materials	x				x						
Fluid Mechanics							x				x
Engineering Economics							x		x		
Heat Transfer							x		x		
Machine Design		x	x								
Fluids Laboratory		x		x							
Engineering Ethics						x		x		x	
Capstone Design		x	x	x		x					x

In general, a curriculum map is organized in a table form with courses listed in the first column and student learning outcomes in the first row with cross references between the courses and outcomes. Table 2.1 shows a sample curriculum mapping for a mechanical engineering program.

2.5 Assessment and Continuous Improvement

Along with curriculum mapping, the program also needs to develop an assessment plan and continuous improvement plan. An assessment plan will describe the process and frequency of assessments that will be performed for each outcome. For example, a program may assess learning outcomes a–c in one semester and d–f during the next semester. Using the information provided in Table 2.1, to assess outcomes a–c, the courses offered during that semester with outcomes a–c will be used. Table 2.1 lists seven different courses that can be used to assess the above three outcomes. It is recommended that each outcome be assessed using multiple courses to ensure quality of the assessment process.

Assessment data must be collected by the instructor teaching these courses. The program faculty may use rubrics for each learning outcome to assess students' work with target achievement levels. Development of rubrics requires performance indicators and levels of performances that can be measured. An example of performance indicators for outcomes a–e is presented in Table 2.2. Table 2.3 shows a sample rubric for student learning outcome *a* that also shows five different levels of performance—missing, emerging, developing, practicing, maturing, and mastering. For example, the program faculty may develop performance criteria that more than 70% of the students in an introduction to engineering course are expected to achieve. To assess learning outcomes, the instructor may select works submitted by students such as homework, examination, and assignments and determine the level of performance for each student in the class using the rubrics.

TABLE 2.2

Student Learning Outcomes and Performance Indicators

Student Learning Outcome	Performance Indicators
An ability to apply knowledge of mathematics, science, and engineering	Application of mathematical and scientific principles to engineering problems Engineering interpretation of mathematics and scientific theories and terms An ability to perform mathematical calculations
An ability to design and conduct experiments, as well as to analyze and interpret data	Conduct experiments logically based on related theories while following safety rules An ability to analyze and interpret collected data using theory An ability to operate instrumentation and testing equipment
An ability to design system component or process to meet desired needs within realistic constraints such as economic, environmental, social, political, ethical, health and safety, manufacturability, and sustainability	Set up a proper design strategy and develop solutions Find solutions using computer tools and engineering resources Support design procedure with documentation and references
An ability to function in multidisciplinary teams	Recognize participants' roles in a team setting and fulfill appropriate roles to ensure team success Integrate input from all team members and make decisions in relation to objective criteria Improve communication among teammates, ask for feedback, and use suggestions
An ability to identify, formulate, and solve engineering problems	Application of engineering principles and scientific concepts to solve engineering problems An ability to formulate strategies for solving engineering problems An ability to identify engineering problems

TABLE 2.3

Sample Rubrics for Student Learning Outcome a

	Learning Outcome a: An Ability to Apply Knowledge of Mathematics, Science, and Engineering					
Performance Indicator	0-Missing	1-Emerging	2-Developing	3-Practicing	4-Maturing	5-Mastering
Application of mathematical and scientific principles to engineering problems	Does not understand the connection between mathematical models and scientific principles to solve problems related to engineering		Chooses a mathematical model and scientific principle but has difficulties in model development		Combines mathematical and/or scientific principles to solve problems relevant to engineering	
Engineering interpretation of mathematics and scientific theories and terms	Theories and terms are not interpreted or interpreted incorrectly		Shows nearly complete understanding of calculus and/or linear algebra but difficulties in interpretation of theory to engineering problems		Shows appropriate engineering interpretation of mathematical and scientific terms	
Ability to perform mathematical calculation	Calculations not performed or performed incorrectly and did not obtain result		Minor errors in calculation and obtains incorrect solution		Executes calculations correctly with appropriate unit and evaluates solution	

If the above performance criteria of 70% are met for an outcome, it can be concluded that the outcome has been achieved. If the performance level is lower than 70%, a continuous improvement plan must be developed and implemented to improve performance. This may require changing the instructional method, spending more time on topics in which students experienced difficulty as demonstrated by their work, or changing the syllabus, textbook, and so on. Even when the performance level is achieved at 70%, the program may still develop a continuous improvement plan to improve the performance further.

2.6 High-Impact Educational Practices

The Association of American Colleges and Universities (www.aacu.org) provided a set of 10 teaching and learning practices (high-impact practices) that have been widely tested and have been shown to benefit college students from many backgrounds. These practices are beneficial especially to historically underserved students, who often do not have equitable access to high-impact learning. Engineering students can also benefit from these practices as they are often challenged with the higher expectations from the courses and programs:

1. First-Year Experiences
2. Common Intellectual Experiences
3. Learning Communities
4. Writing-Intensive Courses
5. Collaborative Assignments and Projects
6. Undergraduate Research
7. Diversity/Global Learning
8. Service Learning, Community-Based Learning
9. Internships
10. Capstone Courses and Projects

The above high-impact practices align closely with ABET student learning outcomes. Common intellectual experiences for engineering students can be assessed by ABET outcome (*a*) *an ability to apply knowledge of mathematics, science, and engineering.* The learning communities and collaborative assignments and projects can be assessed by ABET outcome (*d*) *an ability to function on multidisciplinary teams.* The writing intensive courses can be assessed by ABET outcome (*g*) *an ability to communicate effectively.* Diversity/global learning, service learning, and community-based learning can be assessed by ABET outcome (*h*) *the broad education necessary to understand the impact of engineering solutions in a global, economic, environmental, and societal context.* The internship and capstone courses and projects can be assessed by ABET outcome (*c*) *an ability to design a system, component, or process to meet desired needs within realistic constraints such as economic, environmental, social, political, ethical, health and safety, manufacturability, and sustainability.*

Educational research suggests that the high-impact practices mentioned here increase rates of student retention and student engagement. However, on almost all campuses, utilization of active learning practices is unsystematic to the detriment of student learning.

2.6.1 First-Year Seminars and Experiences

Many schools now build into the curriculum first-year seminars or other programs that bring small groups of students together with faculty or staff on a regular basis. The highest quality first-year experiences place a strong emphasis on critical inquiry, frequent writing, information literacy, collaborative learning, and other skills that develop students' intellectual and practical competencies. First-year seminars can also involve students with cutting-edge questions in scholarship and with faculty members' own research.

2.6.2 Common Intellectual Experiences

The older idea of a *core* curriculum has evolved into a variety of modern forms, such as a set of required common courses or a vertically organized general education program that includes advanced integrative studies and/or required participation in a learning community (see Section 2.6.3). These programs often combine broad themes—for example, technology and society, global interdependence—with a variety of curricular and cocurricular options for students.

2.6.3 Learning Communities

The key goals for learning communities are to encourage integration of learning across courses and to involve students with *big questions* that matter beyond the classroom. Students take two or more linked courses as a group and work closely with one another and with their professors. Many learning communities explore a common topic and/or common readings through the lenses of different disciplines. Some deliberately link *liberal arts* and *professional courses*; others feature service learning.

2.6.4 Writing-Intensive Courses

These courses emphasize writing at all levels of instruction and across the curriculum, including final-year projects. Students are encouraged to produce and revise various forms of writing for different audiences in different disciplines. The effectiveness of this repeated practice *across the curriculum* has led to parallel efforts in such areas as quantitative reasoning, oral communication, information literacy, and, on some campuses, ethical inquiry.

2.6.5 Collaborative Assignments and Projects

Collaborative learning combines two key goals: learning to work and solve problems in the company of others, and sharpening one's own understanding by listening seriously to the insights of others, especially those with different backgrounds and life experiences. Approaches range from study groups within a course, to team-based assignments and writing, to cooperative projects and research.

2.6.6 Undergraduate Research

Many colleges and universities are now providing research experiences for students in all disciplines. Undergraduate research, however, has been most prominently used in science disciplines. With strong support from the National Science Foundation and the research community, scientists are reshaping their courses to connect key concepts and questions with students' early and active involvement in systematic investigation and research.

The goal is to involve students with actively contested questions, empirical observation, cutting-edge technologies, and the sense of excitement that comes from working to answer important questions.

2.6.7 Diversity/Global Learning

Many colleges and universities now emphasize courses and programs that help students explore cultures, life experiences, and worldviews different from their own. These studies—which may address US diversity, world cultures, or both—often explore *difficult differences* such as racial, ethnic, and gender inequality, or continuing struggles around the globe for human rights, freedom, and power. Frequently, intercultural studies are augmented by experiential learning in the community and/or by study abroad.

2.6.8 Service Learning and Community-Based Learning

In these programs, field-based *experiential learning* with community partners is an instructional strategy—and often a required part of the course. Students receive direct experience with issues they are studying in the curriculum and with ongoing efforts to analyze and solve problems in the community. A key element in these programs is the opportunity students have to both *apply* what they are learning in real-world settings and *reflect* in a classroom setting on their service experiences. These programs model the idea that giving something back to the community is an important college outcome, and that working with community partners is good preparation for citizenship, work, and life.

2.6.9 Internships

Internships are another increasingly common form of experiential learning. Internships provide students with direct experience in a work setting—usually related to their career interests—and give them the benefit of supervision and coaching from professionals in the field. If the internship is taken for course credit, students complete a project or paper that is approved by a faculty member.

2.6.10 Capstone Courses and Projects

Whether they are called *senior capstones* or some other name, these culminating experiences require students nearing the end of their college years to create a project of some sort that integrates and applies what they've learned. The project might be a research paper, a performance, a portfolio of *best work*, or an exhibit of artwork. Capstones are offered both in departmental programs and, increasingly, in general education.

2.7 Plan of Study

A plan of study is a comprehensive roadmap for success in completing an academic program. Each undergraduate degree program requires completion of a number of courses from different disciplines. Selection of courses can be difficult due to complexities associated with the prerequisites, offering of the courses, days and times the course is offered, and so on.

For example, BSE mechanical engineering program at University of Michigan–Flint requires completion of 129 credit hours (43 courses) from different areas such as math and science, engineering prerequisites, engineering core, engineering laboratories, and general education. Additional requirements include taking the Fundamentals of Engineering (FE) examination, and a grade point average of 2.5, C or better in all engineering courses.

The primary objective of a plan of study is to familiarize students with program and degree requirements and develop a comprehensive course schedule for each semester. This will provide better understanding of course planning, process, and graduation date. A good plan of study will help students graduate in a timely manner by scheduling sequence of courses based on prerequisites and corequisites.

Another objective is to empower students with information about course, curriculum, and degree requirements so that each student can make both short-term and long-term course planning decisions toward completion of degree requirements. A step-by-step procedure to develop a plan of study for BSE mechanical engineering degree at University of Michigan–Flint is described in Sections 2.7.1 through 2.7.7.

2.7.1 Find Your Catalog Year

Determine your catalog year. Catalog year is typically the academic year you are admitted at University of Michigan–Flint. However, the catalog year can be different as students may update or move forward to a different catalog year for different reasons. It is important to know your catalog year as degree requirement can differ from one catalog year to another. To find your catalog year, you are required to log into student information system of the university using their university student ID and password.

2.7.2 Understand Your Program Requirements

Using the information from step one, determine the program requirements for your catalog year. Program requirements are available in the university catalog, department website, or from an advisor. From the catalog website, select the program requirements for BSE mechanical engineering program for your catalog year.

Catalog years are the same as academic years. For example, 2014–2015 catalog year is for Fall 2014 and Winter 2015. If your first semester at UM-Flint is either Fall 2014 or Winter 2015, your catalog year most likely will be 2014–2015 unless step one shows a different catalog year. After selection of the program, the courses and graduation requirements are displayed. Sample Mechanical Engineering program requirements listed in the university catalog are shown below.

```
Prerequisites. (52 credits).

ENG 112 or EHS 120 (3 cr.).
MTH 121, MTH 122, MTH 222, MTH 303 (16 cr.).
CSC 175 (4 cr.).
CHM 260, CHM 261 (4 cr.).
PHY 243, PHY 245 (10 cr.).
EGR 102, EGR 165, EGR 230, EGR 260, EGR 280 (15 cr.).
Requirements. (54 credits).

Core courses (30 credits).
EGR 310, EGR 315, EGR 330, EGR 350, EGR 353, EGR 356, EGR 370, EGR 432,
EGR 465, EGR 466.
```

```
Laboratory courses (3 credits).
EGR 281, EGR 355; one from: EGR 322, EGR 335, EGR 397, EGR 433.
Elective courses (21 credits).
EGR or AUE courses not already listed as a program prerequisite, core, or
laboratory course at the 300 level or above.
Grades of C (2.0) or better in at least seven of the ten core courses.
A cumulative grade point average of 2.5 or better.
Fundamentals of Engineering (FE) license examination (typically taken
during the final term before graduation).
Completion of at least 129 credits and all requirements of the College of
Arts and Sciences (CAS) Bachelor of Science in Engineering degree,
including General Education requirements.
Minimum Credits For the Degree: 129 credits
```

2.7.3 Determine Prerequisite Requirements for Each Required Course

Most courses have prerequisites except for general education courses or first-year courses at 100 or 200 levels. If a course has a prerequisite requirement, the prerequisite course(s) will be listed in the catalog in the course description area. In general, students must complete the prerequisite course with a grade of C or better before they can register for the course. For example, to register for calculus II, a student must complete calculus I with a grade C or better.

2.7.4 Organize All Materials to Develop Your Plan of Study

Before developing your plan of study, you may organize resources required for developing the plan, such as curriculum from the catalog or department website, sample plan of study for the program, and course descriptions listing prerequisite courses. These resources can be requested from the engineering department faculty or academic advisor.

The curriculum for a program may be different for different catalog years as programs may change requirements and courses from year to year. It is important to use the curriculum for the catalog year for each individual student. A catalog year is typically the year the student is first admitted to the university or college unless the student changes to a different catalog year.

2.7.5 Meet with Your Advisor

After preparing a draft plan of study, make an appointment to meet with your advisor to review and finalize the plan of study. Students must be aware of the advising process at their institution by visiting the academic advising center and the department office that manages the program. Most universities offer advising through a professional academic advisor and department level advisor who can also be a program faculty. It is important to understand the differences in advising in the two processes above.

Academic advisors are knowledgeable in all programs offered at the university and have good understanding of general education courses and requirements. They are full-time advisors and available during office hours, including some evenings and weekends. Students can walk in for advising at the academic advising center or student success center. Academic advisors help students during orientation to select courses during their first semester at the university. Because of the wide range of programs offered by a university, academic advisors are not aware of program specific requirements and updates made at the program level. Academic advisors are great resources for first-year and undecided students who have not selected a major degree program.

Advisors at program levels provide specific advising for students to complete degree requirements in a timely manner. Larger programs usually have professional advisors located in the department office, while other programs provide advising through engineering faculty in the department. Programs in which engineering program faculties are responsible for student advising have limited availability during their office hours. Students are required to make appointments to meet with program level or department advisors several days or weeks in advance. Department level advising is critical for students to develop a clear understanding of degree requirements and prerequisites so they can plan courses for completion of their degree in a timely manner. Once a student chooses a major or degree program, they should seek advising from advisors in the department instead of advisors in the academic advising center. In some programs, students are assigned a specific faculty advisor based on their last name, student ID, or class ranks.

2.7.6 Balance Course Load with Work and Life

Academic plan of study should also require students to balance their work and family lives with academic course loads. In general, students are required to study 3–4 hours per week for each credit hour they register. Some engineering courses may take more time depending on students' background in prerequisite courses and the learning strategies they use. For example, a student enrolled in 12 credit hours of classes is expected to study 36–48 hours per week outside the class to succeed in the course. The following guidelines should be used by engineering students who are working while enrolled in courses. Because of a wide range of family structures and situations, each student should consider the time required to meet family needs and obligations with reduced number of credit hours:

1. Full-time students with more than 12 credit hours (12–18 credit hours) of coursework should not work.
2. Full-time students with 12 credit hours of coursework should work maximum 10 hours per week.
3. Students working 10–20 hours per week should register for 6 credit hour.
4. Students working 20–30 hours should register for 3–6 credit hours of coursework each semester.
5. Students working 30–40 hours should register for maximum 3 credit hours each semester.

2.7.7 Take Courses from Different Areas of the Curriculum

Engineering students should register for courses from different areas such as mathematics, physics, engineering, and general education courses. The common mistake in course selection during first year is that the students take only general education courses without taking mathematics, physics, or engineering courses. Such course selection results in significantly longer graduation time and schedule problems during their later semesters.

Typically, engineering programs require 124–130 credit hours for graduation, of which there are 50–70 credit hours in engineering, 12–15 credit hours in math, 9–12 credit hours in science, and 25–35 credit hours in general education courses (English, humanities,

TABLE 2.4

Plan of Study for a Mechanical Engineering Program

Semester 1	Semester 2	Semester 3	Semester 4
Introduction to Engineering	Computer-Aided Design	Engineering Statics	Mechanics of Materials
Calculus I	Calculus II	Engineering Materials	Thermodynamics and Lab
College Physics I	College Physics II	Materials Lab	Calculus III
First-Year Experience	English Composition	Chemistry I and Lab	Computer Programming
Semester 5	**Semester 6**	**Semester 7**	**Semester 8**
Fluid Mechanics and Lab	Heat Transfer	Manufacturing Process	Engineering Elective 4
Electric Circuits and Lab	Machine Design	Engineering Elective 2	Engineering Elective 5
Differential Equation	Engineering Elective 1	Engineering Elective 3	Global Studies
Dynamics	Engineering Elective 2	Humanities 1	Engineering Lab
English Writing	Social Science 1	Fine Arts	Health and Well-Being
Semester 9	**Semester 10**		
Engineering Elective 6	Capstone Design Two		
Engineering Elective 7	Engineering Elective 8		
Capstone Design One	Social Science 2		
Humanities 2			

social science, global study, fine arts, and so on). General education courses usually do not have prerequisite requirements and students are allowed to choose from a wide variety of courses offered every semester. However, engineering, math, and science courses are more constrained with prerequisite course requirements and are not offered every semester.

Because of these constraints, students should register for engineering, math, and science courses starting their first semester so that they can make progress toward graduation. Students are recommended to register for one or two engineering courses, one mathematics course, one science course, and one general education course during each semester starting from the first semester. Enrollment in mathematics and English courses may require scores in placement examination done by admission or academic advising departments. Table 2.4 shows a sample plan of study for a BSE mechanical engineering program.

2.8 Summary

Engineering is a learned profession that requires problem solving and decision making skills. Engineering students need to develop a good understanding of the roles and responsibilities of an engineer and what engineers do. Both interest and abilities are necessary to prepare and succeed as an engineer. Students can prepare themselves from middle school and high school by taking mathematics and science courses. Engineering education across the world follows a set of accreditation standards to ensure quality. ABET is recognized as the leading organization to set standards for engineering education.

To succeed in engineering education, students are required to review and familiarize themselves with the program or degree requirements. A plan of study must be developed during the first semester and followed as a roadmap to success. The complexities associated with prerequisite courses and other requirements can be effectively planned by students to graduate in a timely manner.

Problems

1. List 11 ABET learning outcomes and identify your level of competencies in each of the outcomes.
2. List and describe 10 high-impact practices. Which of the high-impact practices have you used in your degree program at your institution, and how you are planning for the remaining high-impact practices?
3. Develop a plan of study for your degree program and meet with your advisor to finalize the plan of study.
4. List five recommendations to balance course load with work–life situation.
5. List seven steps in development of a plan of study and describe each of the steps briefly.

Bibliography

Abbott, S. E., Guisbond, L., Levy, J., Newby, D., Sommerfeld, M., and Thomas, B. (August 26, 2014). Hidden curriculum. In S. Abbott (Ed.), *The Glossary of Education Reform*, http://edglossary.org/hidden-curriculum.

Accreditation of Engineering and Technology, *2015–2016 criteria for accrediting engineering programs*, Baltimore, MD, Retrieved October 3, 2015, from http://www.abet.org/accreditation/accreditation-criteria/criteria-for-accrediting-engineering-programs-2015-2016/.

Kuh, G. D. and O'Donnell, K. with Case Studies by Sally Reed (2013). *Ensuring Quality & Taking High-Impact Practices to Scale*, AAC&U, Washington, DC. (For information and more resources and research from LEAP, see www.aacu.org/leap.)

3

How to Learn

CHAPTER OBJECTIVE AND STUDENT LEARNING OUTCOMES

After completing this chapter, students will be able to

1. Define the concept of metacognition or awareness of the learning process.
2. Evaluate their study skills and identify areas for improvement.
3. Describe different learning strategies and which strategies work best for them. [ABET outcome i, see Appendix C]
4. Assess and monitor their progress and develop plans to improve.

3.1 Introduction

Development of appropriate learning strategies is critical to success in higher education as a growing number of engineering and technology students are currently overwhelmed by the numerous topics within a course and the amount of work expected from them. An enduring misconception about academic success is that the amount of time spent studying directly correlates to higher performance by a student. Although there may be some underlying truth in the above statement, a number of issues also surface. The higher expectations from students require them to spend a significant amount of time completing their coursework. However, most students are not aware of the process of learning or how to learn effectively. This chapter will describe how a student can become a better learner by developing learning strategies and using them as appropriate for the context. Metacognitive skills include planning, selecting learning strategies, monitoring progress of learning, and changing strategies as needed to achieve desired objectives. Successful students develop these strategies and use them effectively for different courses and at different levels during their academic life.

The amount of material covered in the classroom sets high expectations from students to learn a wide range of topics in different subjects. This can become overwhelming for students as they may not be able to distinguish and prioritize the topics to spend time on. By developing an individual learning strategy, students can focus on topics they need to study and prioritize time devoted to that topic. Figure 3.1 shows the relationship of students' state of mind with questions about what to learn, how to learn, and why to learn a topic. Students must see a purpose and value to develop motivations about why they

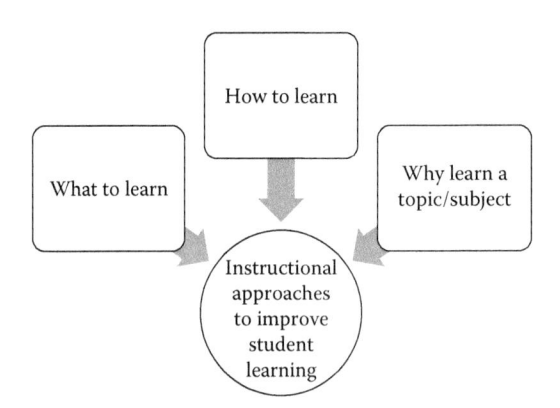

FIGURE 3.1
Students' perception and instructional approaches.

should learn a topic. By considering students' inquiry about the learning process, teachers can develop appropriate instructional approaches to respond to these questions. When instructional approaches respond or align to students' learning needs, the learning process improves.

However, most students do not think about the above three questions. Teachers can embed these questions to develop curiosity and motivate students toward a learning strategy that maximizes their learning process. Metacognition is an important and a useful concept that students can use to become better learners.

3.2 Metacognition

Developing a better understanding of what to learn, how to learn, and which learning strategies are more effective for students can help them become expert learners. There are two types of learners: novice and expert. A novice learner can become an expert learner by using metacognitive strategies.

Novice learners usually neither evaluate their knowledge and comprehension of the topic nor examine the quality of their work. They are classified as surface learners because they do not examine a topic in depth. When these students are asked about a reading assignment, their reply is "I just glanced over the topic." They also do not see any relevance of the topic or try to make connections between the topic and application in their life.

Expert learners continuously evaluate their knowledge and comprehension, why they fail to comprehend, and how they need to redirect their efforts. An expert learner will continue to study the material until he/she develops a deeper understanding of the topic. Expert learners always make connections between topics and their own lives or the world around them. They can distinguish between their level of understanding of different topics and focus their energy on what they do not understand at a higher level (Figure 3.2).

Metacognition: Metacognition refers to the planning, self-regulating, and monitoring of one's cognitive performance. These behaviors are both affective and cognitive in nature. Cognitive issues include understanding about oneself as a learner and the process of learning. Affective components deal with one's emotions, feelings, and beliefs involved in the learning process (Schunk 1996).

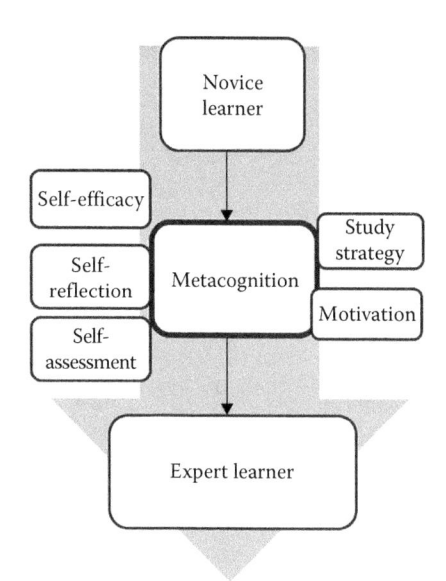

FIGURE 3.2
Use of metacognition to become an expert learner.

Metacognition includes strategies such as problem solving, achievement, motivation, and tactics. The most commonly used techniques are note taking, concept mapping, acquisition procedures, modeling, and so on.

Study strategy: A study strategy is an instructional tool and technique designed to assist students to prepare for content-specific learning. Study strategy instruction provides students with a means to foster analysis for each learning situation in terms of self and context and then develop appropriate actions to improve their own learning (Goetz and Douglas 1991).

3.2.1 Motivation

One of the important elements of learning is motivation as related to what is to be learned. Perseverance, drive, desire, and need determine the direction and efforts toward learning. Motivation affects the amount of time students are willing to devote toward learning. In general, students are motivated to improve their performance because of extrinsic rewards and punishments. Students also work hard for intrinsic reasons, such as finding the subject interesting and enjoyable. Examples of intrinsic motivation factors include challenge, curiosity, competition, reward, and recognition.

Challenges and tasks assigned must be at an appropriate level of difficulty. Easy assignments may bore students and difficult assignments may result in frustration. It is critical to find the balance between these two as students in a class may come from different backgrounds with a wide range of academic preparations. In some courses, prerequisite courses are required so that instructors have a better understanding of the academic preparation and background knowledge possessed by students.

3.2.1.1 Self-Efficacy

Self-efficacy is defined as a personal judgment regarding one's own performance capabilities in a given domain of activity that has diverse effects on activities. As a result, self-efficacy influences such actions as the selection of academic activities, courses, task avoidance, motivation and performance levels, and level of effort expended (Bandura 1986). For example, a student may feel inadequate dealing with quantitative academics such as mathematics. This negative belief may elicit modes of avoidance and thereby block avenues to remedy this negative and anxious feeling about mathematics.

Students with high self-efficacy are more likely to undertake challenging tasks, persist longer, and be more successful than students with low self-efficacy (Harrison et al. 1997). To increase the self-efficacy of students, attributes such as advance organizers, ability formations, attending, help seeking, and higher order of thinking must be integrated in their learning process.

3.2.1.2 Self-Reflection

Self-reflection can be compared to *looking at yourself in a mirror* and seeing through your image—consciously evaluating your strengths and weaknesses in a particular area of interest. It is the ability of individuals to learn more about themselves, their knowledge, behavior, purpose, and essence. For example, students can assess their learning by asking questions about what they already know and what they need to learn to become successful. This can be determined by taking sample examinations on a subject or topic so that strengths and weaknesses are identified.

> Study skills really aren't the point. Learning is about one's relationship with oneself and one's ability to exert the effort, self-control, and critical self-assessment necessary to achieve the best possible results—and about overcoming risk aversion, failure, distractions, and sheer laziness in pursuit of REAL achievement. This is self-regulated learning.
>
> **Linda Nilson**
> *Creating Self-Regulated Learners: Strategies to Strengthen Students Self-Awareness and Learning Skills, Preface, XXVII*

3.2.1.3 Self-Assessment

Self-assessment of students' own abilities, strengths, and weaknesses can motivate them to become a better learner as they become motivated to improve their areas of weakness. Students deliberately engage in the learning process by thinking about what they are learning, why they are learning, how they are learning it, and how effective is their own learning method. These learning strategies depend on students' own abilities and levels of motivation.

Use of self-assessment methods and their success depend on goal setting and guided practice with assessment tools. Goal setting can help students evaluate their current level of performance and their progress when they have a target level of performance to measure the gap. This increases the motivation to learn as they have self-defined, specific, measureable, and relevant learning goals. Instructors must help students in developing short-term, realistic measurable, and attainable goals. For example, a student may set a goal to earn an A in a computer-aided design course as the student's grade in previous courses is at a grade level of B. The instructor and student may evaluate how the student will improve his/her study habit, effort, and strategies to improve performance levels. The student may consider studying topics before they are presented in class, working out more problems than assigned, actively participating in classroom discussion, working in teams, taking sample tests before the exam to identify areas of strength and weaknesses, seeking input from the instructor about topics student needs to understand better, and so on. The student will then continuously monitor his/her progress through homework grades, midterm exams, and quizzes to see whether the performance is on target for the grade of A.

3.3 Types of Knowledge

Awareness of three types of knowledge is essential to develop students' metacognitive abilities: declarative, procedural, and conditional.

1. *Declarative knowledge* is the fact that can be expressed in written or spoken forms. An example of declarative knowledge is knowing the formula for calculating force (force = mass × acceleration).
2. *Procedural knowledge* is the knowledge of a step-by-step process of performing a task, that is, how to do something. For example, how to calculate velocity if you know speed and time. If a car is moving at 70 km per hour, how much distance will the car travel in 3 hours and 15 minutes?
3. *Conditional knowledge* is the knowledge about when to use a procedure, skill, or strategy; why a procedure works and under what conditions; and why a procedure is better than others. For example, engineering economics problems involving selection of different alternatives can be solved using formulas, interest factor tables, or spreadsheets. Some problems can be solved better using formulas than interest factor tables.

3.4 Study Skills/Metacognitive Learning Strategies

Metacognitive strategies help students become better learners. Among various strategies, the following were found to be more effective and useful to students. Teachers can also teach these to their students so that appropriate strategies can be used based on the content and context of the learning environment.

1. Organize/plan:
    a. Set specific, measurable, attainable, realistic, and timely (SMART) goals.
    b. Develop a plan to perform the activities of tasks to accomplish the goals.
2. Manage learning:
    a. Determine how you learn most effectively.
    b. Create environments that help you learn.
    c. Focus attention on topics to be learned and practice learning topics.
3. Monitor progress:
    a. Check your understanding of the subject.
    b. Check progress on your level of understanding and comprehension of the material.
    c. Continue your study until you achieve a higher level of understanding of the subject.
4. Evaluate:
    a. Self-assess your level of accomplishment of your learning goals.
    b. Assess how effective the learning strategies were in accomplishing your goals.

5. Use prior knowledge:
    a. Explore what you already know that is related to the subject or topic.
    b. Relate your prior knowledge to a new topic by relevance and making associations.
6. Make inferences:
    a. Develop strategies to infer from the materials by deep learning.
    b. Understand the underlying meaning by reading between the lines.
7. Make predictions:
    a. Predict new information that may appear.
    b. Make assumptions about what may happen.
8. Apply knowledge:
    a. Apply what you have learned to new situations.
9. Organize knowledge:
    a. Develop metaphors or create an image to understand or represent information.
    b. Develop mental rules and patterns to remember the material learned.
10. Take notes:
    a. Write down important information.
    b. Highlight information in the book.
    c. Paraphrase information in your own words in writing that helps you understand the subject.
11. Use graphic organizer:
    a. Develop a visual representation of the subject or topic using concept maps, flowcharts, Venn diagrams, charts, and so on.
    b. Understand the relationship between multiple concepts and topics in graphical form.
12. Summarize:
    a. Summarize what you have learned in mental, oral, and written forms using keywords.
    b. Rank your level of understanding of different topics as low, medium, or high.
    c. Identify topics with low levels of understanding.
13. Improve:
    a. Focus on topics with low levels of understanding and high levels of importance.
    b. Repeat topics until higher levels of understanding develop.
    c. Assess level of learning and continue the process to next topics.

3.5 Learning Styles

Every student learns differently as he/she has dominant learning styles and techniques. Some students may use different learning styles in different subjects and circumstances. Learning styles can be developed and less dominant styles can be improved as different subjects may be learned more effectively using different learning styles. Most students do not know or understand their own learning styles. By recognizing individual learning styles, students can use these techniques more effectively and may improve their speed and learning quality.

Seven different learning styles have been recognized by researchers and are described as follows:

1. *Visual (spatial)*: Students prefer learning by using picture, images, graphs and charts, and spatial understanding.
2. *Aural (auditory–musical)*: Students prefer learning using sound and music.
3. *Verbal (linguistic)*: Students prefer using words, texts, and sentences, both in speech and writing.
4. *Physical (kinesthetic)*: Students prefer using their body, hands, and sense of touch and feeling.
5. *Logical (mathematical)*: Students prefer using logic, reasoning, mathematics, and systems.
6. *Social (interpersonal)*: Students prefer to learn in groups, team, or with others.
7. *Solitary (intrapersonal)*: Students prefer to work alone and use self-study.

Each learning style uses different parts of the brain. Students can remember and learn more by involving different parts. Most engineering students prefer visual, physical, logical, and solitary styles as these are emphasized in the subjects they take during their study. Engineering students must be trained to develop social learning styles so that they can effectively work in teams to solve complex problems.

3.6 Teaching Styles

As individual students have learning style preferences, teachers also have teaching style preferences based on their personality, background, and institutional context. It is important that both teachers and students understand different teaching styles as alignment of student learning styles, and teaching styles are important for a positive learning environment. Although individual students and teachers have a dominant style of preference, flexibility toward other styles may enhance the overall learning experience. Sometimes, courses and topics can be better learned through specific learning and teaching styles.

Teaching styles can be classified as four different categories: formal authority, demonstrator or personal model, facilitator, and delegator. There are advantages and disadvantages associated with the above styles as described below:

1. *Formal authority*: This teaching style focuses on content and is primarily teacher centered. Teachers feel responsible for controlling and providing the course materials and contents. Students can expect to receive formal instruction and materials from the teacher. Teachers with this style are not interested in student participation in the class and communicating with students outside the classroom. They believe that students should be at the receiving end of the instruction and do not value input from students in the learning process.

2. *Demonstrator or personal model*: Teachers with this style emphasize demonstration and modeling in the classroom and run teacher-centered classes. After a classroom demonstration of skills and processes, they act as a role model or coach to students in developing these skills. They will work through a problem in the class and encourage students to independently solve similar problems using the methods demonstrated.

   This teaching style encourages participation by students in learning the materials presented in the classroom. Students are expected to ask for help to improve their learning and take responsibility for their own learning.

3. *Facilitator*: This style of teaching focuses on activity-based and student-centered learning. Students are expected to take responsibility and initiative for completing the assigned tasks. This style is suitable for independent learners who can actively participate and work in teams with other students. Facilitators require team work, group assignments, and projects that require students to apply knowledge learned in the subject in new situations in creative ways.

4. *Delegator*: Delegators provide a high level of responsibility on individual students or group of students for learning. Delegators believe that empowering students in the learning process may enhance effectiveness in learning. Students typically work individually or in groups with high levels of motivation to accomplish assigned tasks in a timely manner. Teachers also assign independent study or special topics outside the classroom or textbook and require students to develop solutions through research or investigation.

3.7 Time Management

Time management is probably the most challenging task for a student as most students are challenged with balancing their workload for class assignments, homework, jobs, family, and personal time. In general, students are not aware of how they use their time, and do not know how to organize and prioritize their activities to succeed in their courses. The first step in developing time management skills is to investigate and make a list of activities they perform regularly or how they spend their time each day. The following seven steps may help students develop good time management strategies and study habits:

Step 1: Make a list of activities you do every day, including both academic and non-academic tasks. Develop and plan a block of time for study, break, dinner, and so on. Find out the most productive time of the day for yourself such as morning, afternoon, and night (e.g., 8:00 a.m.–10:00 a.m.: prepare for engineering design class, 10:30 a.m.–11:45 a.m.: attend design class, 12:00 p.m.–1:00 p.m.: lunch, 1:00 p.m.–2:15 p.m.: calculus class, 2:30 p.m–3:45 p.m.: physics class, 4:00 p.m.–5:00 p.m.: group study, 5:00 p.m.–6:00 p.m.: dinner, 6:00 p.m.–7:30 p.m.: exercise in gym or play games, 8:00 p.m.–11:00 p.m.: complete homework, reading assignment, study for tomorrow's class, etc.).

Step 2: Identify a dedicated space for study that is free from distractions, such as cell phone, email, or text messaging, where you can concentrate on studying only. In addition to your regular study space, there should also be a backup space such as quiet study areas in the library, department study rooms, and a coffee house. The study space must be carefully chosen so that you can maximize your efficiency in the study of your subject.

Step 3: *Weekly review* of your assignments, notes, and calendar for next week. Select a specific day and time for weekly review such as Saturday night when you will review the exam dates, assignment due dates, deadlines for projects, and group assignments.

Step 4: *Prioritize* assignments, homework, and exam preparation activities. Check the syllabus and additional information provided by the instructor regarding items due next week and work on them. Prioritize based on complexity, due date, and amount of estimated time required for the task. For example, exam preparation should be given higher priority than assignments as exams carry more points for the class and usually take more time to prepare. Assignments that are difficult, take longer, and are due earlier in the week should be priority. Students have a tendency to avoid subjects or assignments that are difficult or time consuming. This behavior must be changed as it may create severe problems with not being able to complete the assigned work in a timely manner resulting in poor performance in the class. Table 3.1 shows a suggested way to prioritize tasks each week.

Step 5: *Accomplish tasks* in a timely manner so that when you start working on an assignment, you can finish it and start the next one. Stopping at the middle of a task and switching between assignments may refrain you from completing the required tasks and could result in poor quality work. If you are experiencing difficulty with an assignment, seek help from outside resources, such as peers, tutoring center, teaching assistant, and course instructor. If you do not complete a task, you did not accomplish much. Make a schedule with time you may need to spend for an assignment and track your progress every hour to see if you are on schedule. You may need to postpone other activities until you have completed your assignment. Do not let low-priority or nonacademic activities distract you from completing your assignment.

Step 6: *Utilize resources* available to help you with your assignment and other tasks. You may have a friend in the class who understands the subject better, there may be a tutor available, or you can use reference books, Internet, and course instructor to help you. Your institution may have a writing center to help you with writing

TABLE 3.1

Weekly Priority of Academic Activities

Type of Activity	Priority
Examination	1
Quiz	2
Homework/assignment (difficult and/or require longer time/due early in the week)	3(a)
Homework/assignment (moderate level of difficulty and/or require moderate time/due middle of the week)	3(b)
Homework/assignment (less difficult/or require less time/due later in the week)	3(c)
Reading assignment for next week	4
Review all remaining tasks for the semester	5
Research/work on the final paper or project	6

assignments, and engineering, mathematics, or physics tutoring centers located in the department or centrally located in the academic advising center. Investigate and become familiar with the resources that can help you succeed. Even if you do not need help, it is a good idea to visit these centers and ask them to check your assignment for improvement. Most of these centers are operated by senior students who took the class before and may provide information about how to study the materials efficiently and suggest topics to focus for examination preparation.

Step 7: Review notes after the class and before the next class to prepare for learning. Reviewing notes will help you identify the topics you may not understand clearly and need to ask the instructor about during the next class. Asking questions in class demonstrates your interest in the subject and your involvement in the learning process. By asking questions, you are not only helping yourself, but also helping other students in the class and the instructor as the instructor needs to know which topics to cover during the next class.

3.8 Bloom's Taxonomy

Bloom's taxonomy of the cognitive domain describes what educators want students to learn in a hierarchical arrangement from lower level of complexity to higher level of complexity. The levels are developed in such a way so that one level (lower level) must be mastered before reaching the next level (higher level). The original six levels were ordered as knowledge, comprehension, application, analysis, synthesis, and evaluation. Knowledge is the lowest level and evaluation is the highest level of learning. The taxonomy was updated by cognitive psychologist Lorin Anderson in the 1990s to reflect relevance to the twenty-first century by changing from nouns to verbs associated with each level. The levels in the new taxonomy are remembering, understanding, applying, analyzing, evaluating, and creating as presented in Figure 3.3. The original and new taxonomies are presented in Table 3.2 with verbs and behaviors for each of the six levels.

Bloom's taxonomy can be effectively used in the learning process by using the verbs in assessing the level of understanding. Typically, first- and second-year courses focus on

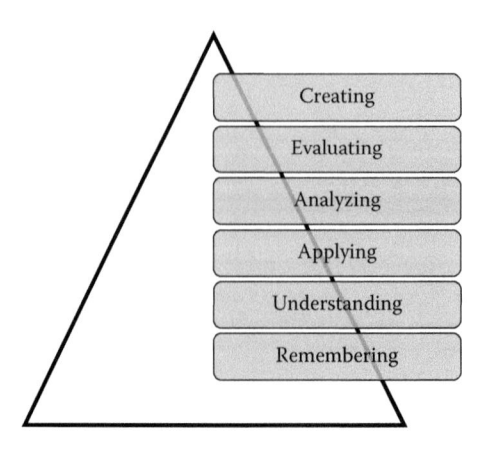

FIGURE 3.3
Bloom's taxonomy (new).

TABLE 3.2
Blooms Taxonomy with Verbs and Behaviors

Learning Levels New (Original)	Definitions	Verbs	Students' Behavior
Remembering (Knowledge)	Can the student recall or remember the information?	List, Define, Name, State, Recall	The student will define all ABET learning outcomes
Understanding (Comprehension)	Can the student explain ideas or concepts?	Explain, Describe, Discuss, Identify, Paraphrase, Classify	The student will explain five ABET learning outcomes
Applying (Application)	Can the student use the information in a different way than learned?	Choose, Demonstrate, Illustrate, Interpret, Operate, Schedule	The student will develop a plan of study for completion of all degree requirements
Analyzing (Analysis)	Can the student distinguish between different parts or topics learned?	Compare, Contrast, Differentiate, Distinguish, Test	The student will compare and contrast the cognitive and affective domains
Evaluating (Synthesis)	Can the student justify a stand or decision?	Argue, Appraise, Defend, Select, Evaluate, Support	The student will design a classification scheme for writing educational objectives that combine the cognitive, affective, and psychomotor domains
Creating	Can the student create a new product or view?	Design, Develop, Formulate, Recommend, Assemble, Create	The student will judge the effectiveness of writing objectives using Bloom's taxonomy

lower levels such as remembering and understanding. As the student progresses toward upper level courses, higher levels of understanding such as analyzing, evaluating, and creating must be emphasized.

3.9 Learning Pyramid

Lecture has been used as the primary mode of instruction at different levels from elementary, high school, and higher education institutions. The lecture model has been used in the last few centuries where the teacher plays an authoritative role as the repository of knowledge. The role of the teacher in the lecture model is viewed as a one-way delivery of knowledge where students are the receivers of knowledge. Lower levels of participation in the classroom with little or no interaction between the teacher and students result in lower effectiveness in the learning process.

The learning model requires a change from a lecture-based model to an active learning model appropriate to prepare students for the competitive global environment. In addition to knowledge gained through lecture, current students need skills in their discipline as well as teamwork, communication, and solving contemporary social and global issues.

Lectures were based on passive reception of knowledge as part of a didactic (instructional) and prescriptive instructional model. To prepare students for future challenges, the learning model needs transformational changes toward activity-based, participative, problem-based, and life-long learning models.

The learning pyramid in Figure 3.4 shows a list of alternate methods to lecture that have higher student retention rates. The learning pyramid shows the highest levels of retention of knowledge by teaching others, practice doing, and discussion. These can be accomplished through group study and discussions outside the classroom. Teachers can incorporate group assignments and discussions in the syllabus to encourage students to use these methods. Discussion can also be facilitated by asynchronous, online discussion questions posted by the teacher where each student needs to participate and respond to the questions related to the subject or topic learned.

Lectures are essential components for in-class learning as teachers typically present new materials to students. Integrating audiovisual, concept demonstrations, and discussion during class can greatly improve student learning and retention. One effective method used by teachers is using the above techniques between lectures. For example, the class may have 15–20 minutes of lecture on a topic followed by demonstration, discussion, and then lecture of another topic. At the end of the class, assessment of student learning may provide valuable information of the level of understanding of the material by students.

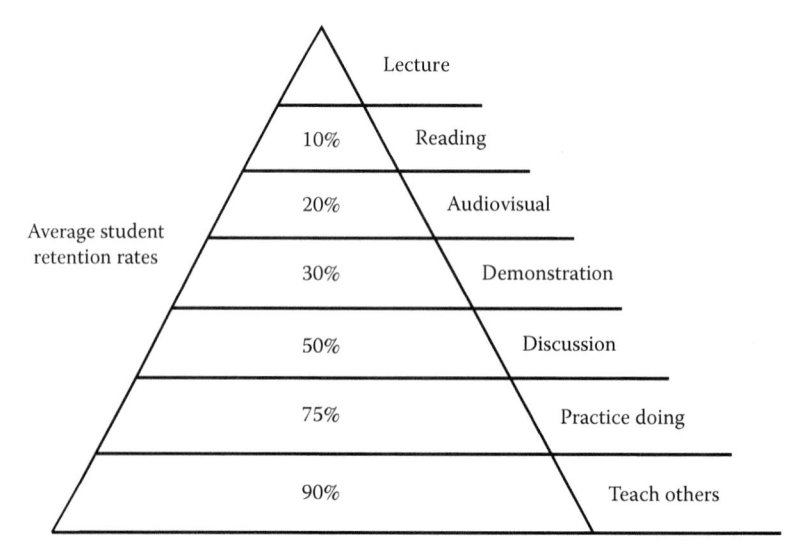

FIGURE 3.4
Learning pyramid. (Courtesy of National Training Laboratories, Bethel, ME.)

3.10 Skill-Based Learning

As engineering and engineering technology professions have globalized during the twenty-first century, employers are demanding increased level of skills from engineering graduates. Some skills required to succeed in engineering jobs cannot be effectively learned through a lecture-based model. Skills are better learned through an instructional approach that requires one-to-one learning by teachers, peers, and tutors. The DEDICT (Demonstrate, Explain, Demonstrate slowly and repeatedly, Imitate, Coach, and Test) approach has been used successfully by military and coaches to train sports teams since the prehistoric time.

The DEDICT method requires development of a set of tools such as forms, worksheets, checklists, flowcharts, concept maps, action matrix, and research reports. In this method, the teacher demonstrates the outcome each student should be able to achieve at the end of class. After the demonstration, the teacher explains why the skill, as well as each step in the process, is important. The demonstration is repeated slowly to allow each student to pay close attention to the details of each step so that they can grasp the steps. Students are asked to practice the steps and imitate the instructor until they feel confident, adequate, and expert. The teacher continues to coach the students to fine-tune their skills until all students develop expertise and high levels of confidence in their skill. The final step in the process is to evaluate whether the students can demonstrate their skills at the same level as the teacher. At this stage, students become self-learners and correct their individual weakness in the skills to become experts.

Although skill-based learning has been widely used in technology and other programs, the DEDICT method can be adapted to undergraduate engineering education to develop skills in problem solving and decision making.

3.11 Summary

This chapter describes different learning approaches, strategies, and tools that can improve student learning. The concept of metacognition and how different metacognition strategies can be used to become an expert learner are described in detail. A novice learner can become an expert learner by using study strategies, motivation, self-efficacy, self-reflection, and self-assessment tools. Student motivation can be improved by developing answers of what to learn, why to learn the topic, and how to learn the topic. Awareness of three types of knowledge—declarative, procedural, and conditional—is essential as they may be used based on the content of the topic.

A list of learning strategies such as organize, monitor progress, evaluate, and take notes is presented to help students become a better learner. Seven different learning styles (visual, aural, verbal, physical, logical, social, and solitary) and four different teaching styles (formal authority, demonstrator, facilitator, and delegator) are described. Both students and teachers need to be flexible from their preferred styles as alignment of learning styles to teaching styles based on the topic. A learning pyramid with alternate teaching styles to lecture is included at the end the chapter. These alternate styles appear to be more effective in student retention of knowledge.

Problems

1. Describe the differences between a novice learner and an expert learner.
2. Explain and illustrate how a novice learner can become an expert learner.
3. List 10 metacognitive learning strategies with a brief description of each of them.
4. Evaluate the learning strategies from the 10 strategies listed in the chapter. Which strategies have been effective for you and which new strategies will you use to improve your academic performance?
5. List seven learning styles. Which one is your dominant learning style? Which style do you think you need to learn for different subjects?
6. Define seven time management steps for academic success. Which of them did you use in the past? Which steps are you planning to try from this point forward?
7. Differentiate between declarative, procedural, and conditional knowledge with examples from your own experience (do not use the examples from the chapter).
8. Compare different levels of Bloom's taxonomy and differentiate among the levels with examples.

References

Bandura, A. (1986). *Social Foundations of Thoughts and Actions: A Social Cognitive Theory*, Prentice Hall, Englewood Cliffs, NJ.

Goetz, E. T. and Douglas, J. P. (1991). The role of students' perceptions of study strategy and personal attributes in strategy use. *Reading Psychology*, 12(3), 199–217.

Harrison, A. W., Ranier, P. K., Hochwarter, W. A., and Thomson, K. R. (1997). Testing self-efficacy—Performance linkage of social cognitive theory. *The Journal of Social Psychology*, 137(1), 79–87.

Huitt, W. (2011). Bloom et al.'s taxonomy of the cognitive domain, *Educational Psychology Interactive*, Valdosta State University, Valdosta, GA. Retrieved from http://www.edpsycinteractive.org/topics/cognition/bloom.html [pdf]. Dated October 8, 2015.

Peirce, W. (2004). Metacognition: Study strategies, monitoring, and motivation. A greatly expanded text version of a workshop, Prince George's Community College, Largo, MD, November 17.

Rampp, L. C. and Guffey, J. S. (1999). The impact of metacognition training on academic self-efficacy of selected underachieving college students, PhD dissertation, Arkansas State University, Jonesboro, AR.

Schunk, D. H. (1996). *Learning Theories: An Educational Perspective*, Merrill, Englewood Cliffs, NJ.

Seven learning styles. Retrieved from http://www.learning-styles-online.com/overview/. Dated May 24, 2015.

The National Academies of Sciences, Engineering, and Medicine. (2000). *How People Learn: Expanded Edition*, National Academies Press, Washington, DC.

Time management. Retrieved from http://www.studygs.net/timman.htm. Dated May 23, 2015.

Section II

Engineering Fundamentals

4

Computer-Aided Design

CHAPTER OBJECTIVE AND STUDENT LEARNING OUTCOMES

After completing this chapter, students will be able to

1. Define fundamental concepts and terminologies used in computer-aided design.
2. Create engineering drawings using computer-aided design software. [ABET outcome g, see Appendix C]
3. Demonstrate competencies with solid modeling, different views, units, and drawing standards. [ABET outcome j, see Appendix C]
4. Modify, update, and analyze drawings using different tools and techniques. [ABET outcome j, see Appendix C]

4.1 Introduction

Computer-aided design (CAD) is the use of computer programs to develop two- or three-dimensional (2D or 3D) graphical representations of physical objects (Figure 4.1). CAD software is used by architects, engineers, drafters, artists, and others to create precision drawings or technical illustrations. The software performs calculations for determining an optimum shape and size for a variety of products and industrial design applications. CAD is also used throughout the engineering process from conceptual design and layout of products, through strength and dynamic analysis of assemblies, to the definition of manufacturing methods. This allows an engineer to both interactively and automatically analyze design variants, to find the optimal design for manufacturing while minimizing the use of physical prototypes.

4.2 Advantages of CAD

CAD software may be used to reduce product development costs, increase productivity, improve product quality, and save time. The objects and features developed in a CAD

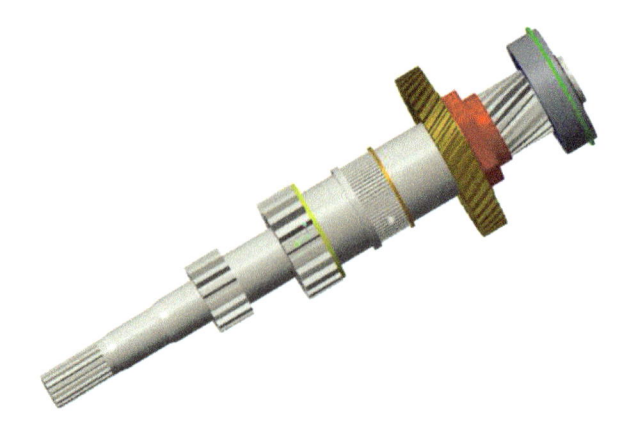

FIGURE 4.1
CAD of a power transmission shaft.

original part were created. You can consider this as being a *perfect world* representation of the component.

If a design engineer develops an assembly of a part, the manufacturing engineer or user needs to assemble it according to the design engineer's intent. Assembling the part from an arbitrary point without carefully considering how the part will be used or assembled may pose challenges during assembly, operation, and maintenance of the part.

3D modeling software may be parametric or direct (nonparametric). *Parametric* modeling enhances seamless update of the overall design and makes it easier to reuse parts and assemblies as well. During the modeling process, when a parameter of a part, such as a dimension, is altered, the entire model will be updated incorporating the change made to a part in the assembly. Parametric modeling helps establish an association between parts, drawings, and assemblies and helps with the manufacturing process. The associativity between parts and their drawings makes the modeling process robust. All the dimensions related to the part are translated in the drawings of the parts through a set of relationships and constraints. Additionally, changing a dimension of a feature such as a hole diameter can be performed from the model tree, which keeps track of every step the engineer performed during the modeling process. Therefore, a change in a hole diameter or the shape of an extruded feature will be immediately reflected in the drawing as well as the entire assembly. Nonparametric modeling, on the other hand, is dependent on the geometry of a form. Rules and constraints are not influential and the nonparametric approach does not store the history of modeling steps. Some of the benefits of CAD are listed as follows:

Visualization: Ability to visualize the final product, subassemblies, and constituent parts in a CAD system speeds the design process. The CAD software helps the designer in visualizing the final product that is to be manufactured. The product can also be animated for better representation of how the functionality of the product is going to perform, thus helping the designer to immediately make modifications if required.

Error: CAD software offers greater accuracy, thereby reducing errors and leading to better design. CAD improves the quality of the design. With CAD software, designing professionals are offered a large number of tools that help in carrying out a thorough analysis of the part using simulations. Among the most common types of simulation, testing for response to stress and modeling the process by which a part might be manufactured or the dynamic relationships among system parts. In stress tests, model surfaces are shown by a grid or mesh that distorts as the part comes under simulated physical or thermal stress. Additionally, simulation allows an engineer to visually inspect the machining process, and catch costly tool gauges and collisions before the parts reach the computer numerical control machine. Simulation also provides detailed information about the toolpath, cycle times, part deviation analysis, the ability to create simulation presentations, and so on.

Database: CAD software helps in better documentation of the design, fewer drawing errors, and greater legibility. The documentation of designing includes geometries and dimensions of the product, its subassemblies and its components, material specification for the components, bill of materials (BOMs) for the components, and so on. The modeled parts, drawings, and assemblies must be stored in a secure place (database). All the data used for designing can easily be saved and used for the future reference; thus, certain components do not have to be redesigned. Similarly, the drawings can also be saved, and any number of copies can be printed or electronically shared with the involved parties anytime required.

4.3 CAD Software

Although various 3D modeling CAD softwares are available for commercial and educational use, such as Creo Parametric, AutoCAD, SolidWorks, 3D Studio Max, and Maya, we will be using Google's SketchUp Pro for demonstration purposes in this chapter.

4.3.1 Installing Google SketchUp Pro

- Type www.sketchup.com/download into your address bar to see the download page.
- Select "Educational Use" from the drop-down menu.
- Enter your email address, first name, last name, and school name.
- Select platform (notice I used Windows' operating system).
- Select "SketchUp Pro 2015" (or equivalent) to select the software to download.
- Accept the license agreement and hit "Download" (Figure 4.2).
- Locate the downloaded file (usually it is downloaded to your "Downloads" folder) and click "Run" to start the extraction process (Figure 4.3).
- Once the extraction is complete, accept the License Agreement and click "Next" (Figure 4.4).
- Select a desired location to install the software (such as c: drive) and click "Install."

4.3.2 Starting SketchUp Pro

To start SketchUp Pro, double click the program icon on the desktop or find SketchUp Pro from the installed programs in the start menu. Starting the program will take you to the welcome screen, where you can select templates for the 3D modeling exercise. Choose template (click on the first selection and then "Start using SketchUp") (Figures 4.5 and 4.6).

The working space in Figure 4.7 contains various tools on the toolbar at the top of the screen and a large work area, where all of your sketching and modeling will take place. Note that

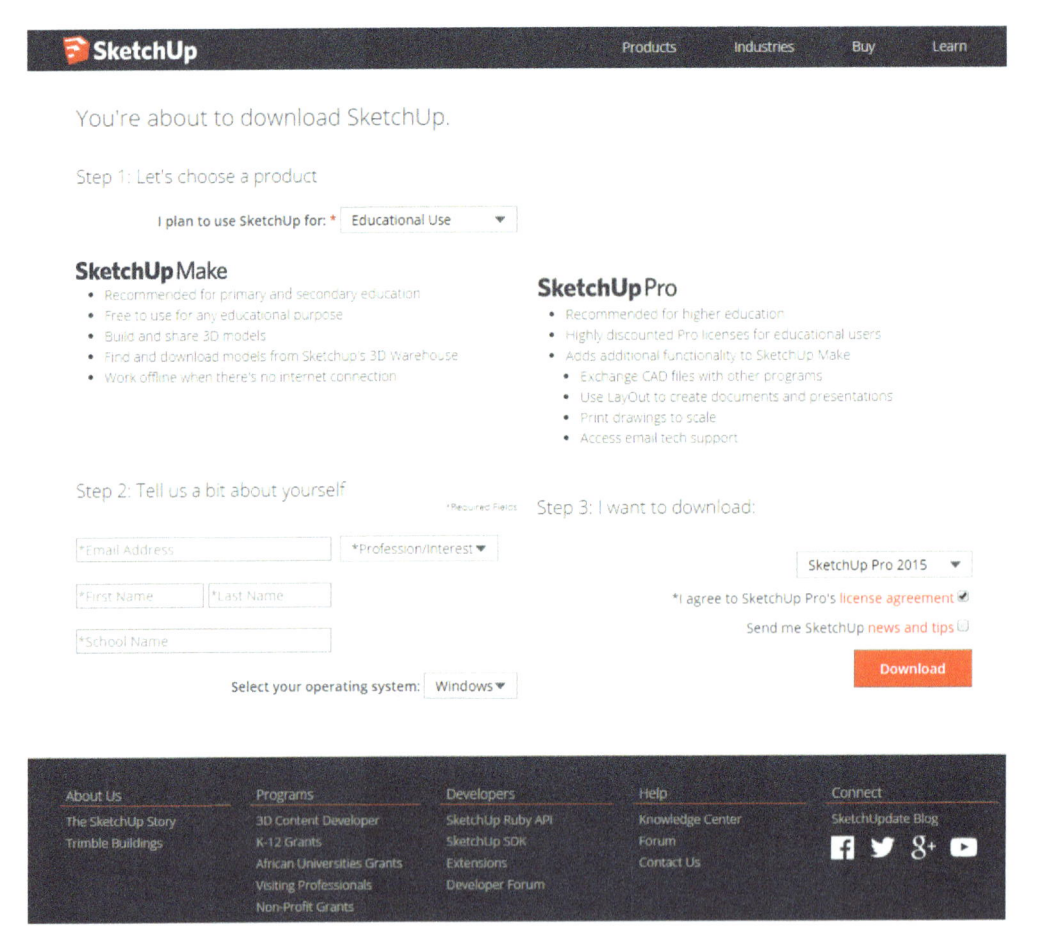

FIGURE 4.2
Download screen of Google SketchUp Pro 2015.

the number of tools on your interface might be different than the one below. You can always add/remove the tools by customizing the toolbar. To customize SketchUp Pro Toolbar, simply right click on any gray area and select the tool set you are interested in (Figure 4.8).

4.3.3 Toolbar

SketchUp Pro Toolbar contains the most frequently used tools (Figure 4.9). The number and type of tools might change during the modeling process because of the design intent. The toolbar can be customized by right clicking on the empty area on the toolbar and selecting desired tool sets. Some of the most frequently used tools for the Bracket tutorial are as follows:

Principal Toolbar: The buttons on the Principal Toolbar activate the Select tool, Make Component, Paint Bucket tool, and Eraser tool (Figure 4.10).

Edit Toolbar: The Edit Toolbar contains geometry modification tools. The tools on this toolbar are the Move tool, Push/Pull tool, Rotate tool, Follow Me tool, Scale tool, and Offset tool (Figure 4.11).

Computer-Aided Design   61

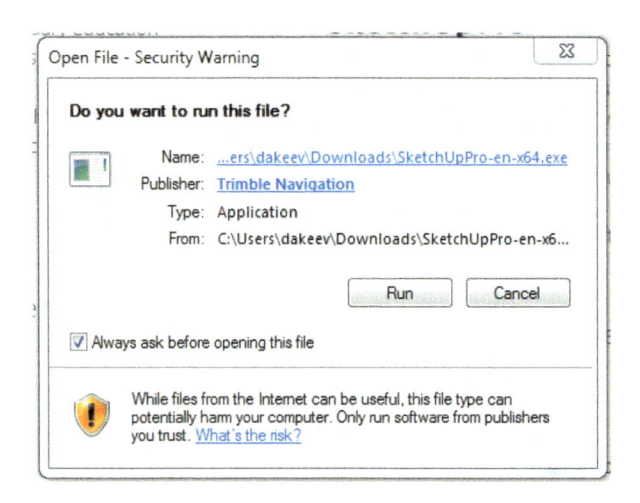

FIGURE 4.3
Google SketchUp installation (step 1).

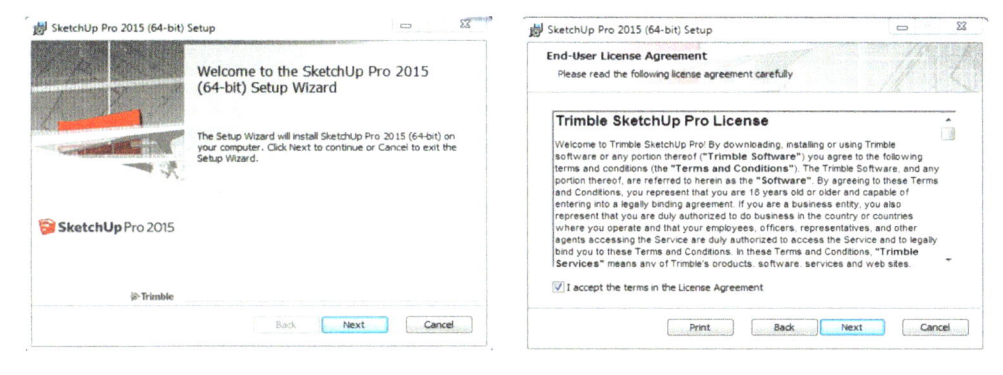

FIGURE 4.4
Google SketchUp installation (step 2).

Camera Toolbar: The buttons on the Camera Toolbar activate the Orbit tool, Pan tool, Zoom tool, Zoom Window tool, Previous, Next, Zoom Extents tool, Camera tool, Position Camera tool, and Look Around tool (Figure 4.12).

To activate necessary tools for this chapter, we can activate "Large Toolset" that contains almost all of the tools we will be using. To activate "Large Toolset," right click on the toolbar→Check "Large Toolset."

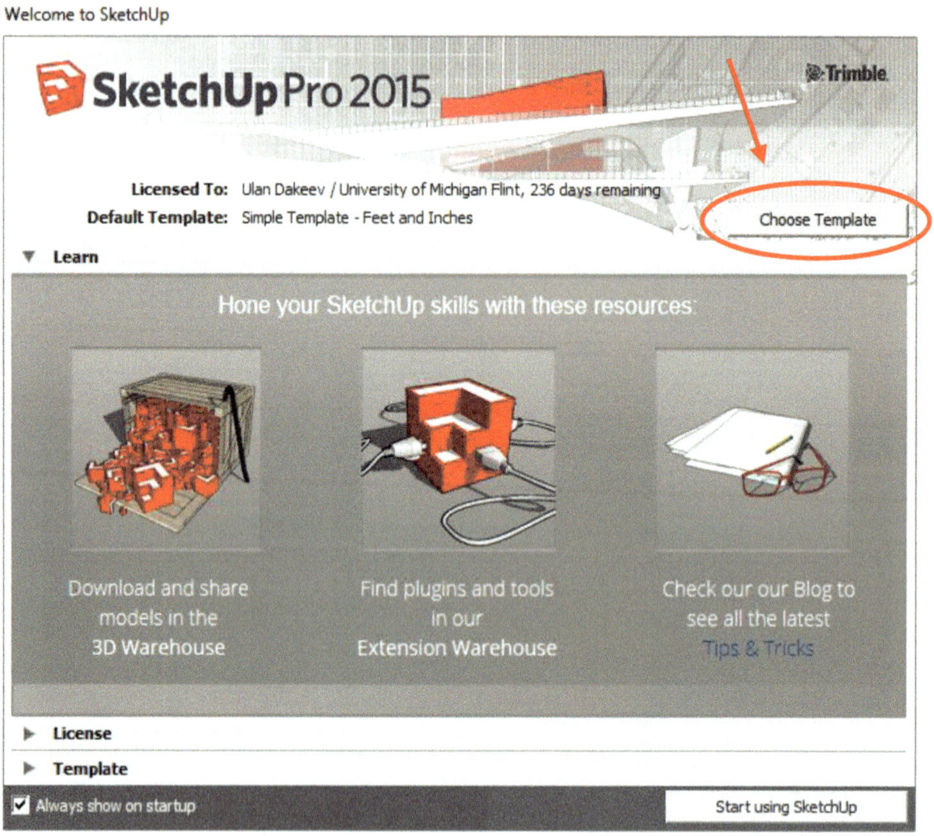

FIGURE 4.5
Selection of templates for drawing.

4.3.4 Dynamic Viewing Function

3D Dynamic Rotation-Middle Mouse Button: Press down the middle mouse button (MMB) in the working area. Drag the mouse on the screen to rotate the model about the screen.

Pan Shift + Middle Mouse Button

Zoom Mouse Wheel

4.3.5 Abbreviations

LMB (Left Mouse Button—Left Mouse Button Click)

Left Mouse Button is used for most operations, such as selecting menus and icons, or picking graphic entities.

MMB (Middle Mouse Button—Press Mouse Wheel)

MMB is often used to zoom (wheel), rotate (press and hold), and pan (shift + hold MMB).

RMB (Right Mouse Button—Right Mouse Button Click)

Right Mouse Button is used to call additional options.

Computer-Aided Design

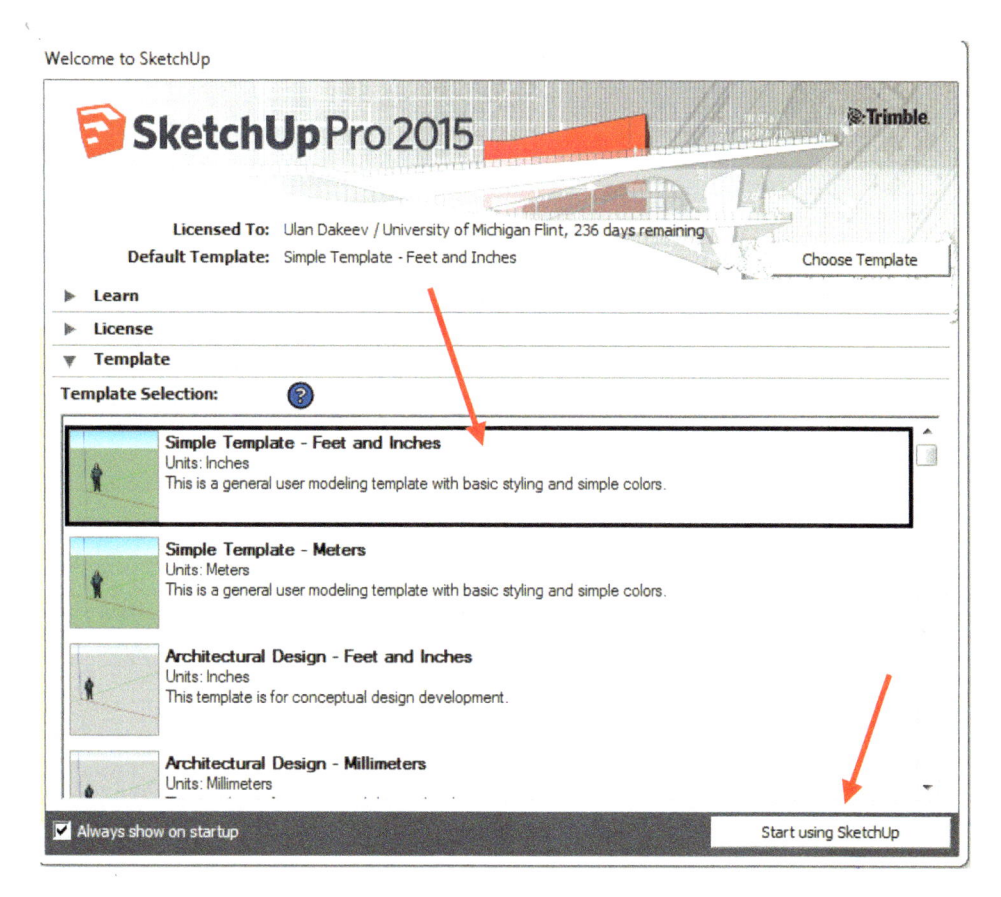

FIGURE 4.6
Selection of template with units (feet and inches).

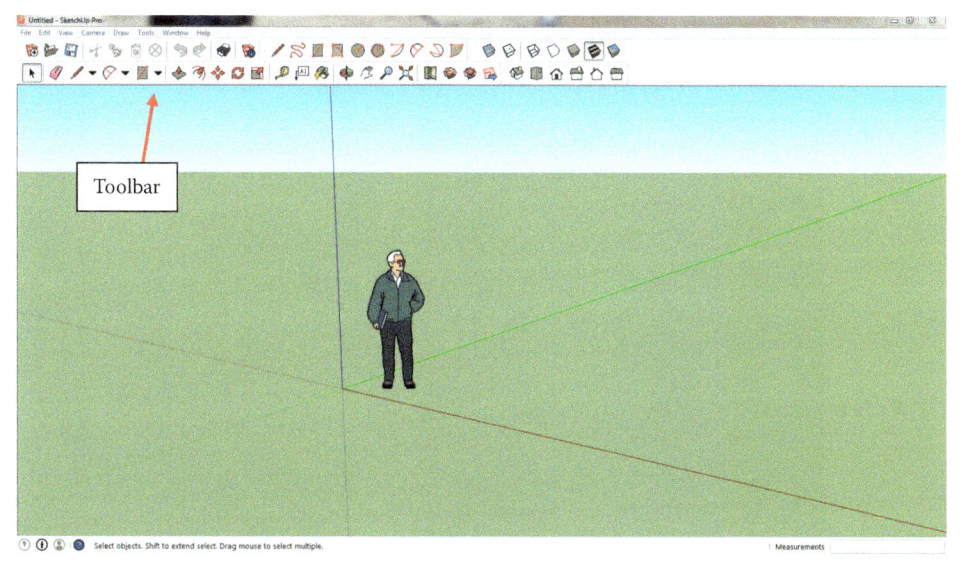

FIGURE 4.7
SketchUp graphical user interface.

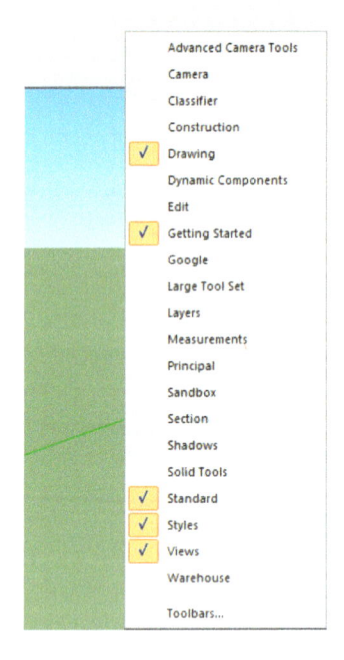

FIGURE 4.8
Available toolbars in SketchUp Pro.

FIGURE 4.9
SketchUp Pro Toolbar.

FIGURE 4.10
Principal Toolbar.

FIGURE 4.11
Edit Toolbar.

FIGURE 4.12
Camera Toolbar.

Computer-Aided Design

4.3.6 Sketch and Extrusion

After properly downloading the software and becoming familiar with the toolbars, a drawing can be created. Different CAD programs offer several powerful tools so that there is always room to learn depending on the complexity of the program. However, the 3D modeling technique is similar across platforms and starts with 2D sketching in the baseline. The Extrusion process is a technique of creating objects with a fixed cross-sectional profile (Sketch). A material is pushed or pulled through using the Push/Pull tool in SketchUp Pro. A step-by-step tutorial is provided below to become familiar with the drawing creation process.

4.3.7 Modeling 3D Object (Bracket Tutorial)

In this step-by-step tutorial, we will develop a 3D representation of a bracket that assembles to the front brake system of a motorcycle. The general approach to modeling the bracket is to start with a 2D sketch on an x–y plane and extrude it in the z-direction (Figure 4.13):

- Select the Human Object on the working area→press "Delete" on your keyboard.
- Right click on the Toolbar and ensure the "Views" tool set has a checkmark (Figure 4.8).

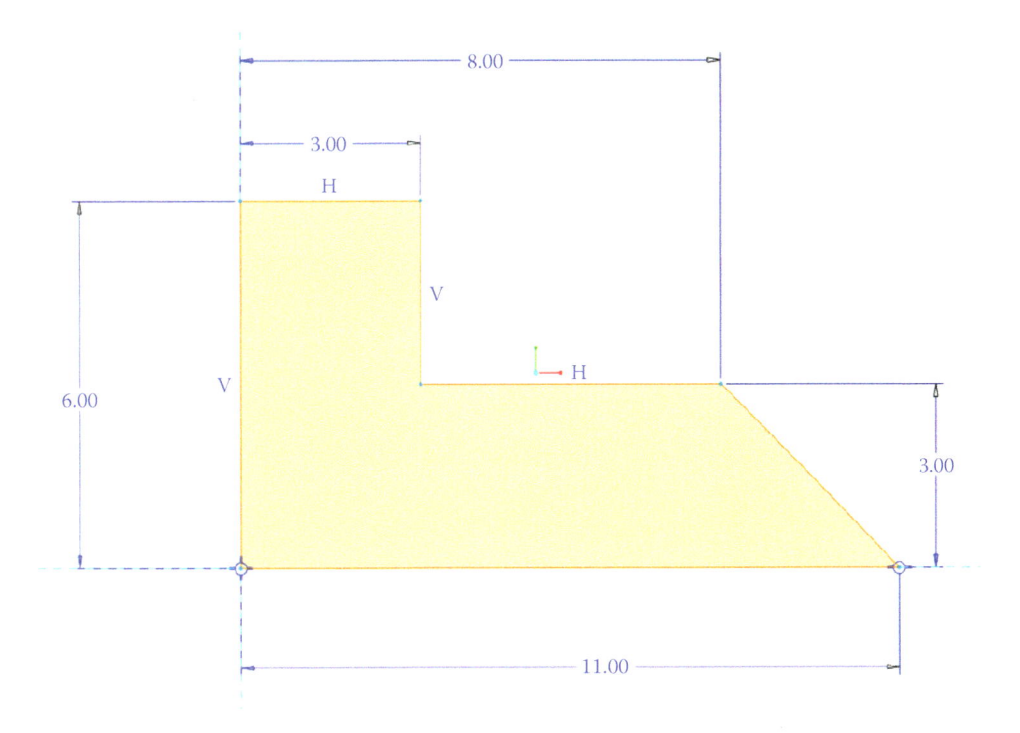

FIGURE 4.13
Two-dimensional sketch of the bracket.

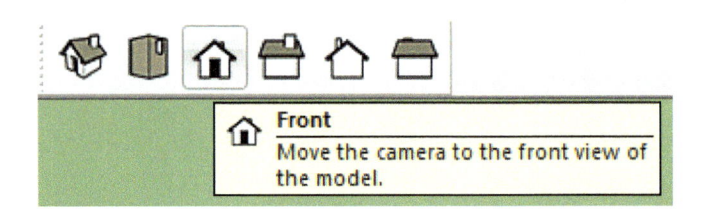

FIGURE 4.14
Setup for the Front view.

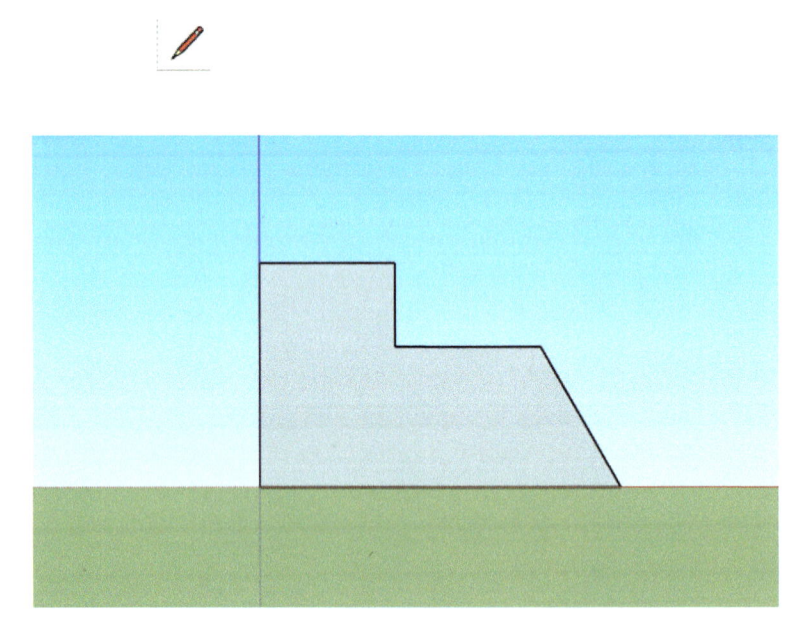

FIGURE 4.15
Bracket Tutorial-1.

- Select the "Front" view to orient the working area (Front orientation will be active) (Figure 4.14).
- Select the "Line" tool→Make a sketch (as we make each line we will be entering the correct values for the length) (Figure 4.15).
- With the "Line" tool selected, click on the origin, move your cursor up vertically and enter 6. Your vertical line will measure 6 inches (Figure 4.16).
- Move your cursor right (horizontally) and enter 3 to define the length. You will have two lines, vertical and horizontal, of 6 and 3 inches, respectively.
- Because the given sketch does not tell us how far down we need to sketch a vertical line, we need to calculate the distance. Because the entire height is 6 inches and the distance between two horizontal parallel lines is 3 inches, we can bravely conclude that the length of the vertical line pointing downward should be 3 inches (Figure 4.17).
- Horizontal line can now be calculated similarly: 8 − 3 = 5 (Figure 4.18).

Computer-Aided Design 67

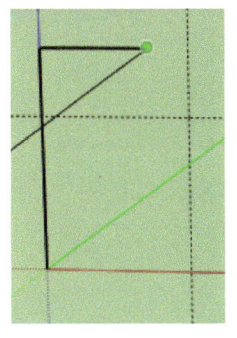

FIGURE 4.16
Bracket Tutorial-2.

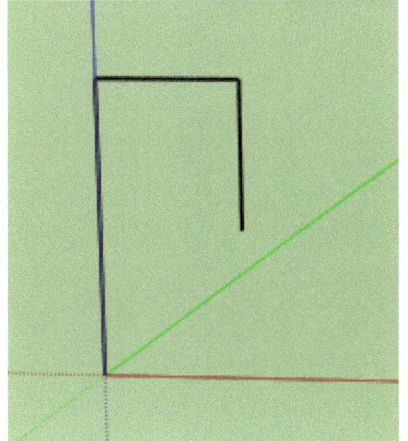

FIGURE 4.17
Bracket Tutorial-3.

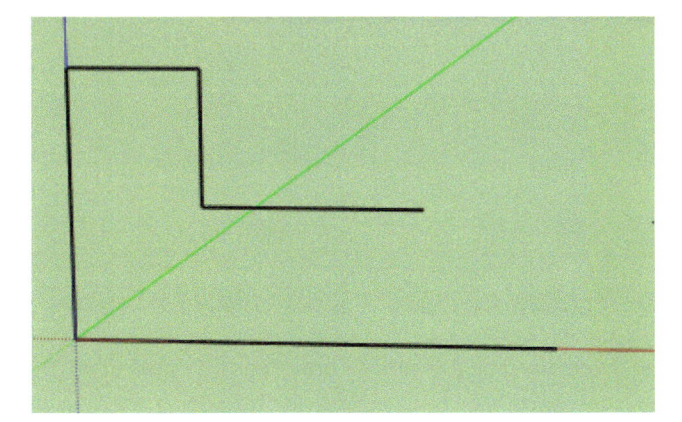

FIGURE 4.18
Bracket Tutorial-4.

- Because we do not know the angle to sketch the slanted line, we can start a new line from the origin of 11 inches and connect two end points of the parallel lines. Note that to deselect the "Line" tool you need to click on the "Select" tool and activate the "Line" tool again to resume the sketch.
- Connect two end points. Notice the content of the shape is shaded (Figure 4.19).

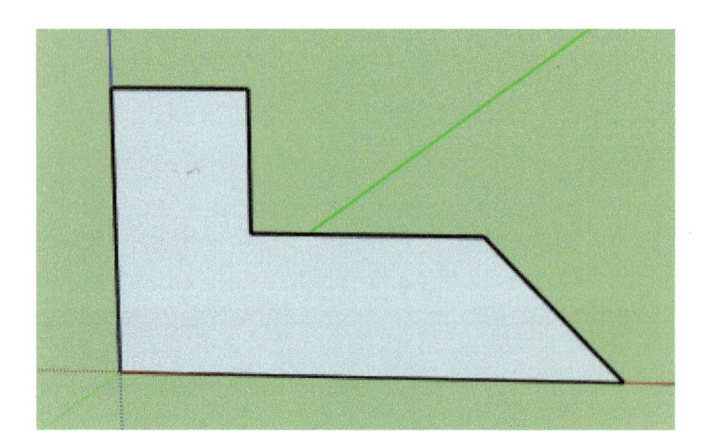

FIGURE 4.19
Bracket Tutorial-5.

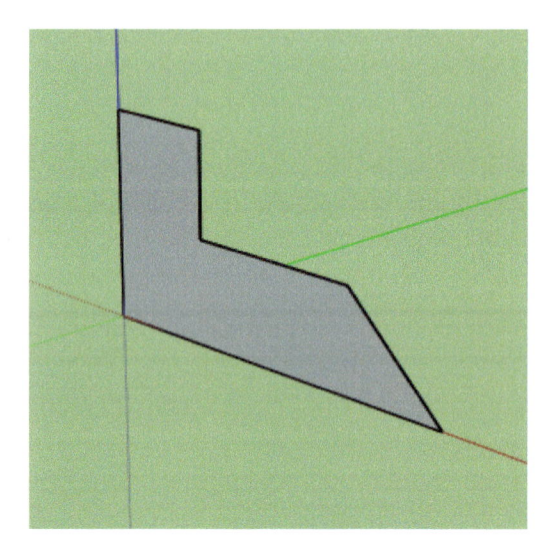

FIGURE 4.20
Bracket Tutorial-6.

- Select the "Iso" view on the "View" toolset to orient the geometry to Isometric view (Figure 4.20).

- Select the "Push/Pull" tool and select the shaded portion of your geometry to pull 5 inches (Figure 4.21).

4.3.8 Adding Dimensions to the Model

At this point, the motorcycle bracket part is completed in its 3D form. To ensure that we modeled the part with the correct dimensions, we will measure all the necessary driving dimensions.

Computer-Aided Design

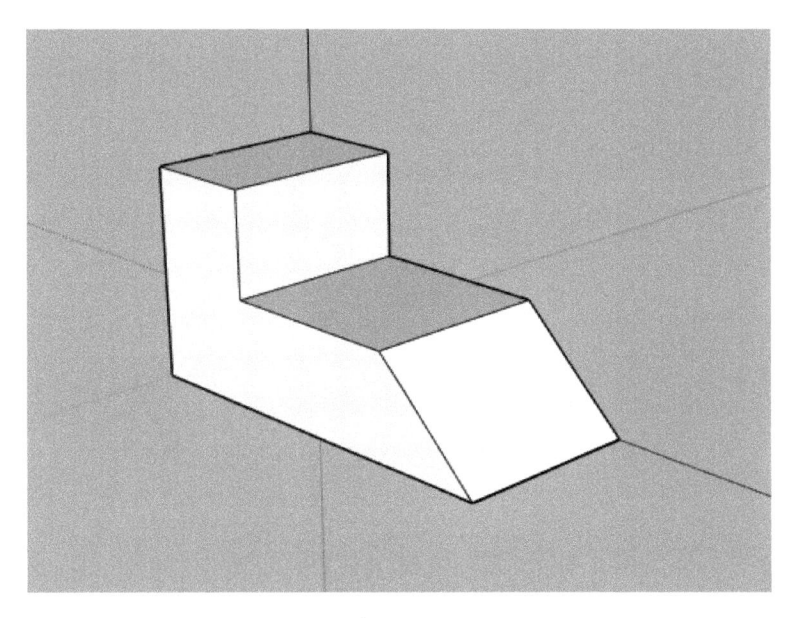

FIGURE 4.21
Bracket Tutorial-7.

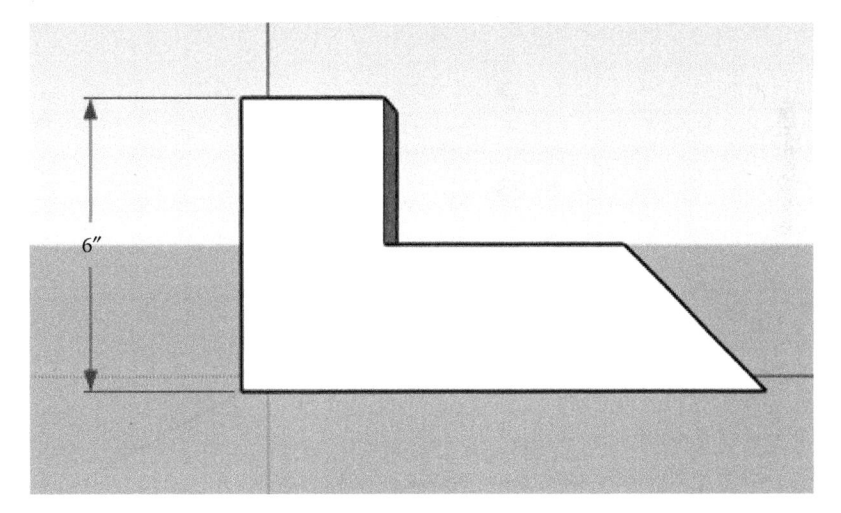

FIGURE 4.22
Bracket Tutorial-8.

- Orient back to the Front view.
- Select the "Dimension" tool from the "Large Toolset" and select two end points of the vertical line to define the height of the bracket. Continue this step until all dimensions are defined (Figures 4.22 and 4.23).

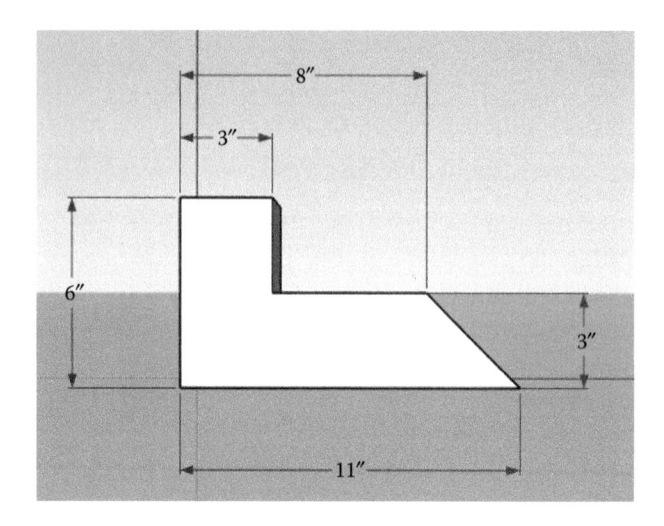

FIGURE 4.23
Bracket Tutorial-9.

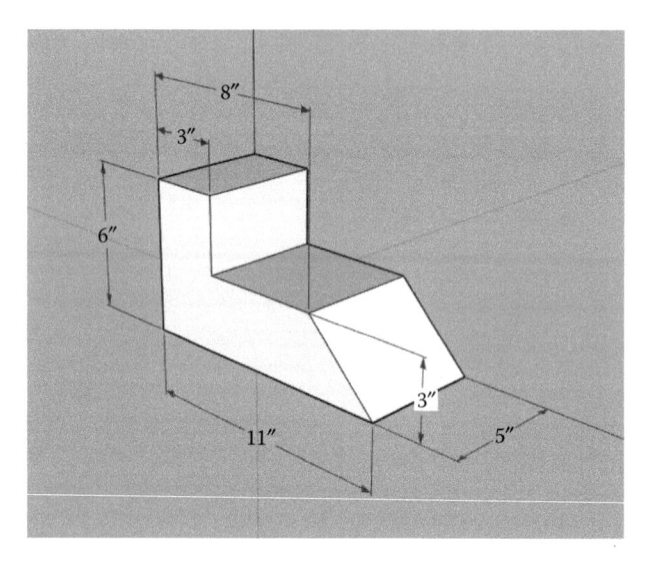

FIGURE 4.24
Orientation to the isometric view to define the thickness of the part.

To define 3-inch height on the right, select two parallel line edges and pull the dimension value out (Figure 4.24).

4.4 Drawings in SketchUp Pro

The majority of applications in manufacturing facilities require the use of 2D (orthogonal) drawings. Using the solid model as the starting point for a design, solid modeling tools can easily generate all the necessary orthogonal views. In this section, the general procedure of creating multiview drawings from solid models is discussed.

Computer-Aided Design

4.4.1 Part Tutorial

In this tutorial, we will be modeling a part in millimeters and adding round features to the part. Additionally, a 2D drawing will be developed for this model (Figure 4.25).

- Start SketchUp Pro unless it is already running.
- Choose a metric template (second from top)→Start using SketchUp. Because the default metric units are going to be in meters, we will need to modify to millimeters.
- Click Window→Model Info. From Format, change m (meters) to mm (millimeters) and close.
- Erase the Human Object and orient to the Front view.
- With the Line tool selected, sketch all sides with given dimensions (Figure 4.12).
- If the line cannot be seen, click "Zoom Extents" to fit the line to screen (Figure 4.26).
- Select the "Protractor" tool to define the angle guide. Procedure: Click the end point of the bottom line→select any point along the same line→move mouse up to define angle (enter 45 or track the "angle" box to show 45 degrees). A 45 degree guide will appear (Figure 4.27).

With the Line tool, follow the 45 degree guide. Do not worry if you exceed the necessary height; you can make another horizontal line from the end of the vertical to connect the slanted line and erase the excessive line segment.

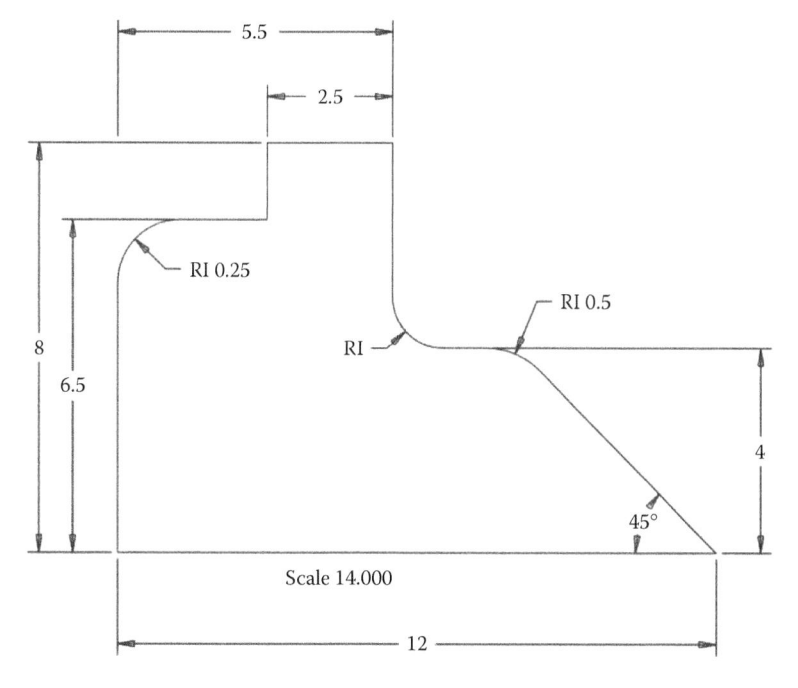

FIGURE 4.25
Definition of dimensions.

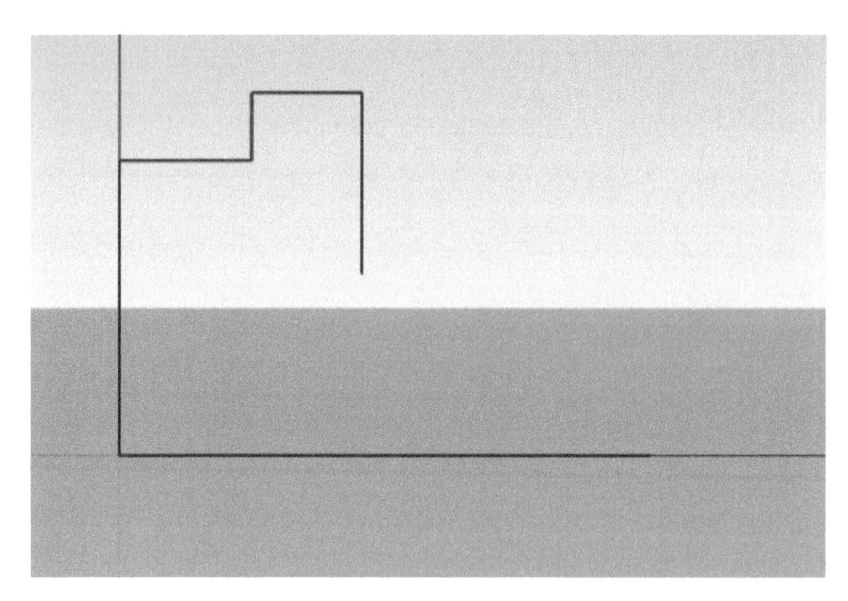

FIGURE 4.26
Part Tutorial-1.

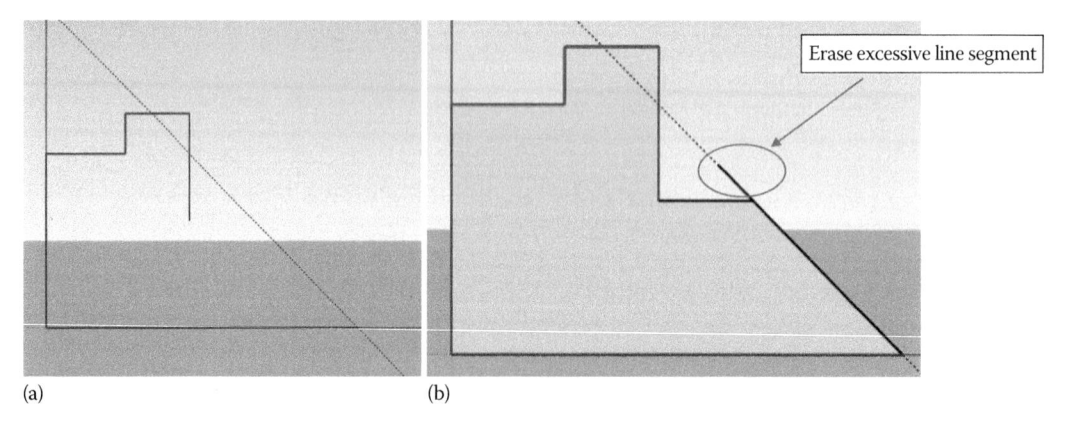

FIGURE 4.27
(a) Part Tutorial-2; (b) Part Tutorial-3.

4.4.1.1 Adding Rounds

- Select the "2 Point Arc" tool→Select two adjacent edges until the arc changes to a different color→Type 1.25 to define radius (Figure 4.28).
- Repeat the step to make the other two arcs (Figure 4.29).
- Orient to isometric view→add 5 mm thickness (Figure 4.30).

Computer-Aided Design

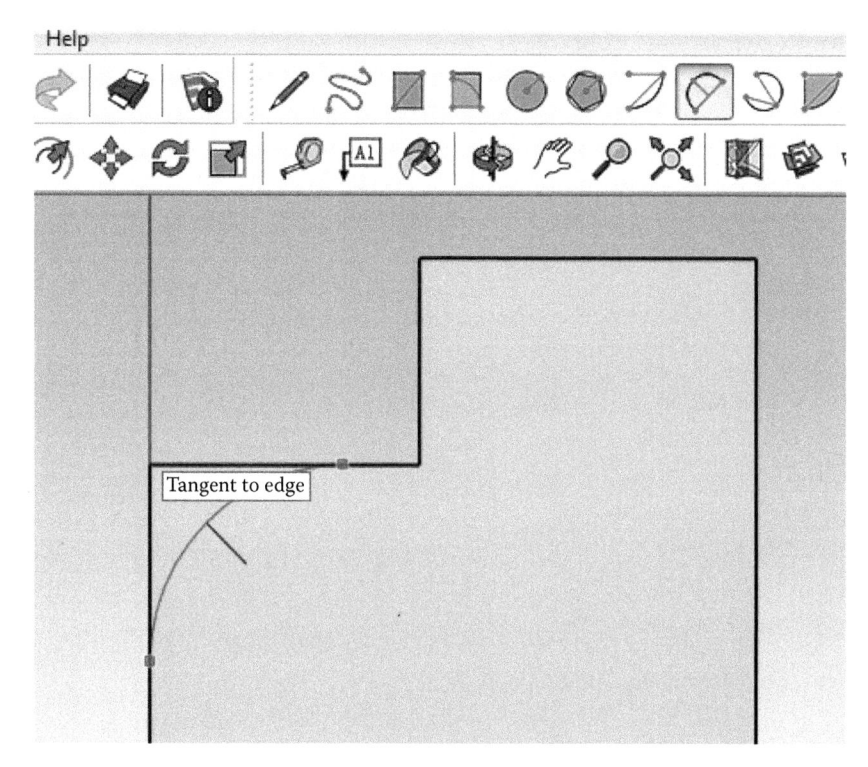

FIGURE 4.28
Part Tutorial-4.

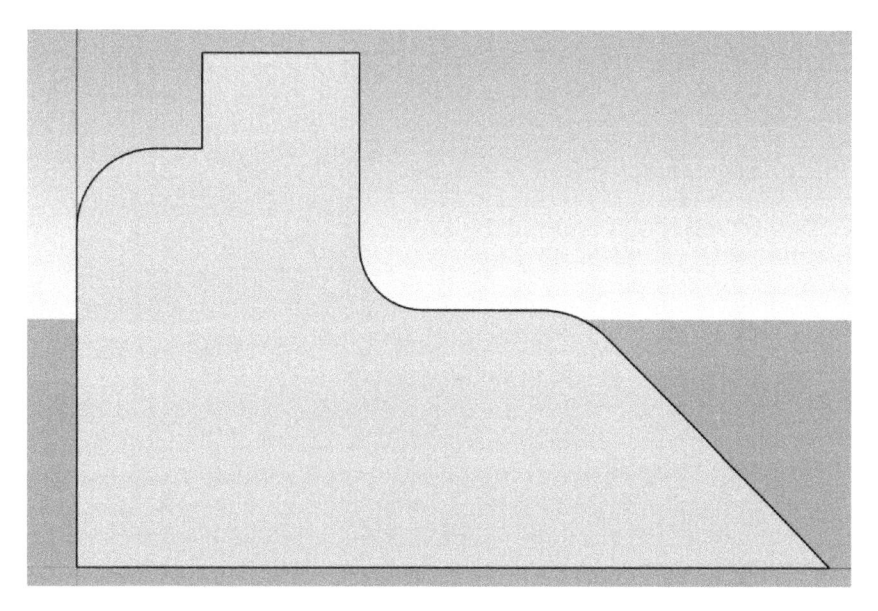

FIGURE 4.29
Part Tutorial-5.

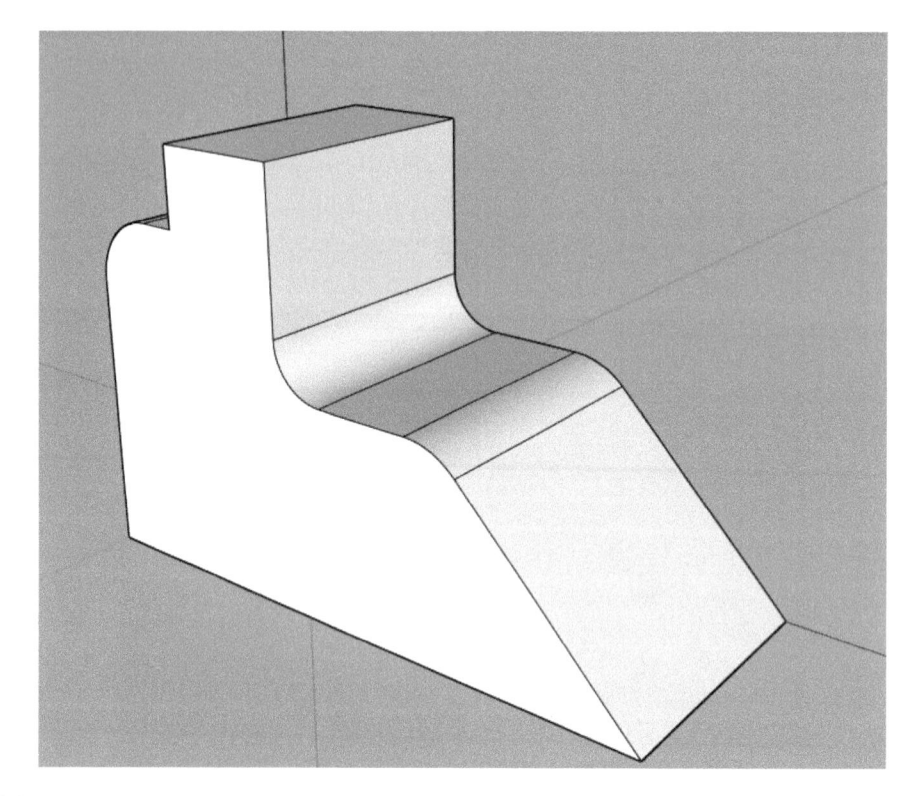

FIGURE 4.30
Part drawing (isometric view).

4.5 Orthographic Drawings in CAD

2D drawings of parts provide important information to the manufacturer but it may be inadequate. In order to manufacture a part, it is important to develop a 3D model and provide all necessary dimensions on the drawing. These models are developed with 2D sketches and projected 2D views of an isometric (3D) view called orthogonal views. In Figure 4.31, you can see the projection views of a simple triangular 3D object. These orthogonal views illustrate how the orientation of a 3D object to that particular side would look on a 2D plane. Figure 4.32 illustrates how different orthogonal views represent the end product from all the necessary sides (top, side, and front).

You can visualize a 3D part in orthogonal views with Figure 4.33. As illustrated in the figure, you can see the projected 2D views of an object from different viewpoints.

Similarly, cylindrical objects might be projected as a simple trapezoidal shape on one projection wall, while on the other one (top or bottom views) circular outline of the shapes will be projected. Orthographic projection is used as an unambiguous and accurate way of providing information, primarily for manufacturing and detail design.

When a designer works with an engineering drawing, they must be familiar with the precise meaning of the various drawing simplifications and terminology as specified in

Computer-Aided Design

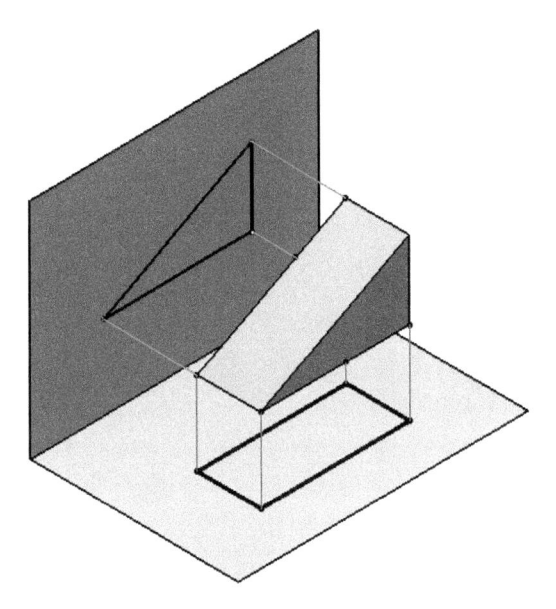

FIGURE 4.31
Orthogonal view projections.

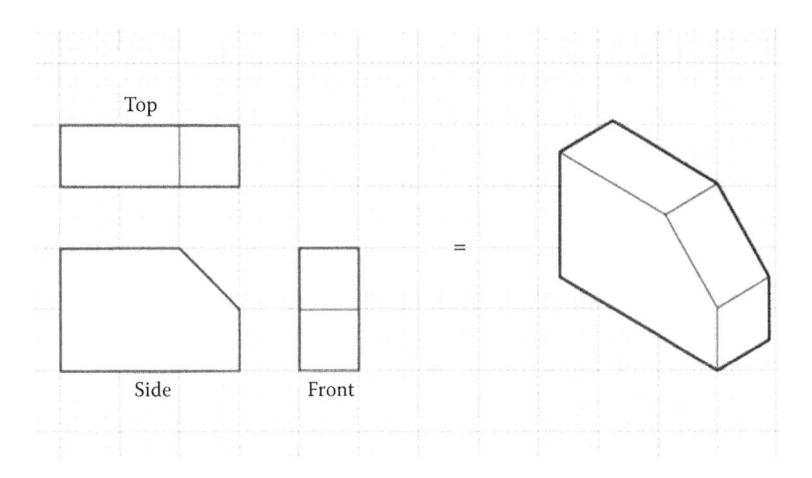

FIGURE 4.32
Orthogonal views.

the relevant standards. Standards are developed by private companies or by internationally recognized institutions. Two such international standards are

British Standard Institution: **BS 8888**.
American National Standards Institute: **Y14 series**.

Following the relevant standards bring all engineers around the globe on the same page. Numerous manufacturing organizations work internationally. They might have a

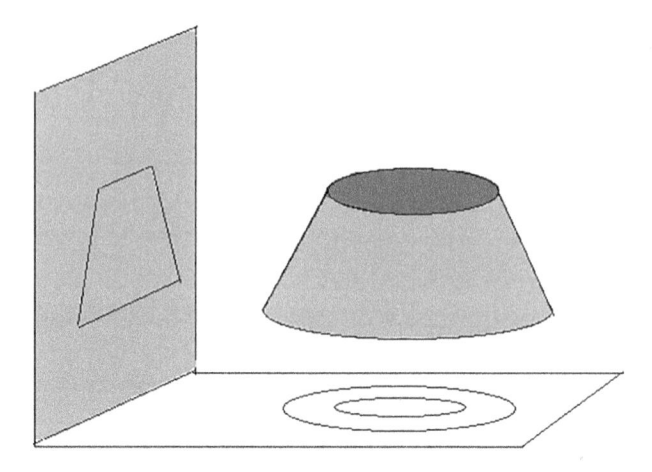

FIGURE 4.33
Orthogonal projections.

manufacturing plant in one location and an assembly plant in another country or state. 2D drawings are the communication tools between involved parties. Through the use of specific symbols (such as geometric dimensions and tolerances), suppliers and customers manufacture the exact product as specified in the drawings. The appearance of an engineering drawing varies depending on when, where, and for what it was produced. However, the general procedure for developing an engineering drawing from a solid model is fairly well defined. Because 3D modeling is not limited with modeling the necessary part only, engineers often use virtual models of parts in the assemblies as well. The assemblies pursue the same intent, which is to illustrate how various parts (probably from different suppliers, or older and modern) function. Assembly drawings usually do not include the dimensions of the individual parts, but they illustrate the general outlook of the assembled parts. Two important items need to be placed in the assembly drawings: (1) representation of entire assembly in an isometric mode, and (2) BOMs. The isometric representation of the assembly needs to be in an *exploded* view, which shows the relationships of mating parts and their constraints. BOMs represent individual part names, their quantities, and indexed number in the balloon on the exploded isometric view. Other items placed in the assembly drawing might vary by organization. Each company may choose what exactly they want to include to represent their drawings. However, three items usually are always present in any drawing by Y14.5 standard: (1) *Title Block*: It contains the name of the drawing, the engineer's name (the person who developed the drawing), company's name, and date. (2) *Revision Block*: It is located on the top-right corner of the drawing. This block contains the revision of the drawing and the date, symbol illustrating the change on current revision, and the zone where the symbol is located (Figure 4.34).

If a drawing size is A or B, the zone information may not be necessary unlike larger sizes such as C and E. Placing the zone and the symbol illustrating the location of the change will help the reviewer of the drawing to locate the change. (3) *Part information*: All necessary views (orthogonal, detailed, sectioned, auxiliary, etc.), geometric dimensions and tolerances (GD&T), and general notes about the parts communicating the material, heat treatment information, and primer or coating to be applied are placed (Figures 4.35 through 4.37).

Computer-Aided Design

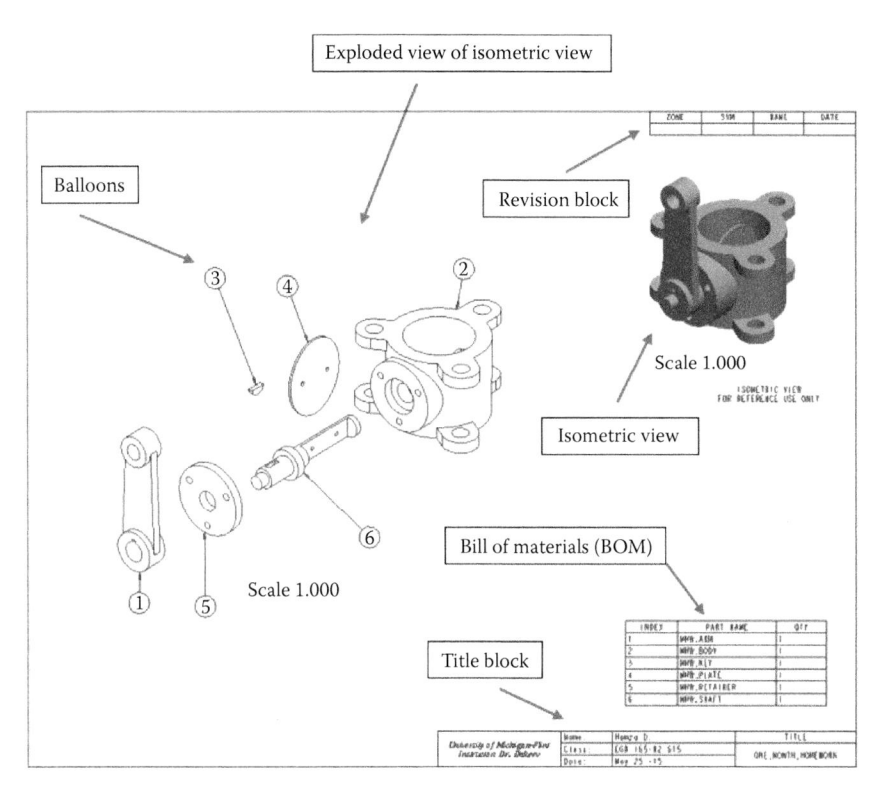

FIGURE 4.34
Example of assembly drawing.

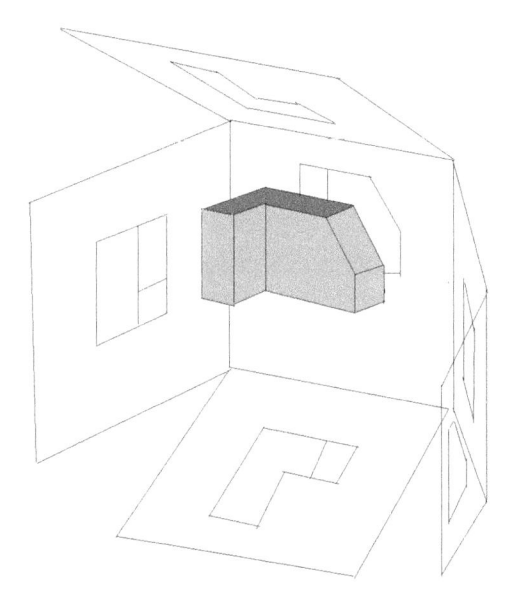

FIGURE 4.35
Orthogonal View Creation-Step 1.

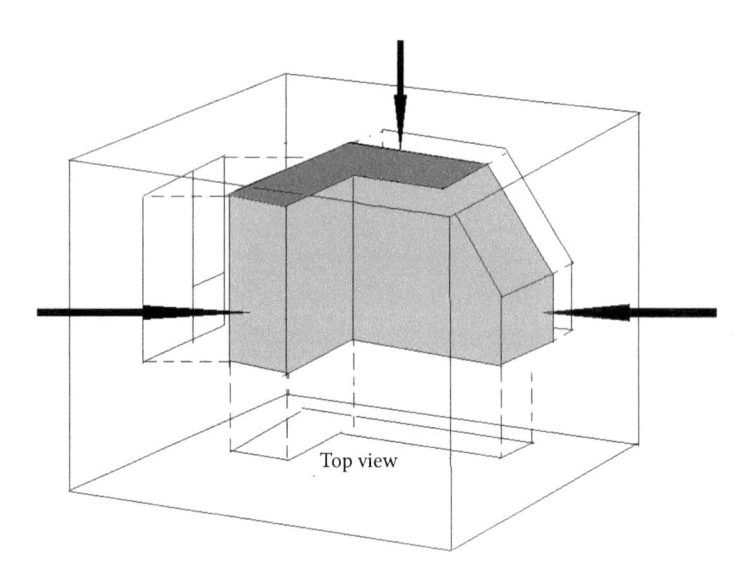

FIGURE 4.36
Orthogonal View Creation-Step 2.

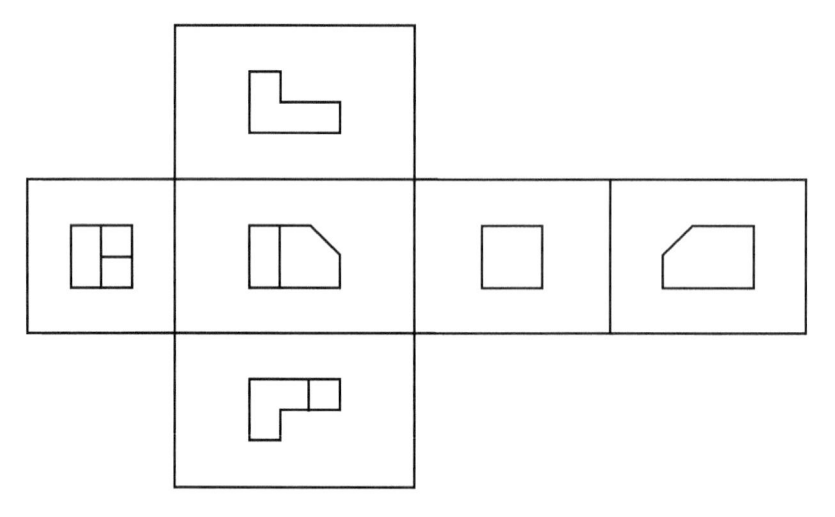

FIGURE 4.37
Orthogonal View Creation-Step 3.

4.6 Preparing Model for LayOut

SketchUp Pro lets users save orientation in terms of styles and transfer to LayOut mode to define dimensions in orthogonal views. In this section, we will prepare the modeled 3D part for the drawing and export to SketchUp Pro LayOut.
Adding views

- Select "Window"→Scenes (Scenes dialog box will open).
- Click (+) button to add a new scene (Figures 4.14 and 4.38).

Computer-Aided Design 79

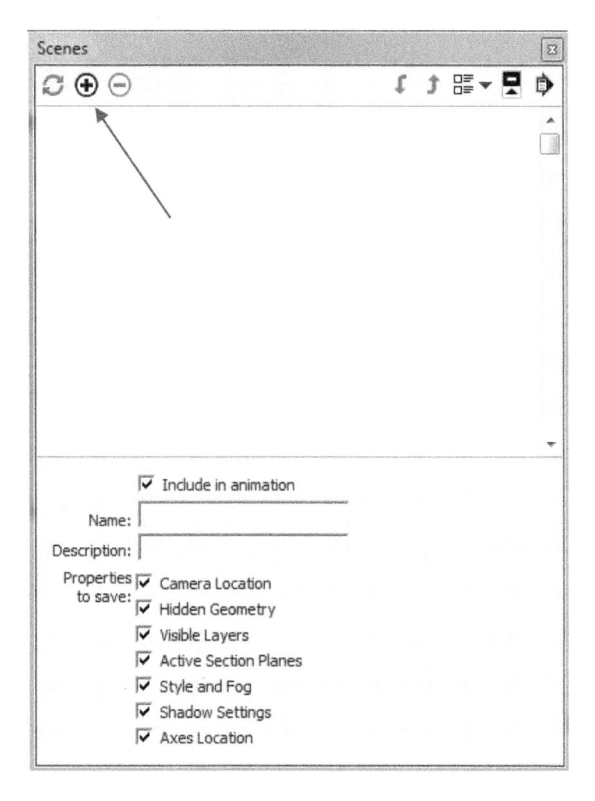

FIGURE 4.38
Adding Scenes (views)-1.

- If additional options are not displayed, rename Scene 1 (right click to rename).
- Add three scenes, renaming them Isometric, Front, and Top, respectively (Figure 4.39).
- Close the Scene dialog box.
- Select "Front Scene" on the top left of the working area→click Front orientation from the toolbar.
- Right click "Front Scene"→Update (Front orientation will be updated in this Scene).
- Select "Top Scene"→Orient model to Top (View)→Update (Figures 4.15 and 4.40).
- Select each of the Scene tabs to ensure that all orientations are saved accordingly.
- Save your work.

4.6.1 Opening Model in LayOut

Once the model is completed and the scenes are prepared for the 2D drawing mode, we can send the part to LayOut.

- Select File Menu→Send to LayOut.
- From the Templates dialog box, scroll down to select A3 Landscape with no grid (Figure 4.41).
- Click "Open" (the isometric view of the model will open in the LayOut mode) (Figure 4.42).

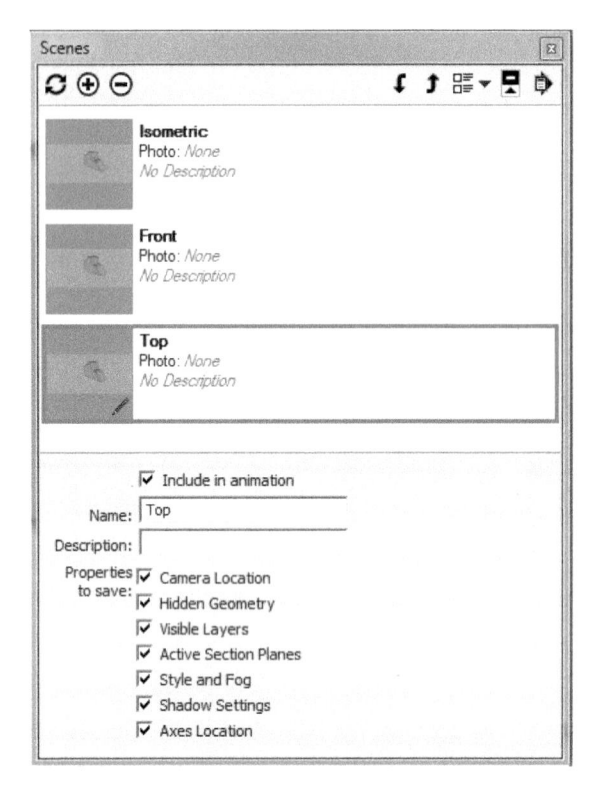

FIGURE 4.39
Adding Scenes (views)-2.

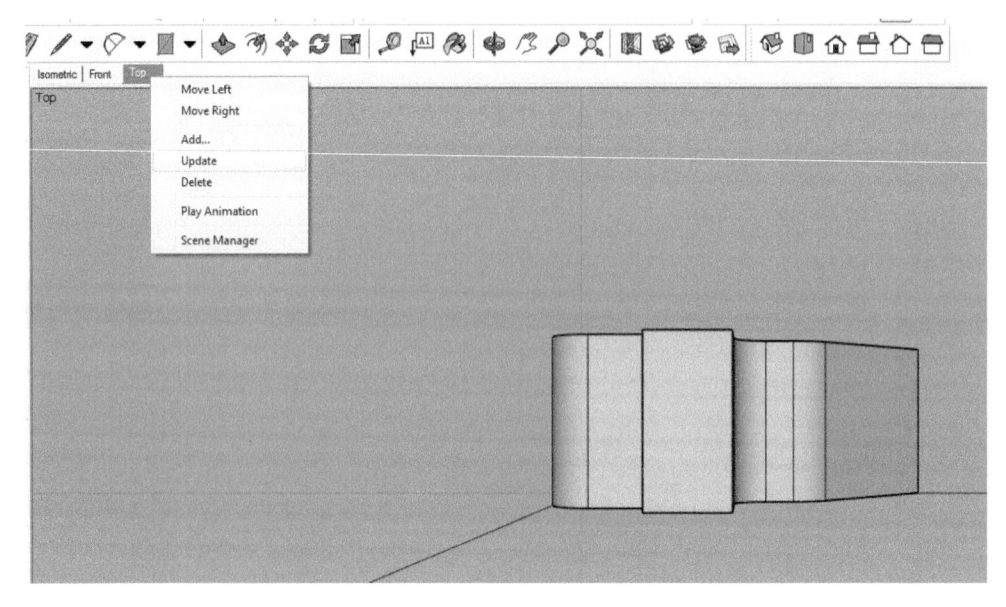

FIGURE 4.40
Adding Scenes (views)-3.

Computer-Aided Design 81

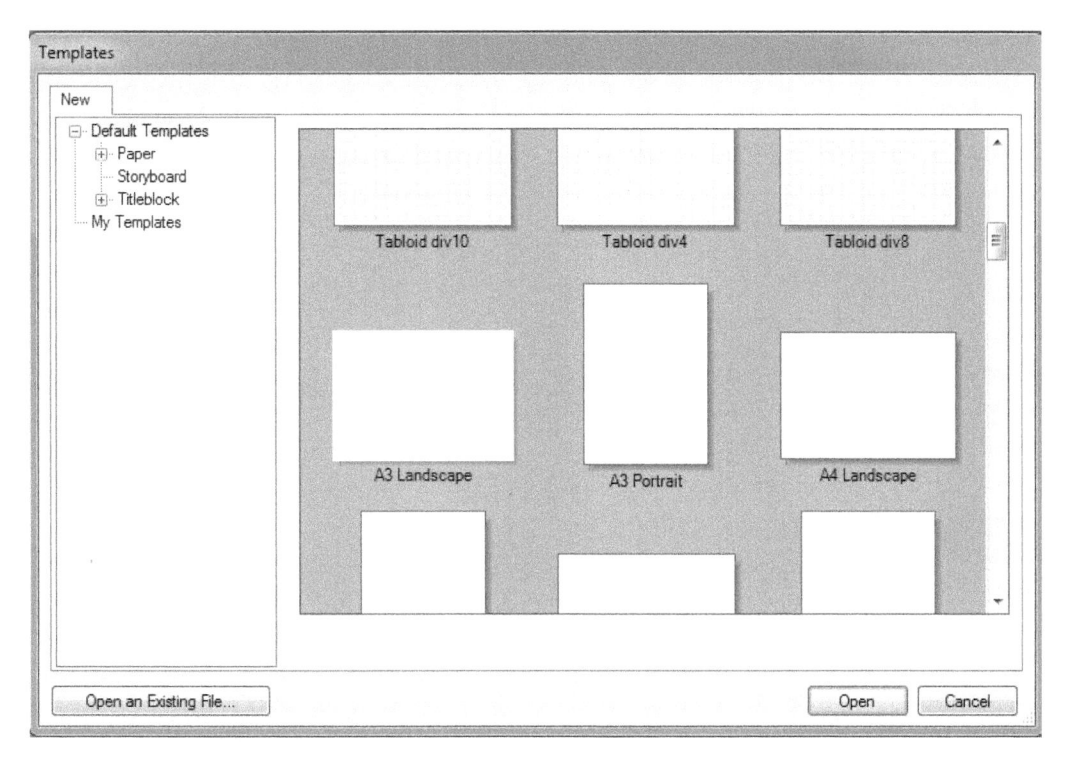

FIGURE 4.41
Select Template (A3-Landscape).

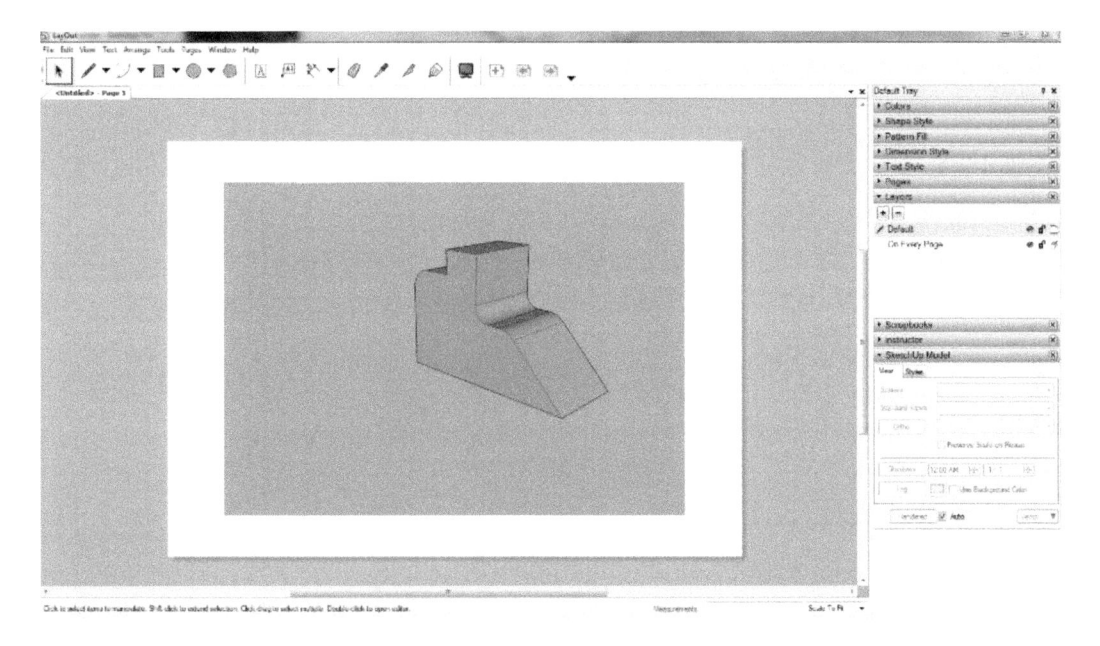

FIGURE 4.42
Update orientation for view.

We have successfully transferred the model into LayOut, where we will be developing orthogonal views and dimensions for print. To ensure that the model and the drawing are seamless, we will need to make some adjustments in LayOut. Note that LayOut placed the isometric scene right at the center of the A3 Landscape paper.

- Select the scene (left click).
- On the right panel of the LayOut window, expand the "SketchUp Model" tab (Figure 4.43).
- Change Raster to Vector (bottom-right corner of SketchUp Model Panel).
- Resize the Isometric Scene by dragging from the bottom-right corner (Figure 4.44).
- On the SketchUp Model Panel, click on the "Ortho" button (scale will change).
- From the drop-down menu, change scale to 1 mm: 200 mm (1:200).
- Double click Isometric Scene→RMB→Zoom Extents→LMB outside the scene (Figure 4.45).

The object is refitted to the window.

- Select "Scene"→On the SketchUp Model panel. Click Scenes drop-down menu→select "Front" (Figure 4.46).
- Repeat Zoom Extent to fit the Front orientation to the Scene (Figure 4.47).
- Select the "Styles" tab on the SketchUp Model panel→click Style Collection drop-down menu→Select Styles (Figure 4.48).
- Scroll down to locate Straight Lines Style collection→Double click→Straight Lines 04pix (Figure 4.49).

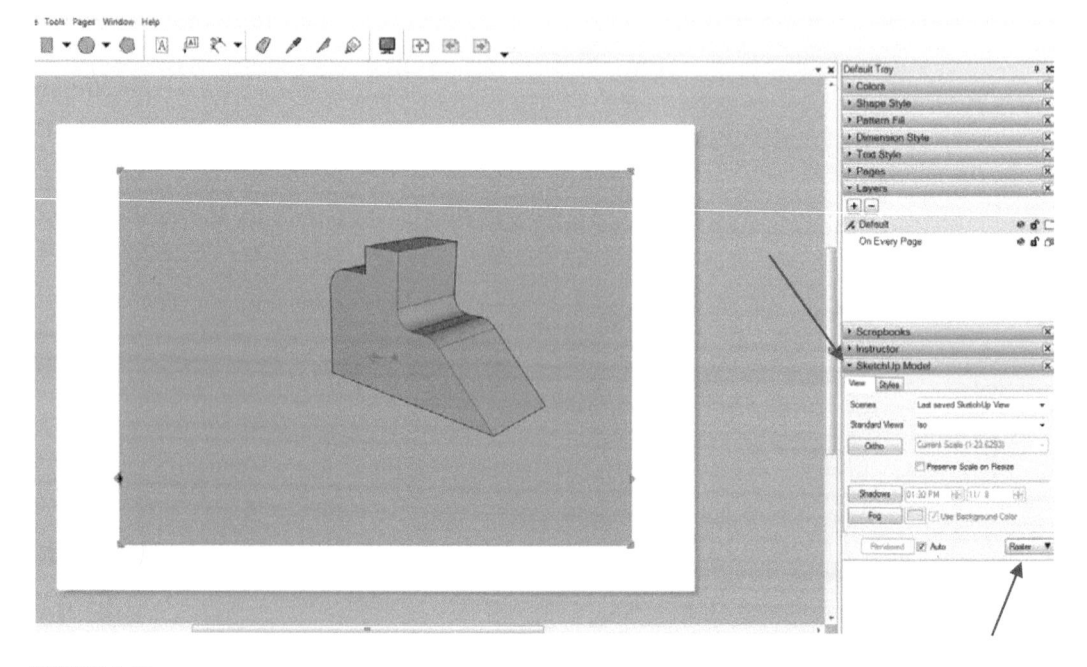

FIGURE 4.43
Orthogonal View Creation-1.

Computer-Aided Design

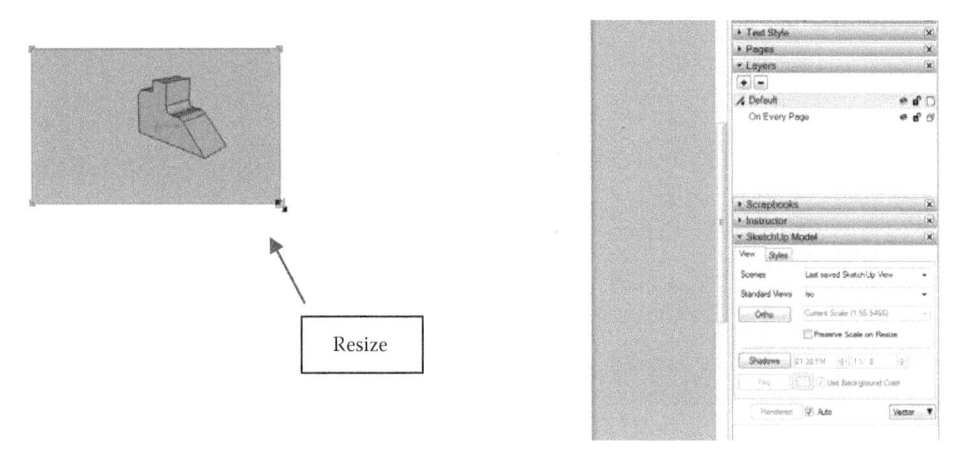

FIGURE 4.44
Orthogonal View Creation-2.

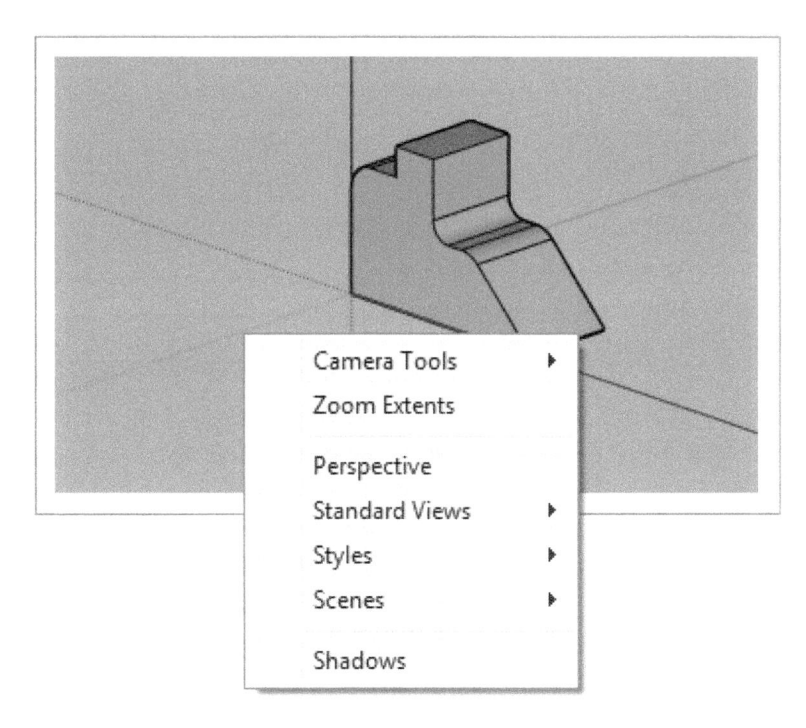

FIGURE 4.45
Orthogonal View Creation-3.

Notice the Front Scene has changed to Straight Lines with no color. Next, we will insert Top and Isometric views.

- Select "Front Scene"→hold the Ctrl (Command Mac) button→drag your cursor down (this will duplicate the Front Scene) (Figure 4.50).
- Select the "View" tab on the SketchUp Model panel→Change Standard views from Front to Top.

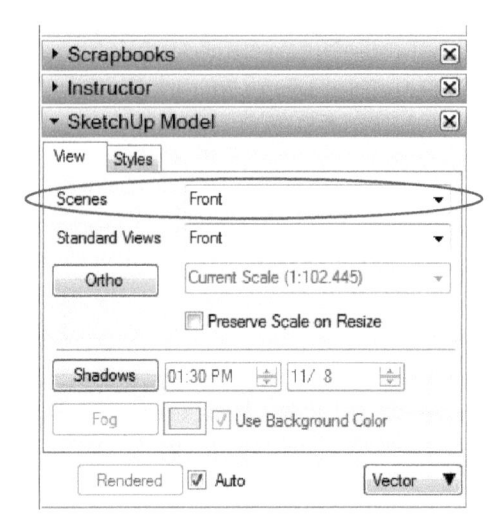

FIGURE 4.46
Orthogonal View Creation-4.

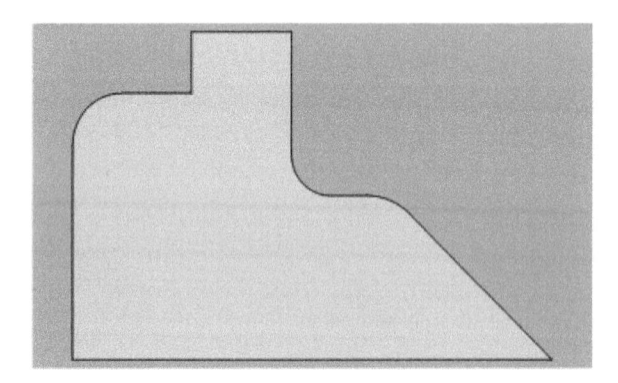

FIGURE 4.47
Orthogonal View Creation-5.

- Repeat the previous steps to duplicate any of the scenes for the Isometric view (You may need Zoom Extents to fit the Isometric Scene into the window.) (Figures 4.51 and 4.52).

4.7 Dimensioning Orthogonal View

We have successfully completed adding all the necessary views for the Angle Bracket model. We needed two orthogonal views (Front and Top) to define the dimensions of the part. The isometric view of the part communicates how the end product should look. Drawings are the important communication tools between engineers in various

Computer-Aided Design

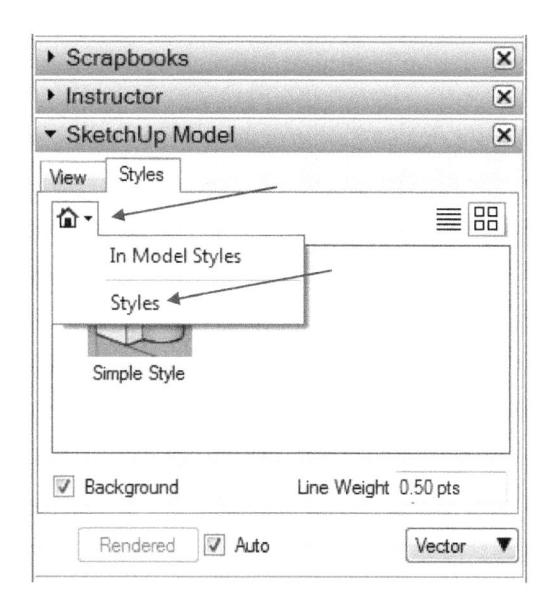

FIGURE 4.48
Orthogonal View Creation-6.

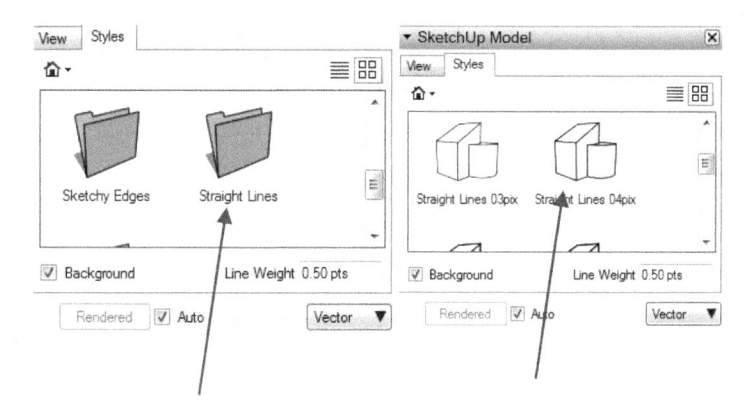

FIGURE 4.49
Orthogonal View Creation-7.

departments. All departments may not have the necessary software to open the 3D model. Therefore, providing 2D drawings to related individuals will help them understand how the manufactured model will look in terms of shape and size. Next, we will define all the necessary dimensions in the drawing.

- On the LayOut Toolbar (on top of the screen) select the "Dimension" tool.
- Define all linear dimensions. The Dimension tool works the same as it did in the Model screen. Select two points to define the distance between them.
- To define angle, select a vertex point and any of the lines, hold Ctrl (Cmd Mac), and select an adjacent line.

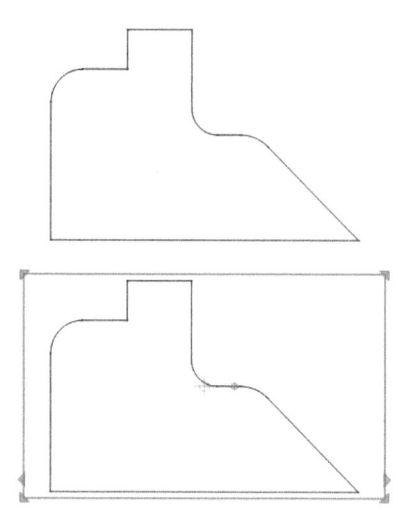

FIGURE 4.50
Orthogonal View Creation-8.

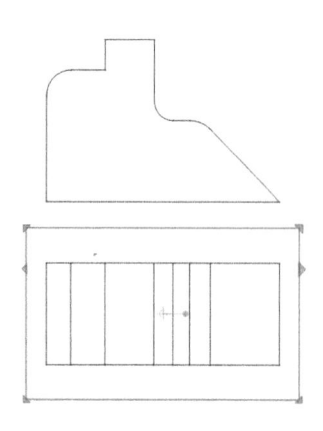

FIGURE 4.51
Orthogonal View Creation-9.

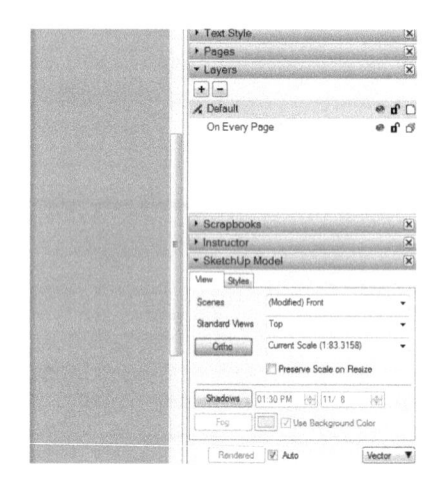

- To define the radius of rounded corners, select the Dimension tool, click the round feature→move toward the center of the arc. Once the center is located, define the distance from the edge to the center (radius).

Title Block: Every drawing needs to have a Title Block that carries important information about the part and the manufacturer. For simplicity purposes, we will have a simple Title Block with name, date, and title of the drawing. To draw the Title Block, select the "Rectangle" tool on the toolbar (Figure 4.53).

Computer-Aided Design

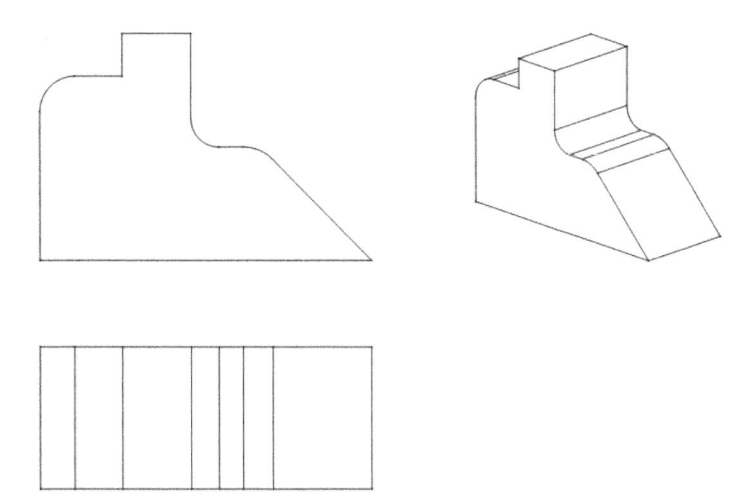

FIGURE 4.52
Orthogonal View Creation-10.

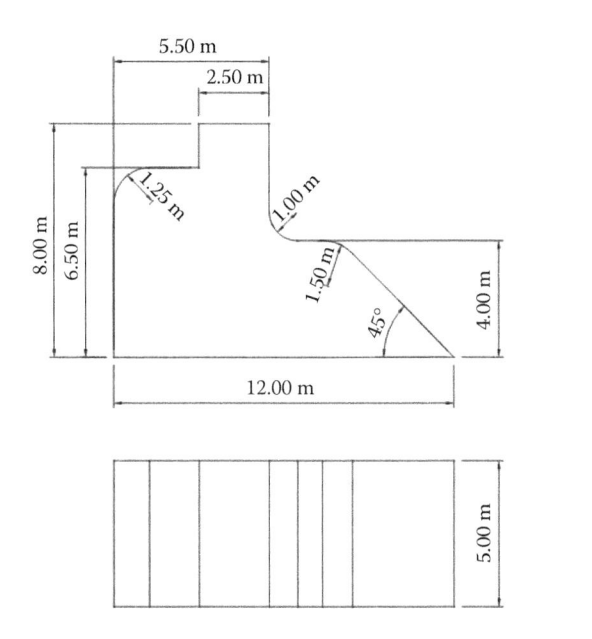

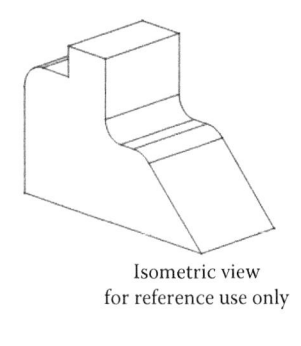

Isometric view
for reference use only

Name	John Smith	Title
Date	5-9-2015	Angle bracket

FIGURE 4.53
Drawing with Orthogonal Views and Dimensions.

4.8 L-Bracket Design

The following example will show you how to develop a 3D object with Boolean tools (Figure 4.54).

- Start SketchUp Pro with Metric Template→Ensure Millimeter Units are defined.
- Orient to Front view→13 by 60 mm Rectangle→Zoom Extents to fit window.
- Sketch a 60 by 19 mm rectangle starting from the left edge of the previous rectangle.
- Delete the horizontal line that crosses the first rectangle (Figures 4.19 and 4.55).
- Select the "Move" tool→Select first rectangle→Hold Ctrl and drag mouse to the right (Figure 4.56).
- Type 47 to define the distance. Note the length of a rectangle is 13 mm. Therefore, to ensure the 60 mm distance, we need to define 47 mm.
- Erase all lines crossing the geometry leaving only the U-shaped object.
- Sketch additional rectangle with 25 mm by 13 mm length and height. Note that the 25 mm length is needed because the entire length of the object is 115 mm (115 mm – 83 mm = 25 mm). This 25 mm rectangle will serve as the center point of the outer rounded feature, whereas 83 mm is the location of the hole center.
- Extrude the sketched geometry to 60 mm with the Push/Pull tool (Figure 4.57).
- Make two guidelines from both edges to define the location of the center rectangle for removing material with the Tape Measure tool.

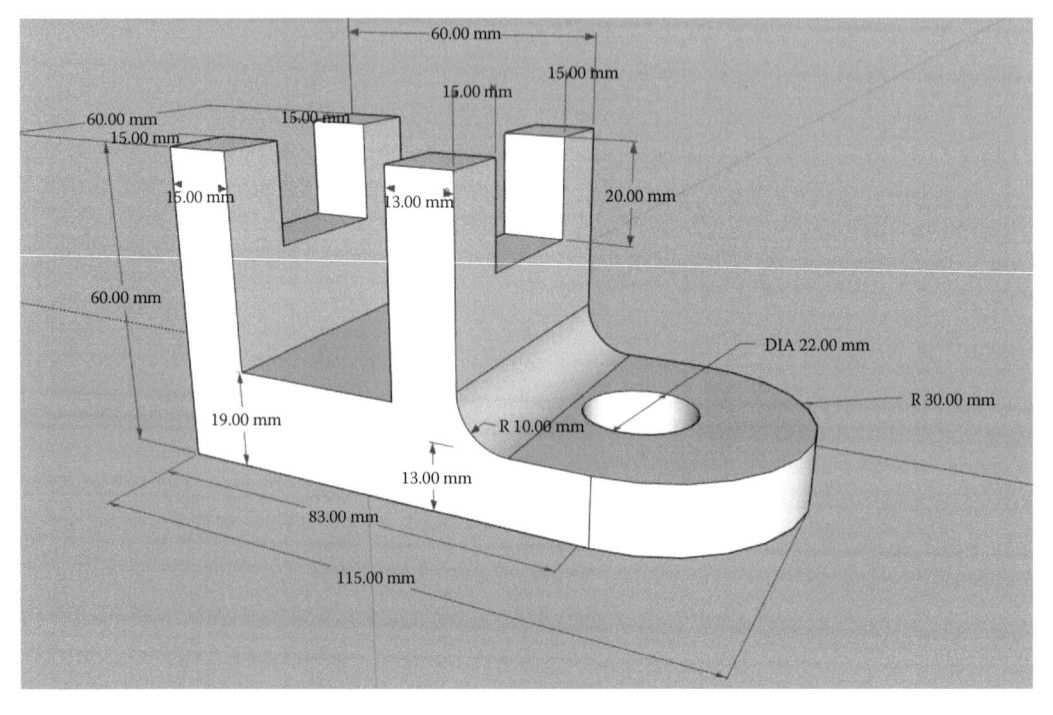

FIGURE 4.54
L-Bracket with dimensions.

Computer-Aided Design

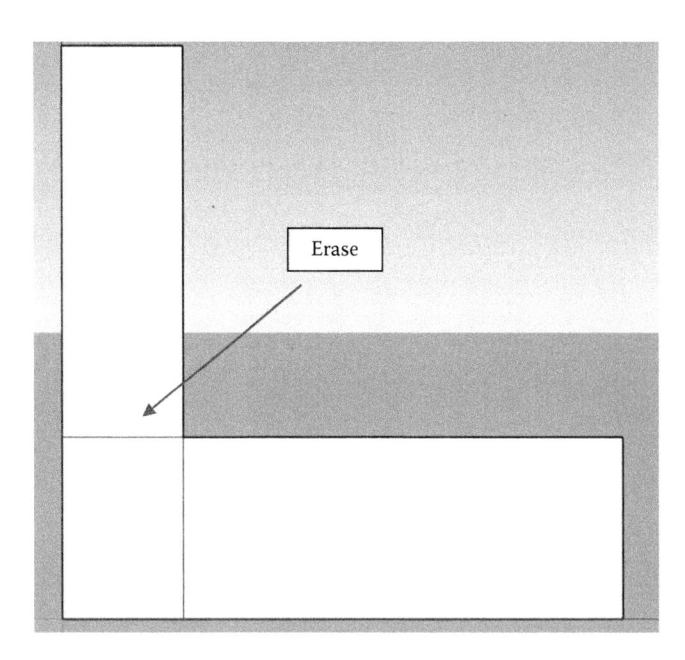

FIGURE 4.55
L-Bracket Tutorial-1.

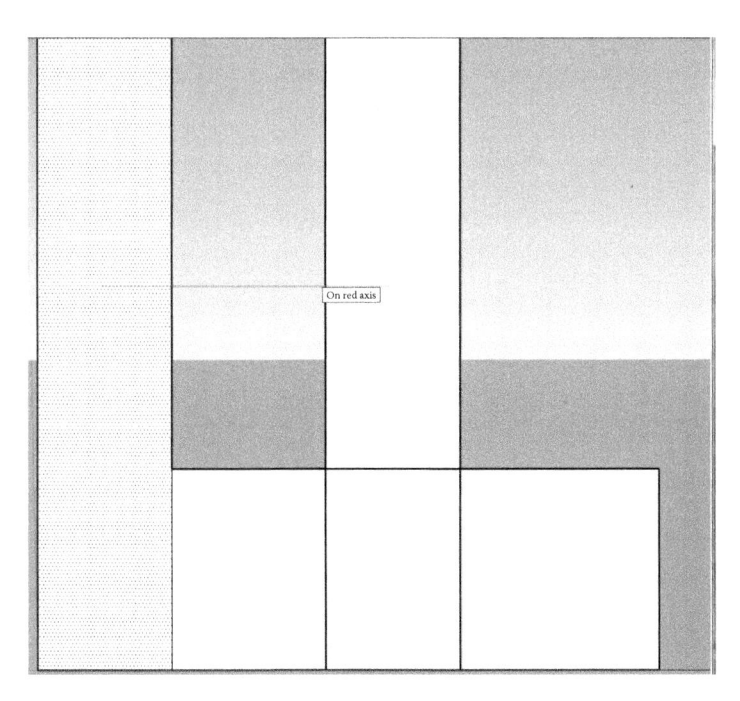

FIGURE 4.56
L-Bracket Tutorial-2.

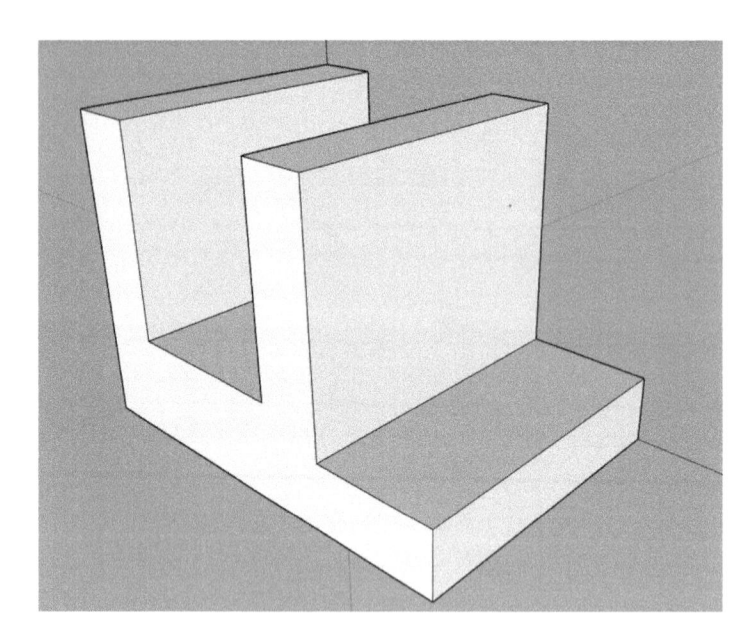

FIGURE 4.57
L-Bracket Tutorial-3.

- Sketch a 30 mm by 20 mm rectangle→Remove material with the Push/Pull tool. Snapping to the opposite edge will remove the material (Figures 4.58 and 4.59).
- Repeat the above steps to remove material from the back as well (Figure 4.60).
- Sketch a 2 Point Arc with 30 mm radius→Extrude to 13 mm→Erase crossing lines.
- Sketch a guide for the middle hole→Define 23 mm from the edge. This way the center of the hole will be located exactly 83 mm from the back edge.

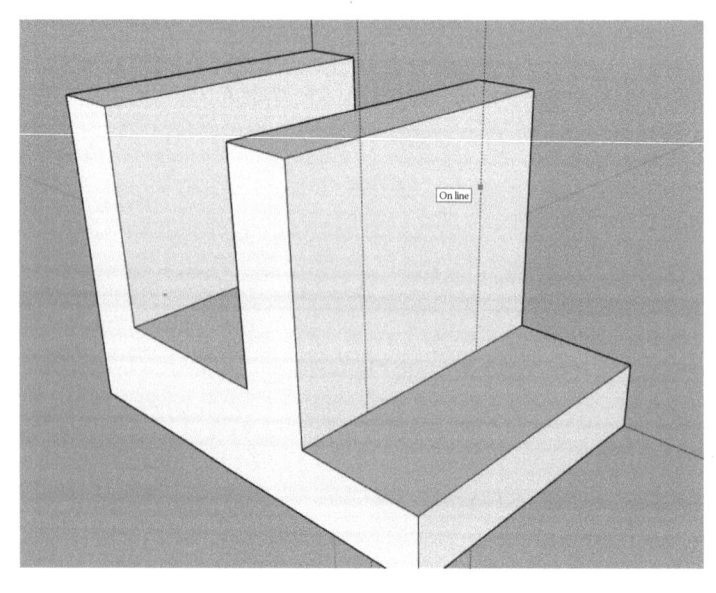

FIGURE 4.58
L-Bracket Tutorial-4.

Computer-Aided Design

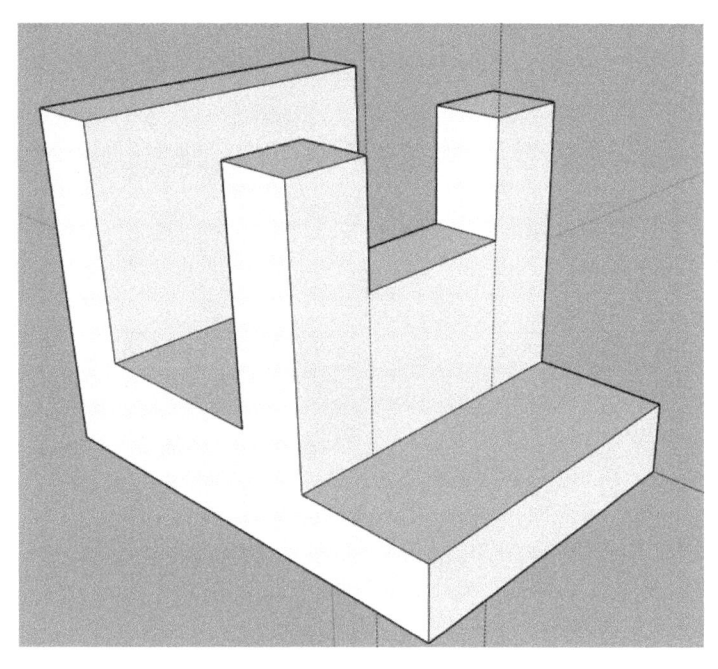

FIGURE 4.59
L-Bracket Tutorial-5.

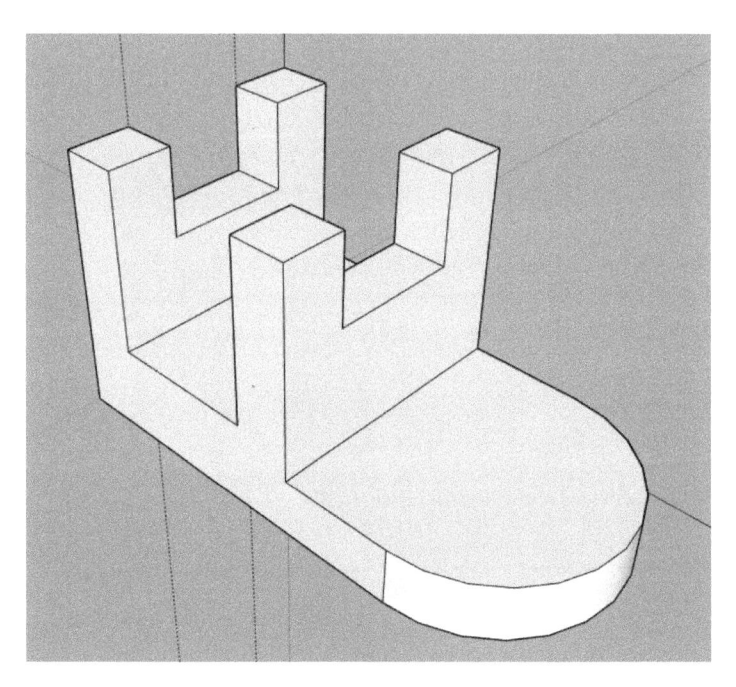

FIGURE 4.60
L-Bracket Tutorial-6.

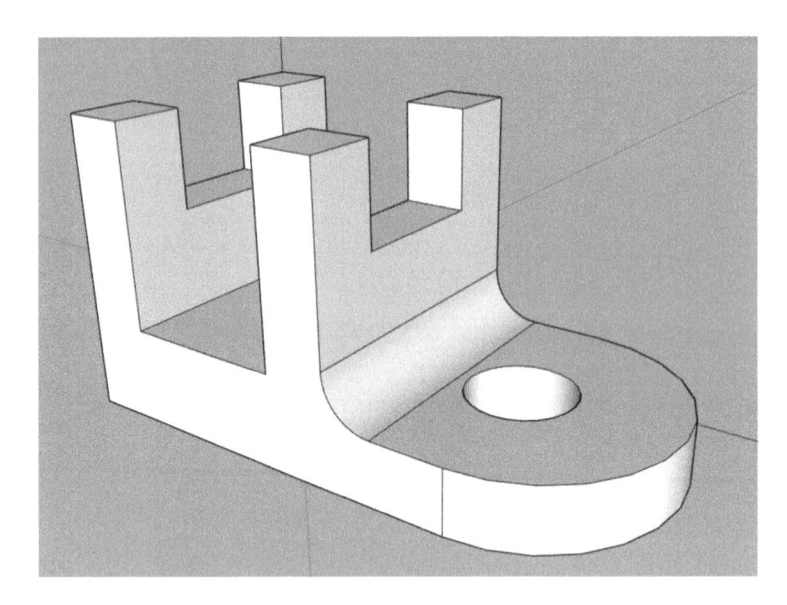

FIGURE 4.61
L-Bracket Tutorial-7.

- Sketch a circle at the midpoint of the guide→Extrude to the bottom face. You may need to rotate the object to select the bottom face to remove the circular hole.
- Orient to Front view→Sketch 2 Point Arc for the Rounded feature→Extrude Arc to the opposite side of L-Support.

Note that 2 Point Arc will need to turn magenta for round feature.

- Remove crossing lines from the geometry (Figure 4.61).
- Select the "Dimension" tool to define all necessary dimensions.
- To define the 83 mm distance from the back edge of the part: Select Back Edge→Select Center of the Hole→Pull out to define the dimension (Figure 4.62).

ABET Program Outcomes

The relationship of CAD Chapter with ABET-related Program Outcomes addresses the following:

- (g) Communicate effectively: (high coverage). The course teaches graphical communication language that engineers practice. The outcome is assessed through assignments and projects requiring engineering drawings to be generated.

Computer-Aided Design

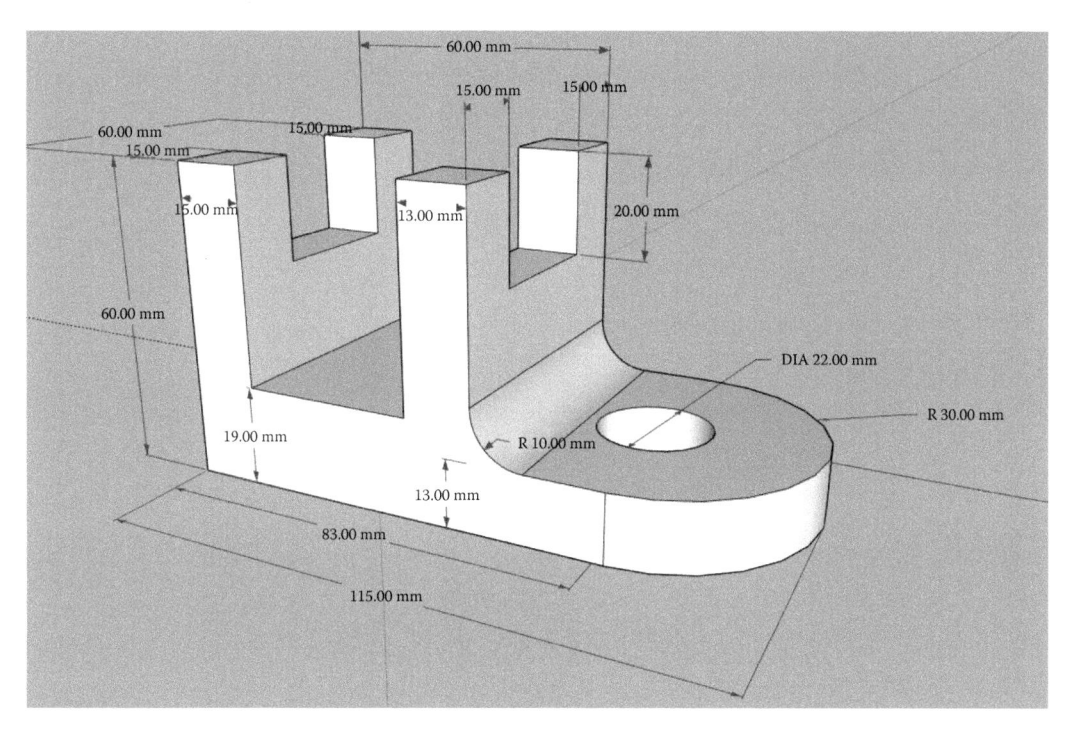

FIGURE 4.62
Completed L-Bracket with dimensions.

- (j) Use the techniques, skills, and modern engineering tools necessary for engineering practice: (high coverage). The course teaches students freehand drawing skills, graphical techniques, as well as use of modern CAD software. These skills are assessed through assignments, exams, and the final project.

Review Questions

1. What is CAD and who uses CAD software?
2. List two benefits of CAD.
3. What is parametric modeling?
4. List three frequently used types of CAD software. State major differences between them and SketchUp Pro (Research Online).
5. What is the difference between Sketch and Extrusion?
6. What is the difference between SketchUp Modeling and LayOut?

7. What are guidelines and how does the Tape Measure tool help make them?
8. Identify the following quick-key commands:
   (a) Shift + MMB
   (b) MMB in all directions
   (c) Ctrl + LMB

Problems

Develop 3D models of the following parts:

1. Drawing Problem 1

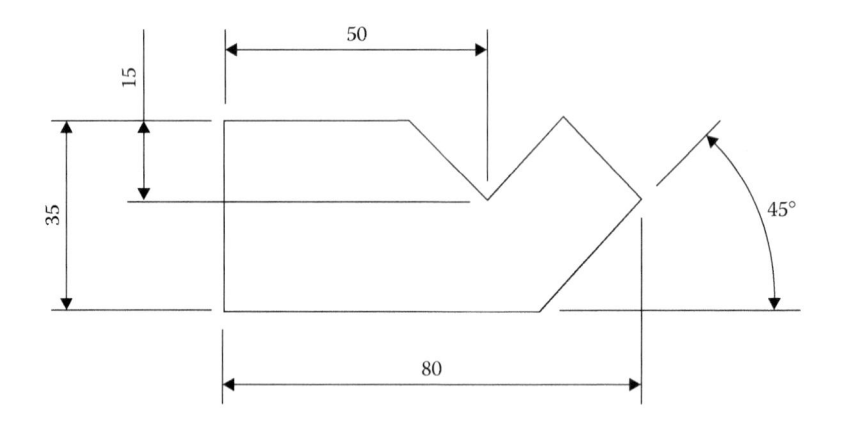

2. Drawing Problem 2

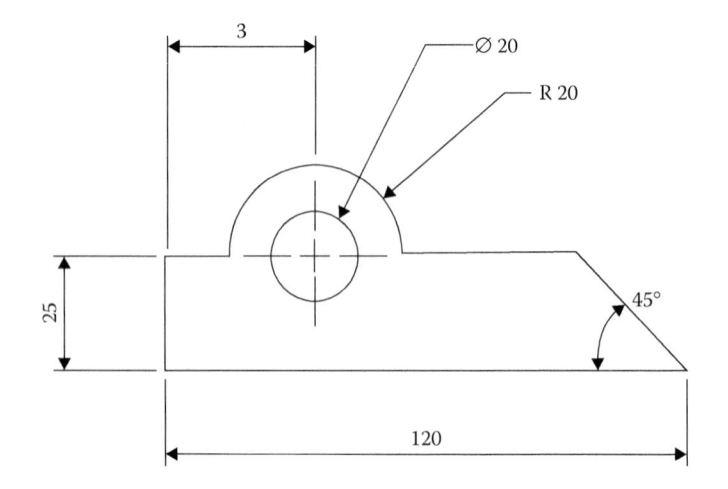

Computer-Aided Design

3. Drawing Problem 3

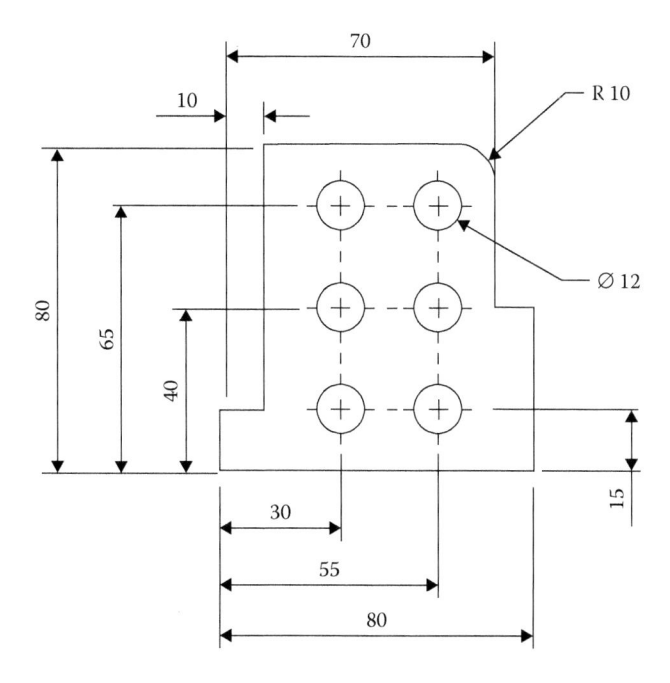

4. Drawing Problem 4

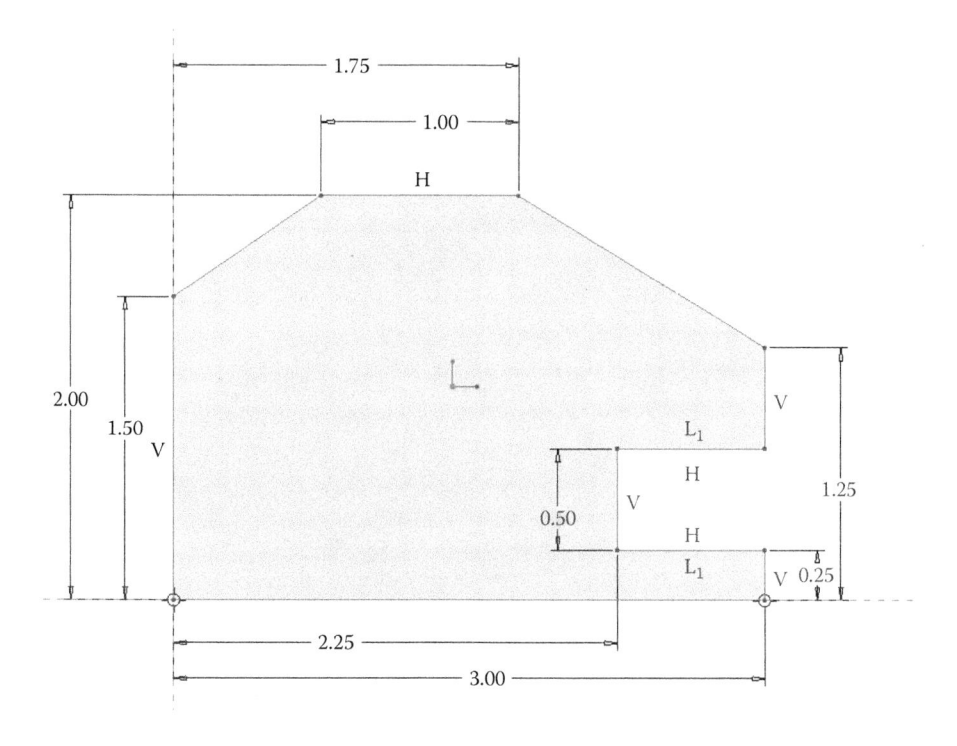

96 Introduction to Engineering

5. Drawing Problem 5

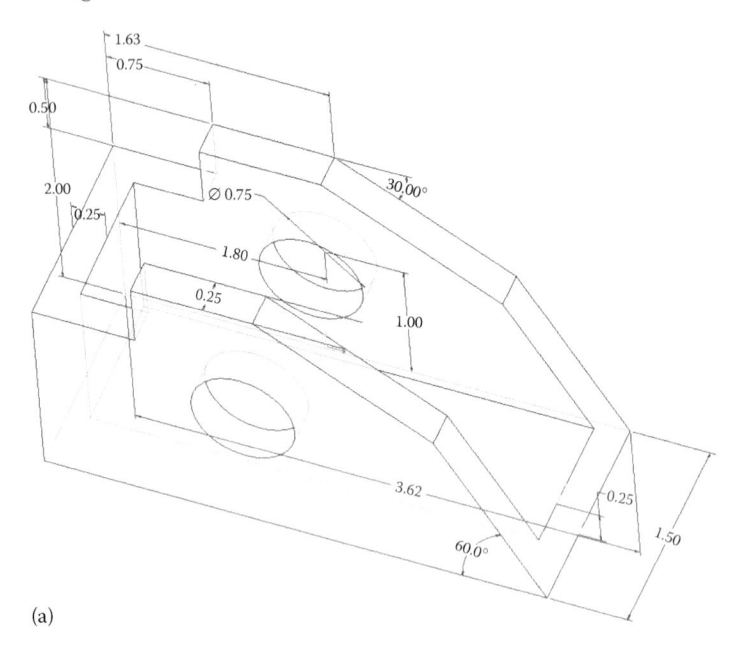

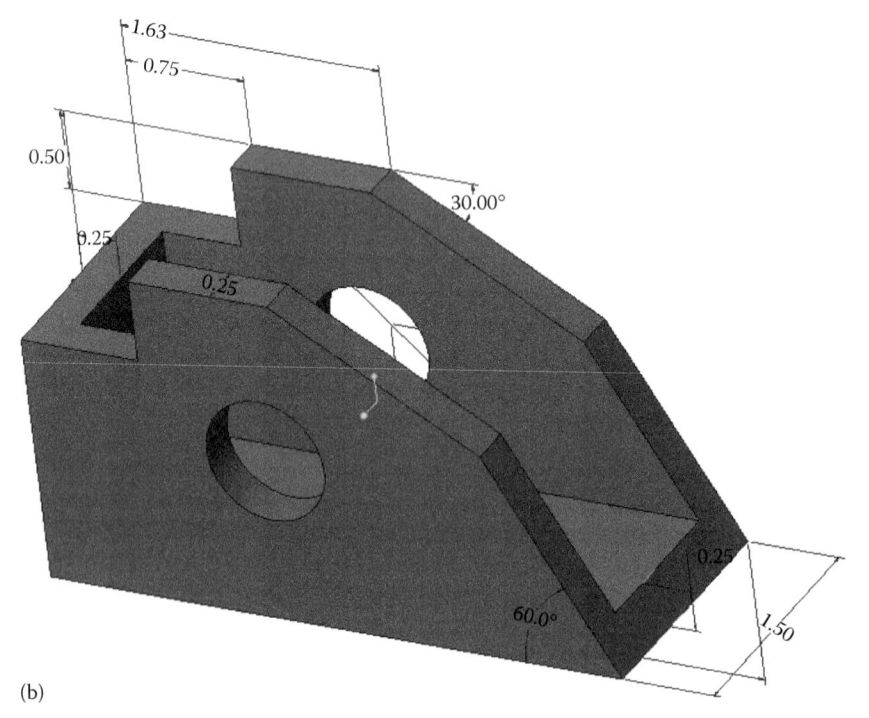

(a)

(b)

Computer-Aided Design

6. Drawing Problem 6: Develop Orthogonal Views (Front, Top, and Side)

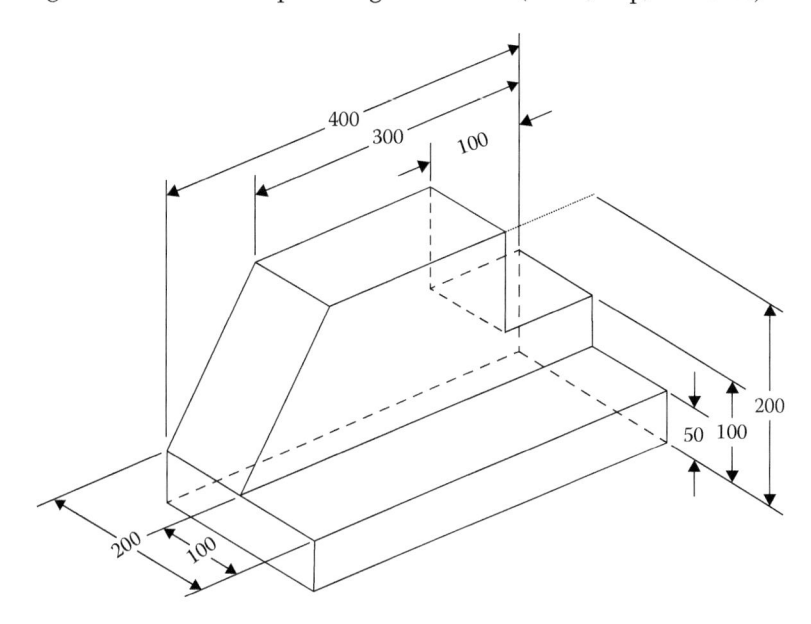

7. Drawing Problem 7: Create Isometric View

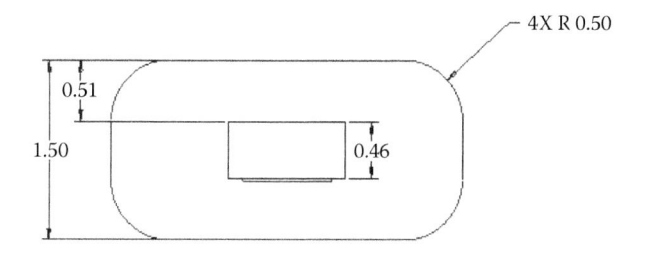

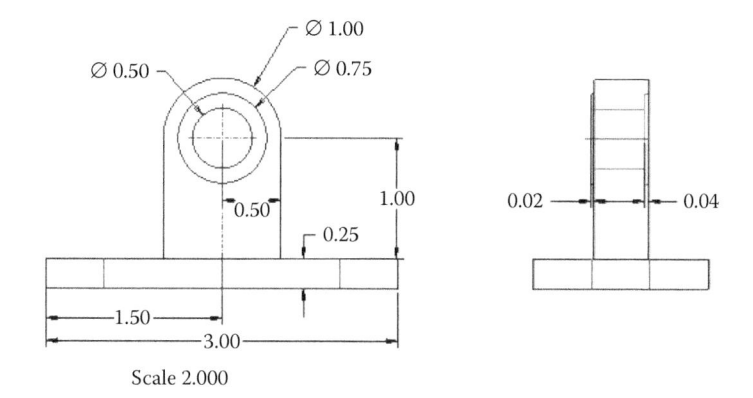

Scale 2.000

5

Statics

CHAPTER OBJECTIVE AND STUDENT LEARNING OUTCOMES

After completing the chapter, students will be able to

1. Define vectors and perform vector addition and subtraction.
2. Draw a free-body diagram for a statically equilibrium system.
3. Calculate forces and moments on a statically equilibrium system. [ABET outcome e, see Appendix C]
4. Calculate centroid and moment of inertia for different shapes. [ABET outcome a, see Appendix C]

5.1 Introduction

As the first branch of engineering mechanics, statics deals with stationary, rigid bodies or bodies that are in static equilibrium. This chapter will provide information about how to replace two or more forces acting on a particle with an equivalent force. Force is an action that one body exerts on another. Examples of force include gravitation, electrostatic, magnetic, and contact influences. Force can be divided into two parts: external and internal. External force is the action on a rigid body. If external force is in a state of unbalance, it will cause the body to move. Internal forces hold parts of a rigid body together. Internal forces can cause deformation but cannot cause motion of a body.

5.2 Scalars and Vectors

Quantity comes in two forms in engineering mechanics. Scalars, such as temperature, time, and mass, have magnitude but no direction. Vectors are quantities that require both magnitude and direction to describe them. Examples of vectors include location in a two- or three-dimensional plane, velocity, acceleration, force, and so on.

A force can be characterized by its point of application, its magnitude, and its direction, so force is always represented as a vector. In this chapter, all vectors will be written with

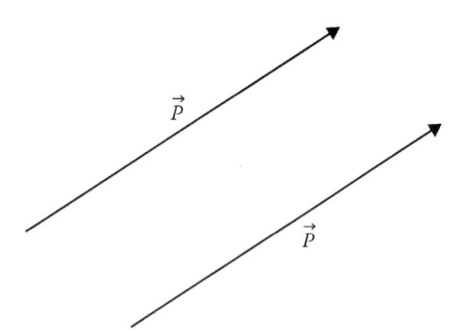

FIGURE 5.1
Vectors with same magnitude and direction.

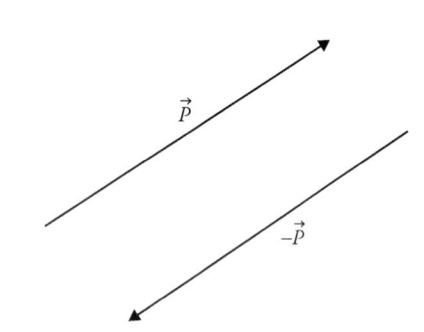

FIGURE 5.2
Vectors with same magnitude and opposite direction.

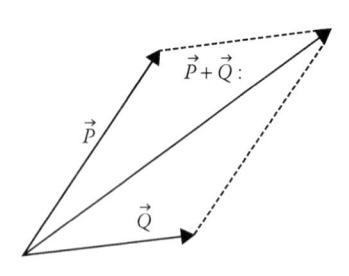

FIGURE 5.3
Addition of vectors using parallelogram law.

a short arrow above the letter, such as $\vec{P}$. A vector can be fixed or bound. Two vectors that have the same magnitude and direction are equal, or equivalent, even if they originate at different points (Figure 5.1). The negative sign of a given vector $\vec{P}$ means that a vector has the same magnitude of $\vec{P}$ but the opposite direction, written as $-\vec{P}$ (Figure 5.2), so $\vec{P} + (-\vec{P}) = 0$.

The sum of two vectors $\vec{P}$ and $\vec{Q}$ can be shown by the parallelogram law and obtained by attaching the two vectors to the same point (Figure 5.3). The addition of two vectors is commutative, and written as

$$\vec{P} + \vec{Q} = \vec{Q} + \vec{P}$$

We can simplify the parallelogram law to another method that determines the sum of two vectors, named the triangle rule. Because the side of the parallelogram opposite $\vec{Q}$ is equal to $\vec{Q}$ in magnitude and direction, half of the parallelogram is a triangle that tells us the magnitude and direction of the added vectors (Figure 5.4).

The subtraction of a vector can be defined as the addition of the corresponding negative vector, written as $\vec{P} - \vec{Q} = \vec{P} + (-\vec{Q})$ (Figure 5.5).

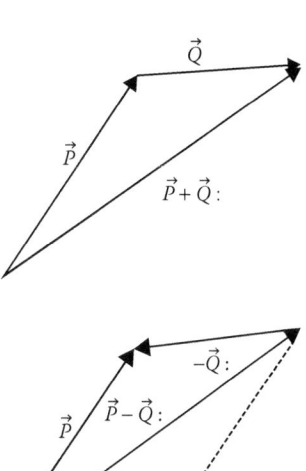

FIGURE 5.4
Addition of vectors using triangles.

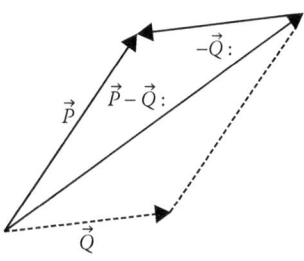

FIGURE 5.5
Subtraction of vectors.

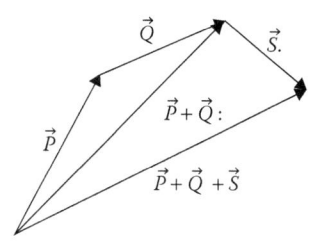

FIGURE 5.6
Addition of three or more vectors.

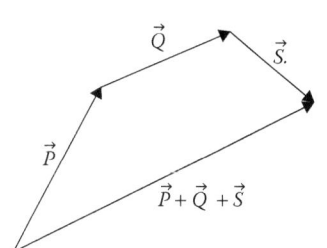

FIGURE 5.7
Addition of vectors using polygon rule.

For the addition of three or more vectors, if all vectors are coplanar, it is easy to use the triangle or polygon rule to solve it. As shown in Figure 5.6, apply the triangle rule to $\vec{P}+\vec{Q}$ first, and then apply again to obtain $\vec{P}+\vec{Q}$ and $\vec{S}$. As shown in Figure 5.7, apply the polygon rule to get the addition of vectors by connecting the tail of the first vector with the tip of the last one.

Two or more forces acting on a particle can be replaced by a single force, the same way as a single force $\vec{F}$ can be replaced by two or more vectors. These forces are called components of the original force. Forces are always represented in terms of a unit vector and force component. A unit vector is a vector of unit length directed along a coordinated axis, denoted by $\vec{i}, \vec{j}$. The components of a two- or three-dimensional force can be found using trigonometry to solve for the sine or cosine of the true angle made by the force vector with the x-, y-, and z-axes.

$$F_y = F \sin \theta$$

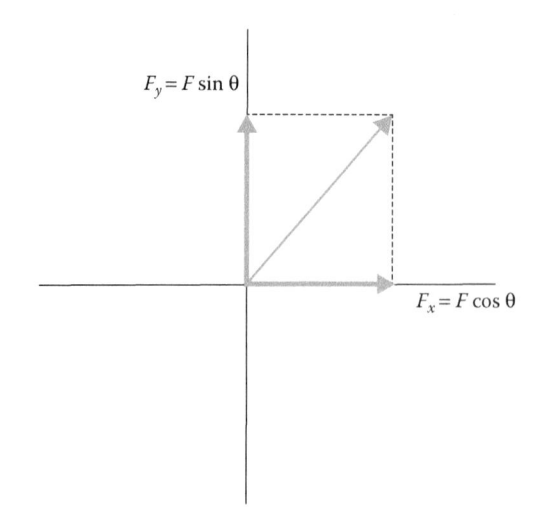

$$F_x = F \cos \theta$$

FIGURE 5.8
Resolving forces in x and y components.

In Figure 5.8, the force $\vec{F}$ has been resolved into components $\vec{F}_x$ and $\vec{F}_y$, which make a rectangle, so $\vec{F}_x$ and $\vec{F}_y$ are called rectangular components, which can also be represented as $\vec{F} = F_x \vec{i} + F_y \vec{j}$.

Denoting by F the magnitude of the force $\vec{F}$ and by θ the angle between $\vec{F}$ and the x-axis, we can express the x- and y-components of $\vec{F}$ as follows:

$$F_x = F \cos \theta; \quad F_y = F \sin \theta$$

Example 5.1

The four forces shown act at point A (Figure 5.9). What is the magnitude of the resultant force?

Solution:

$$\sum F_x = 30 \text{ N} + (60 \text{ N})\cos 30° + (75 \text{ N})\cos 90° + (90 \text{ N})\cos 120° = 36.96 \text{ N}$$

$$\sum F_y = (30 \text{ N})\sin 0° + (60 \text{ N})\sin 30° + 75 \text{ N} + (90 \text{ N})\sin 120° = 182.94 \text{ N}$$

$$R = \sqrt{(36.96)^2 + 182.94^2} = 186.64 \text{ N}$$

Example 5.2

Three forces act at point O (Figure 5.10). Determine the magnitude of the resultant force.

Solution:

$$\sum F_x = (1000 \text{ N})\cos 30° - (750 \text{ N})\cos 15° - (800 \text{ N})\cos 60° = -258.42 \text{ N}$$

Statics

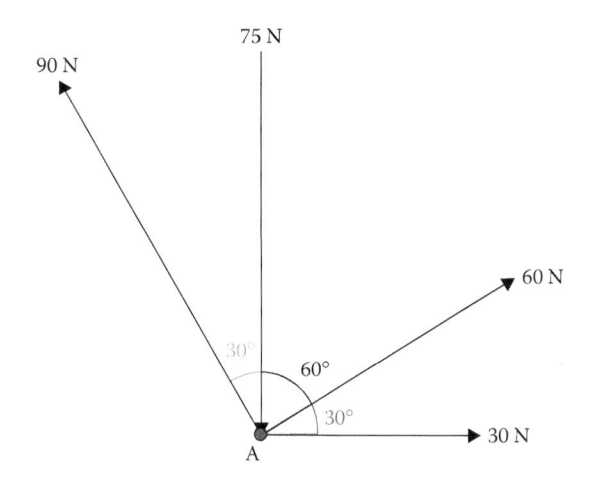

FIGURE 5.9
Resultant of four forces.

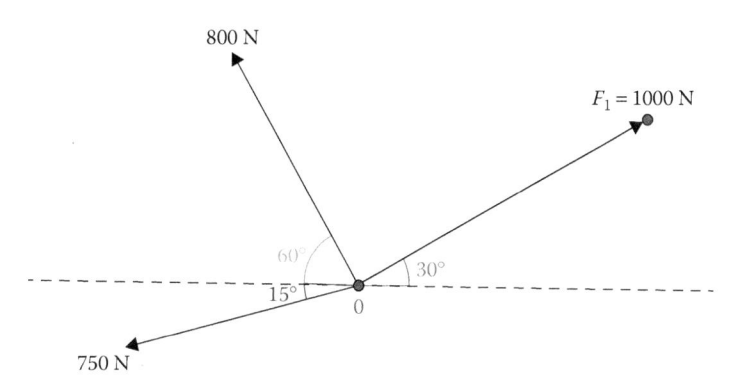

FIGURE 5.10
Resultant of three forces.

$$\sum F_y = (1000 \text{ N})\sin 30° - (750 \text{ N})\sin 15° + 75 \text{ N} + (800 \text{ N})\sin 60° = 998.71 \text{ N}$$

$$R = \sqrt{(-258.42)^2 + 998.71^2} = 1031.6 \text{ N}.$$

5.3 Equilibrium of Forces

Forces that act on rigid bodies can be divided into two groups: external and internal forces. External forces are the actions on a rigid body. Internal forces can cause deformation but never motion of a body. Equilibrium of forces means the resultant of all the external forces acting on a particle is zero (Figure 5.9). The expression of the equilibrium of point A algebraically is $\vec{R} = \Sigma\vec{F} = 0$. Resolving each force $\vec{F}$ into rectangular components, $\Sigma\vec{F}_x = 0$; $\Sigma\vec{F}_y = 0$.

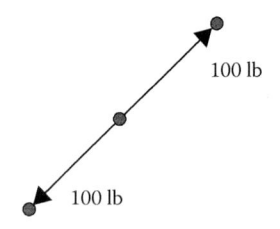

FIGURE 5.11
Equilibrium of forces.

The sum of two-dimensional forces is equal to the sum of the components: $R = \sqrt{F_x^2 + F_y^2}$ (Figure 5.11).

Example 5.3

A 100 kg block rests on an incline. There is no friction between the block and the ramp. The mass of the cable is negligible, and the pulley at point C is frictionless. What is the support force of the block and the smallest mass of block B that will start the 100 kg block moving up the incline? ($g = 9.8$ m/s²) (Figures 5.12 and 5.13).

Solution:

$$\sum F_y = N - m_A g \cos\theta = 0$$

$$N = m_A g \cos\theta = 100 \times 9.8 \times \frac{4}{5} = 784 \text{ kg}$$

FIGURE 5.12
Example 5.3.

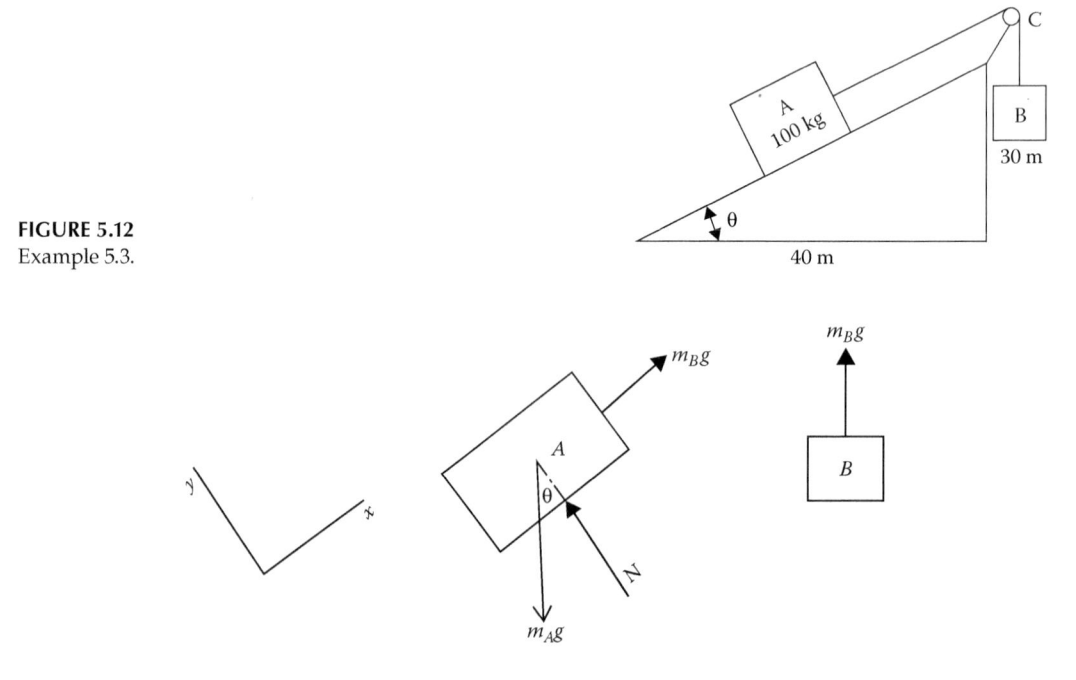

FIGURE 5.13
Free-body diagram of Example 5.3.

Statics

$$\sum F_x = m_B g - m_A g \sin\theta = 0$$

$$m_B = m_A \sin\theta = 100\left(\frac{3}{5}\right) = 60 \text{ kg}$$

Example 5.4

A 500 kg block rests on a frictionless incline. Forces are applied to the block as shown in Figures 5.14 and 5.15. What is the minimum force, P, such that no downward motion occurs?

Solution:

$$\sum F_x = P - mg\sin\theta = 0$$

$$P = mg\sin\theta = (500\,\text{kg})\left(9.81\,\frac{\text{m}}{\text{s}^2}\right)(\sin 30°) = 2452.5 \text{ N}$$

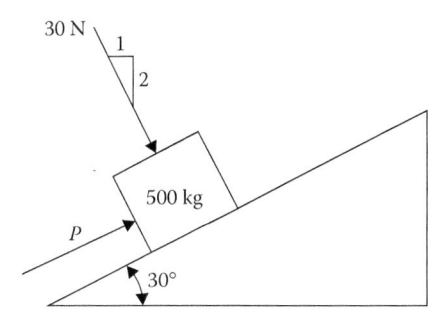

FIGURE 5.14
Example 5.4.

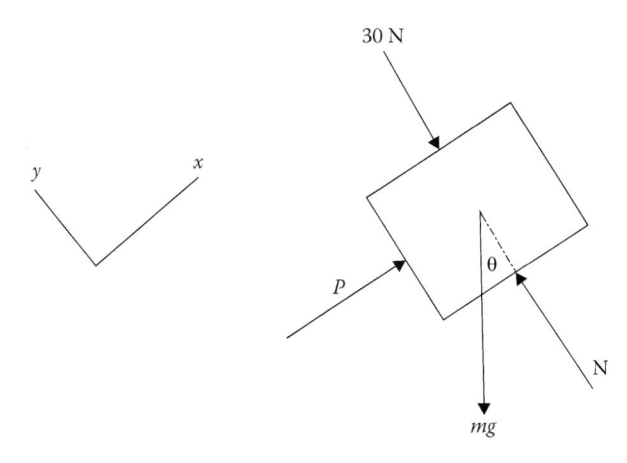

FIGURE 5.15
Free-body diagram of Example 5.4.

5.4 Moment, Couple, Torque

Moment is the tendency of a force to rotate, turn, or twist a rigid body about an actual or assumed pivot point. The moment of $\vec{F}$ about O is written as the vector product of the position vector $\vec{r}$ and the force $\vec{F}$:

$$\vec{M}_o = \vec{r} \times \vec{F}$$

The moment $\vec{M}_o$ must be perpendicular to the plane containing O and the force $\vec{F}$. If the rotation is counterclockwise, then the moment is positive; if the rotation is clockwise, the moment is negative.

According to the angle θ between the lines of action of the position vector $\vec{r}$ and the force $\vec{F}$, the magnitude of the moment of $\vec{F}$ about O is

$$M_o = rF\sin\theta = Fd$$

where d is the perpendicular distance from O to the line of action of $\vec{F}$

Two forces that have the same magnitude, parallel lines of action, and opposite sense are defined as a couple. The sum of a couple is zero, but the sum of the moment of the couple is not zero. The sum of the moments of $\vec{F}$ and $-\vec{F}$ about O is represented by the vector

$$\vec{M} = \vec{r} \times \vec{F}$$

where:
$\vec{r}$ is the vector joining the points of application of the two forces
$\vec{M}$ is the moment of the couple

The magnitude of $\vec{M}$ is

$$M = rF\sin\theta = Fd$$

where d is the perpendicular distance between the lines of action of $\vec{F}$ and $-\vec{F}$.

Torque is another name for moment, but it is used mainly with shafts and other power-transmitting machines.

Example 5.5

The loading shown in Figure 5.16 requires a resisting moment of 40 N m at the support. Calculate the value of force F.

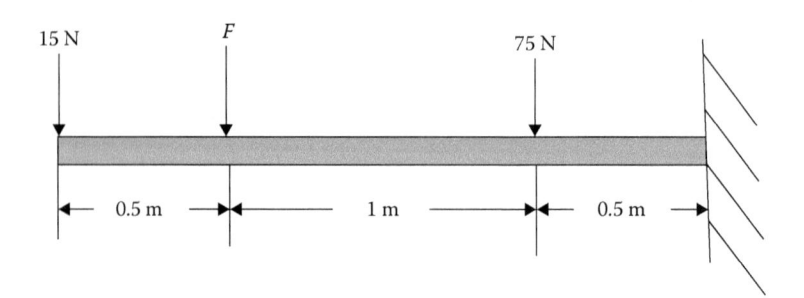

FIGURE 5.16
Cantilever beam with multiple forces of Example 5.5.

Statics

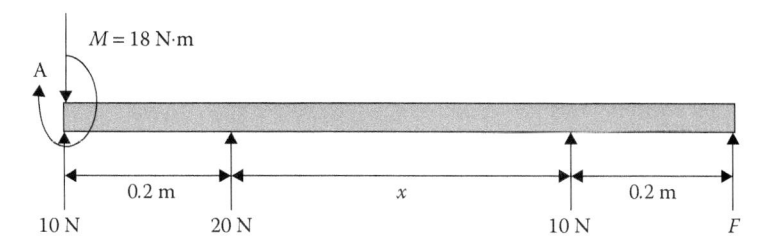

FIGURE 5.17
Simply supported beam with multiple forces of Example 5.6.

Solution:

$$\sum M_A = 0$$

$$= 40 \text{ N m} - (75 \text{ N})(0.5 \text{ m}) - F(1.5 \text{ m}) - (15 \text{ N})(2 \text{ m})$$

$$F = 18.33 \text{ N}$$

Example 5.6

A cantilever beam is acted on by a moment and several concentrated forces, as shown in Figure 5.17. Find the unknown force, F, and distance x, that will maintain equilibrium on the member shown.

Solution:

$$\sum F_y = 0$$

$$= -45 \text{ N} + 10 \text{ N} + 20 \text{ N} + 10 \text{ N} + F = 0$$

$$F = 5 \text{ N}$$

To determine the distance x where the resultant force $F = 5\text{N}$ acts, it is necessary to take moment about a point. As we can ignore the forces acting on the point, where moment is taken, it is convenient to take moment about a point where the maximum number of forces act. In the above-mentioned example, there are two forces acting on point A. Therefore, moment was taken about point A:

$$\sum M_A = 0$$

$$= 18 \text{ N m} - (20 \text{ N})(0.2 \text{ m}) - 10(0.2 \text{ m} + x) - (5 \text{ N})(0.4 \text{ m} + x)$$

$$= 4 + 2 + 10x + 5x = 18$$

$$x = 0.8 \text{ m}$$

5.5 Free-Body Diagram

When solving a problem of the equilibrium of a rigid body, it is necessary to consider all forces acting on the rigid body. The first step is to draw a free-body diagram. When drawing a free-body diagram, all external forces should be located correctly, such that their direction and magnitude are accurate. A clear, accurate free-body diagram is a critical tool to the work of an engineer.

Table 5.1 shows all essential reactions exerted on a rigid body.

Example 5.7

Block B sits freely on the homogeneous bar and experiences a gravitation force of 100 N (Figure 5.18). The homogeneous bar experiences a gravitational force of 50 N. There are

TABLE 5.1

Reaction Forces and Moments of Different Type of Support

Type of Support	Reaction and Moments	Number of Unknowns
Simple, roller, rocker, ball, or frictionless surfaces	Reaction from normal to suface, no moment	1
Cable in tension or link	Reaction in line with cable or link, no moment	1
Frictionless guide or collar	Reaction from normal to rail, no moment	1
Built-in, fixed support	Two reaction components, one moment	3
Frictionless hinge, pin connnection, or rough surface	Reaction in any direction, no moment	2

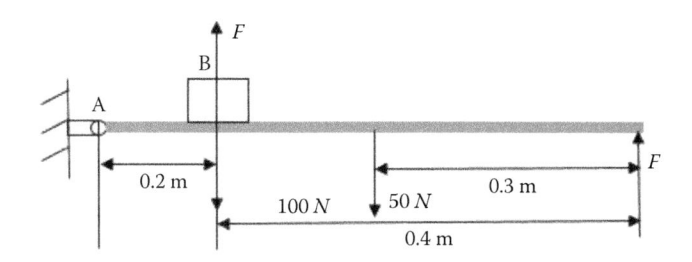

Statics 109

FIGURE 5.18
Equilibrium of force.

two equal support forces, *F*, located as shown. What is the force between the bar and block B? Please draw a free-body diagram to calculate.

Solution:

The free-body diagram of the bar (Figure 5.19):

$$\sum M_A = 0$$

$$= (100\ N - F)(0.2\ m) + (50\ N)(0.4 - 0.3 + 0.2\ m) - F(0.4 + 0.2\ m)$$

$$F = 43.75\ N$$

The free-body diagram of block B (Figure 5.20):

$$\sum F_y = 0$$

$$F - 100\ N + N = 0$$

$$N = 100\ N - 43.75\ N = 56.25\ N$$

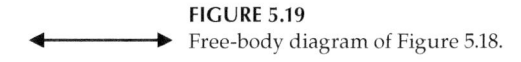

FIGURE 5.19
Free-body diagram of Figure 5.18.

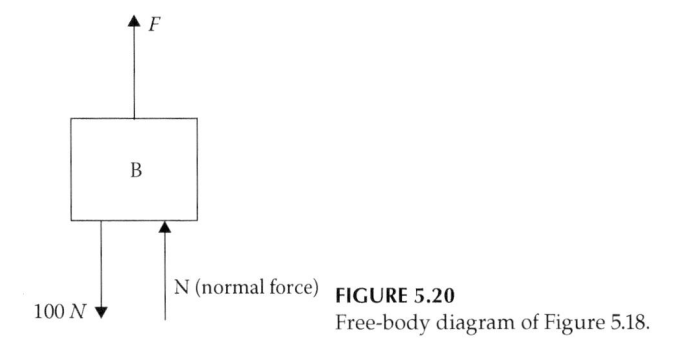

FIGURE 5.20
Free-body diagram of Figure 5.18.

5.6 Centroids and Center of Gravity

Centroid is the geometric center of a shape. In a homogeneous shape, the centroid is the same as the center of gravity. The center of gravity is the point of balance. In a free-body diagram, this is the point where gravity is drawn as an acting force. The weight $\vec{W}$ and the coordinates $\bar{x}, \bar{y}$ of the equations

$$W = \int dW; \quad \bar{x}W = \int x\,dW; \quad \bar{y}W = \int y\,dW$$

define the center of gravity G of a two-dimensional structure.

The coordinates $\bar{x}, \bar{y}$ of the equations

$$\bar{x}A = \int x\,dA; \quad \bar{y}A = \int y\,dA$$

define the center of gravity of a homogeneous plate. The point $(\bar{x}, \bar{y})$ is also known as the centroid C of the plate. Another important term is the first moment of the area, A, which is a geometric property that describes a shape's resistance to bending about a certain axis. The first moment of the area A can be expressed as

$$Q_y = \bar{x}A; \quad Q_x = \bar{y}A$$

Table 5.2 shows the centroid of common 2D shapes and Table 5.3 shows centroids of common lines.

Example 5.8

What are the x- and y-coordinates of the centroid of the area? See Figure 5.21.

Solution:

Divide the area into two rectangles, HCBA and FEDG. Their areas are

$$A_1 = (8 \text{ cm})(3 \text{ cm}) = 24 \text{ cm}^2$$

$$A_2 = (5 \text{ cm})(3 \text{ cm}) = 15 \text{ cm}^2$$

The total area is

$$A = A_1 + A_2 = 24 \text{ cm}^2 + 15 \text{ cm}^2 = 39 \text{ cm}^2$$

By inspection, the x-coordinates of the centroids of the rectangles are $x_{c,1} = x_{c,2} = 4$ cm. This is because of the symmetry of the shape. To prove this, we will still perform the analysis of the x-coordinate of the centroid as follows:

$$\bar{x} = \frac{\int x\,dA}{A} = \frac{\sum x_{c,n} A_n}{A} = \frac{(4 \text{ cm})(24 \text{ cm}^2) + (4 \text{ cm})(15 \text{ cm}^2)}{39 \text{ cm}^2} = 4 \text{ cm}$$

TABLE 5.2
Center of Gravity and Area of Common Geometries

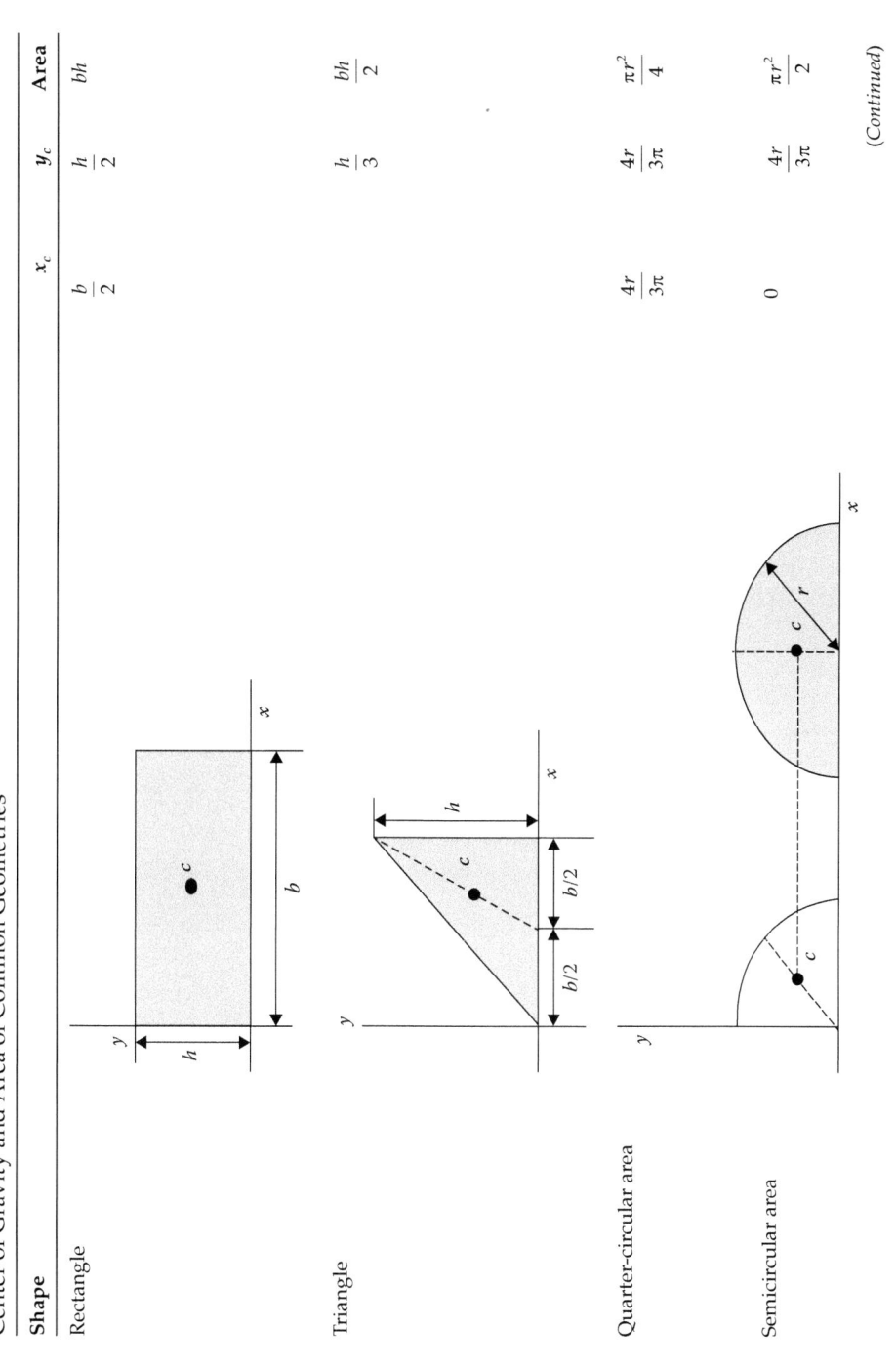

Shape	x_c	y_c	Area
Rectangle	$\dfrac{b}{2}$	$\dfrac{h}{2}$	bh
Triangle		$\dfrac{h}{3}$	$\dfrac{bh}{2}$
Quarter-circular area	$\dfrac{4r}{3\pi}$	$\dfrac{4r}{3\pi}$	$\dfrac{\pi r^2}{4}$
Semicircular area	0	$\dfrac{4r}{3\pi}$	$\dfrac{\pi r^2}{2}$

(Continued)

TABLE 5.2 (Continued)
Center of Gravity and Area of Common Geometries

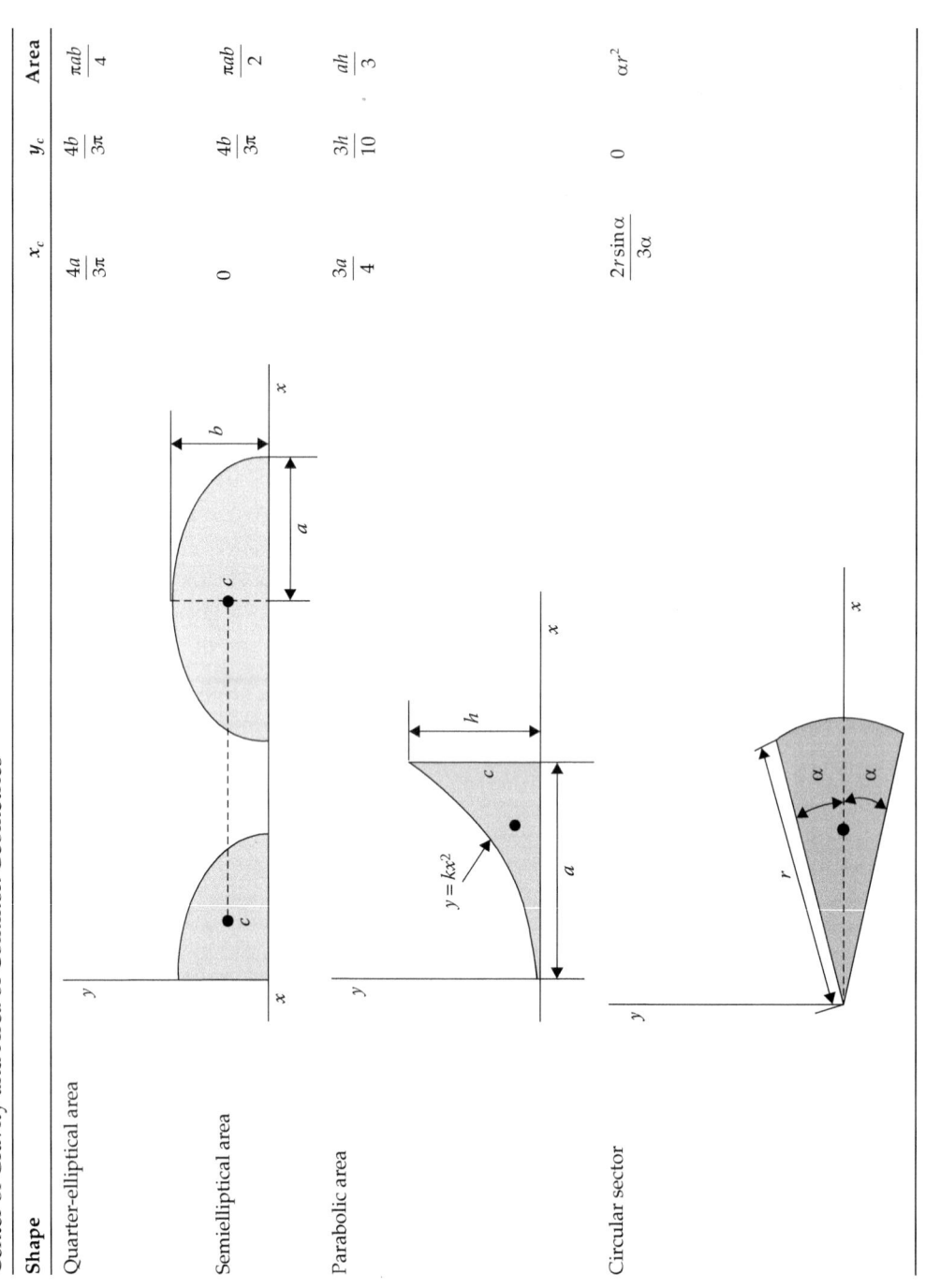

Shape		x_c	y_c	Area
Quarter-elliptical area		$\dfrac{4a}{3\pi}$	$\dfrac{4b}{3\pi}$	$\dfrac{\pi ab}{4}$
Semielliptical area		0	$\dfrac{4b}{3\pi}$	$\dfrac{\pi ab}{2}$
Parabolic area		$\dfrac{3a}{4}$	$\dfrac{3h}{10}$	$\dfrac{ah}{3}$
Circular sector		$\dfrac{2r\sin\alpha}{3\alpha}$	0	αr^2

TABLE 5.3
Center of Gravity and Length of Common Geometries

Shape	x_c	y_c	Length
Quarter-circular arc	$\dfrac{2r}{\pi}$	$\dfrac{2r}{\pi}$	$\dfrac{\pi r}{2}$
Semicircular arc	0	$\dfrac{2r}{\pi}$	πr
Arc of circle	$\dfrac{r\sin\alpha}{\alpha}$	0	$2\alpha r$

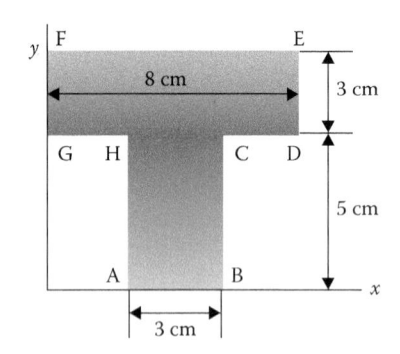

FIGURE 5.21
Centroid of a T-shaped section.

By inspection, the y-coordinates of the centroids of each rectangle are $y_{c,1} = 2.5$ cm; $y_{c,2} = 6.5$ cm. The y-coordinate of the centroids of the total area is

$$\bar{y} = \frac{\int y\, dA}{A} = \frac{\sum y_{c,n} A_n}{A} = \frac{(2.5 \text{ cm})(24 \text{ cm}^2) + (6.5 \text{ cm})(15 \text{ cm}^2)}{39 \text{ cm}^2} = 4.04 \text{ cm}$$

5.7 Moment of Inertia

The equations of the second moment, or moment of inertia, of the beam section with respect to the x-axis and y-axis that is denoted by I_x and I_y are $I_x = \int y^2 dA$ and $I_y = \int x^2 dA$, respectively.

If the moment of inertia is known with respect to one axis, the moment of inertia with respect to a parallel axis can be calculated using the parallel axis theorem. This theorem is used to evaluate the moment of inertia of areas that are composed of two or more basic shapes. This theorem can be expressed as

$$I_{\text{parallel}} = I_c + Ad^2$$

where d is the distance between the centroidal axis and the second, parallel axis.

The equation of polar moment of inertia with respect to the pole O is $J_O = \int r^2 dA$, where r is the distance from O to the element of area dA. However, it is easier to use the perpendicular axis theorem to quickly calculate the polar moment of inertia. The perpendicular axis theorem can be expressed as

$$J = I_x + I_y$$

Table 5.4 shows the moments of inertia of common geometric shapes.

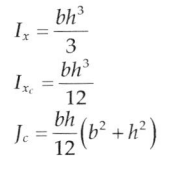

TABLE 5.4
Moment of Inertia of Common Geometries

Shape		I, J
Rectangle	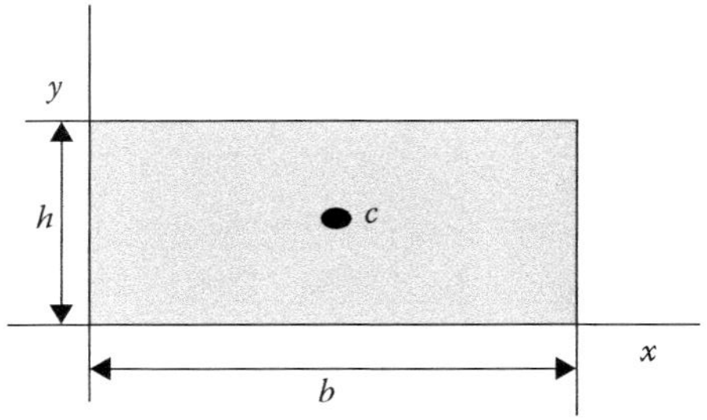	$I_x = \dfrac{bh^3}{3}$ $I_{x_c} = \dfrac{bh^3}{12}$ $J_c = \dfrac{bh}{12}(b^2 + h^2)$
Triangle	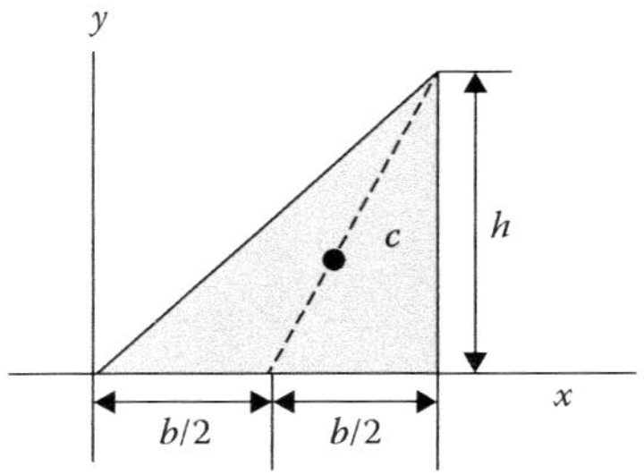	$I_x = \dfrac{bh^3}{12}$ $I_{x_c} = \dfrac{bh^3}{36}$
Circle	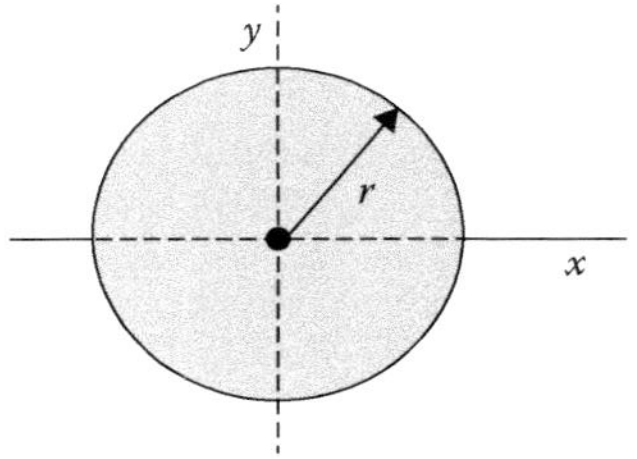	$I_x = I_y = \dfrac{\pi r^4}{4}$ $J_c = \dfrac{\pi r^4}{2}$

(Continued)

TABLE 5.4 (*Continued*)
Moment of Inertia of Common Geometries

Shape		I, J
Quarter-circular area		$I_x = I_y = \dfrac{\pi r^4}{16}$ $J_c = \dfrac{\pi r^4}{8}$
Semicircular area		$I_x = I_y = \dfrac{\pi r^4}{8}$ $I_{x_c} = 0.1098 r^4$ $J_o = \dfrac{\pi r^4}{4}$ $J_c = 0.5025 r^4$
Quarter-elliptical area		$I_x = \dfrac{\pi a b^3}{8}$ $I_y = \dfrac{\pi a^3 b}{8}$ $J_O = \dfrac{\pi a b}{8}(a^2 + b^2)$
Semielliptical area		
Parabolic area		$I_x = \dfrac{a h^3}{21}$ $I_y = \dfrac{3 h a^3}{15}$

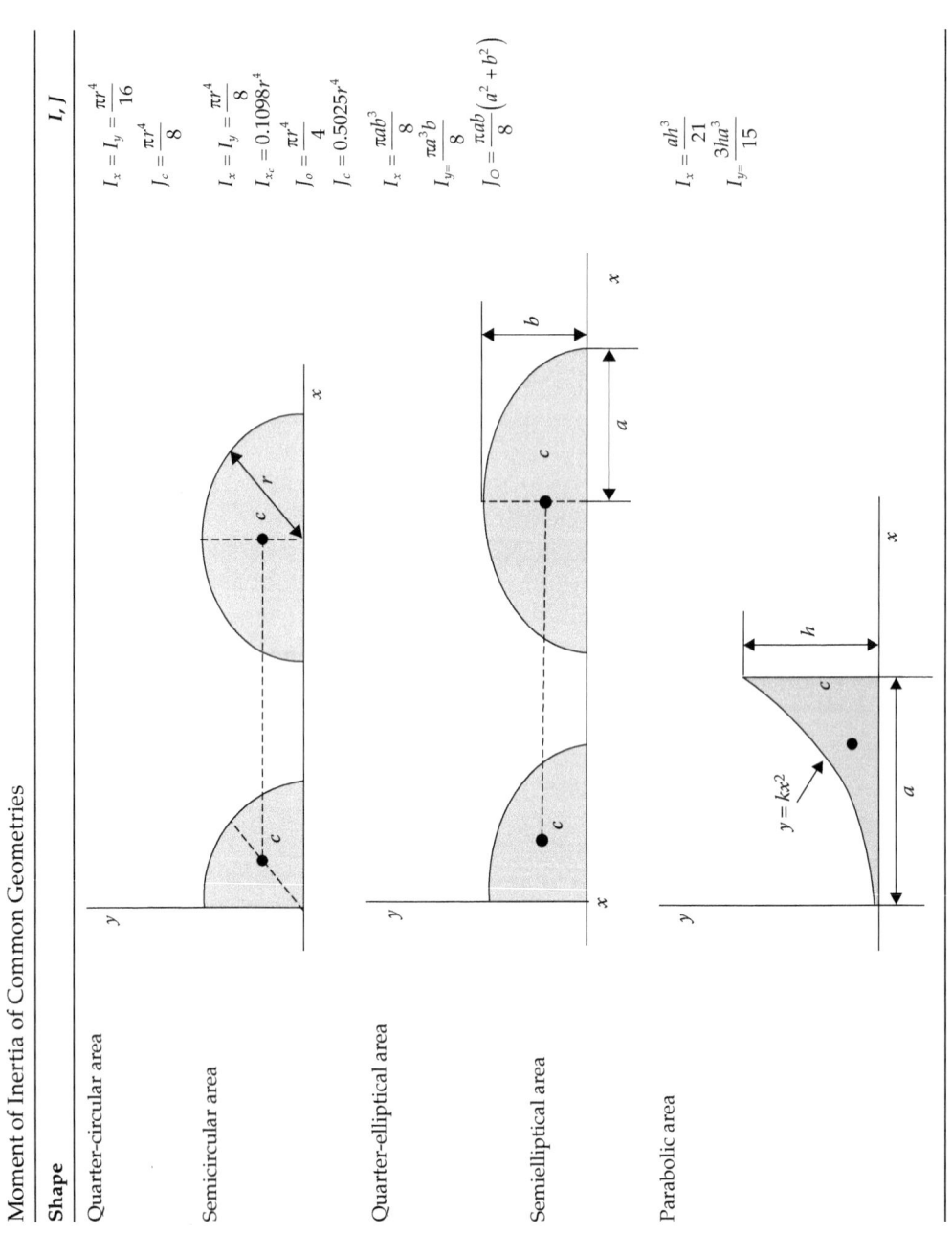

Statics 117

Example 5.9

According to Figure 5.22, please answer the following questions:

a. What is the moment of inertia about the *x*-axis?
b. What is the centroidal moment of inertia with respect to the *x*-axis?
c. What is the centroidal polar moment of inertia?

Solution:

a. The formula for the moment of inertia about an edge of a rectangle is given in Table 5.4.
For rectangle HCBA,

$$I_{x,1} = \frac{bh^3}{3} = \frac{(3 \text{ cm})(5 \text{ cm})^3}{3} = 125 \text{ cm}^4$$

Use the parallel axis theorem to calculate the moment of inertia of rectangle FEDG, $d = 6.5$ cm is the distance from the centroid of FEDG to the *x*-axis.

$$I_{x,2} = \frac{bh^3}{12} + Ad^2 = \frac{(8 \text{ cm})(3 \text{ cm})^3}{12} + (8 \text{ cm})(3 \text{ cm})(6.5 \text{ cm})^2 = 1032 \text{ cm}^4$$

The moment of inertia for the total area is

$$I_x = I_{x,1} + I_{x,2} = 125 \text{ cm}^4 + 1032 \text{ cm}^4 = 1157 \text{ cm}^4$$

b. From Example 5.8, the centroidal *y*-coordinate is 4.04 cm. Use the parallel axis theorem to find the centroidal moment of inertia of each rectangular area:

$$I_{x,c} = \left(I_{c,1} + Ad_1^2\right) + \left(I_{c,2} + Ad_2^2\right)$$

$$= \left[\frac{(3 \text{ cm})(5 \text{ cm})^3}{12} + (15 \text{ cm}^2)(4.04 \text{ cm} - 2.5 \text{ cm})^2\right] +$$

$$\left[\frac{(8 \text{ cm})(3 \text{ cm})^3}{12} + (24 \text{ cm}^2)(6.5 \text{ cm} - 4.04 \text{ cm})^2\right]$$

$$= 66.824 \text{ cm}^4 + 163.2384 \text{ cm}^4 = 230.0624 \text{ cm}^4$$

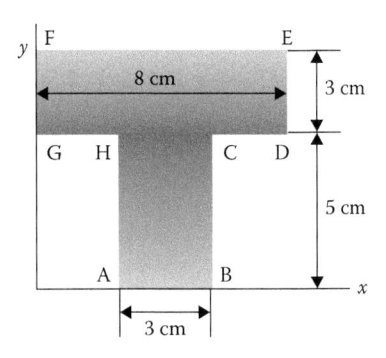

FIGURE 5.22
Moment of inertia of a T-shaped section.

c. The centroidal polar moment of inertia is given by

$$J_c = I_{x,c} + I_{y,c}$$

$$I_{y,c} = \frac{(5\text{ cm})(3\text{ cm})^3}{12} + \frac{(3\text{ cm})(8\text{ cm})^3}{12} = 139.25\text{ cm}^4$$

$$J_c = 230.0624\text{ cm}^4 + 139.25\text{ cm}^4 = 369.3124\text{ cm}^4$$

Problems

1. Resolve the 500 N force into two components, one along line P and the other along line Q (F, P, Q are coplanar) (see Figure 5.23).
2. Three concurrent forces act as shown in Figure 5.24. If the forces are in equilibrium and $F_2 = 15\text{N}$, what is the magnitude of F_1?
3. An inclined force, F, is applied to a block of mass m. What is the minimum coefficient of static friction μ_s between the block and the ramp surface such that no motion occurs? (The friction force is less than or equal to the coefficient of static friction multiplied by the normal force: $F_f \leq \mu_s \times N$.)

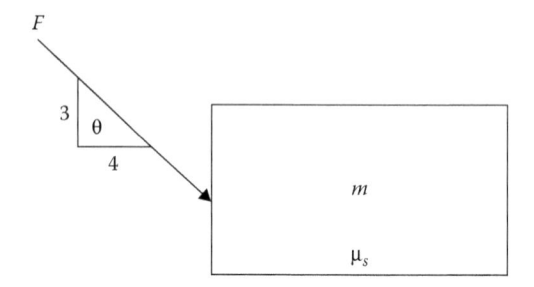

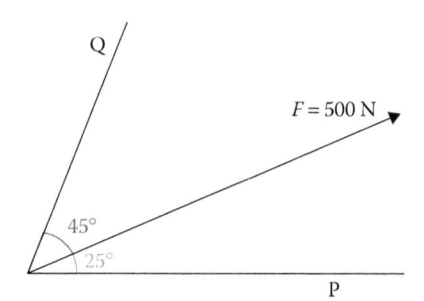

FIGURE 5.23
Resolution of coplanar forces.

Statics

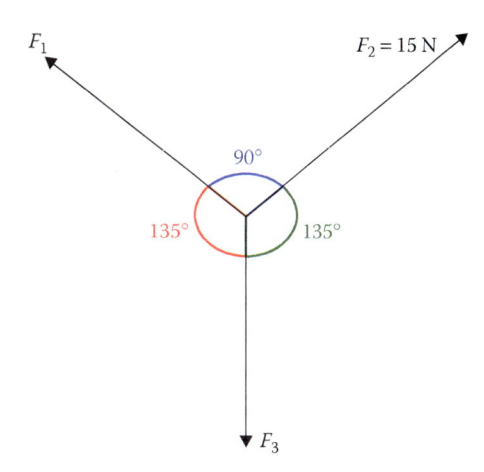

FIGURE 5.24
Resolution of three concurrent forces.

4. Two spheres, one with a mass of 8 kg and the other with a mass of 10 kg, are in equilibrium as shown. If all surface are frictionless, what is the magnitude of the reaction at point B?

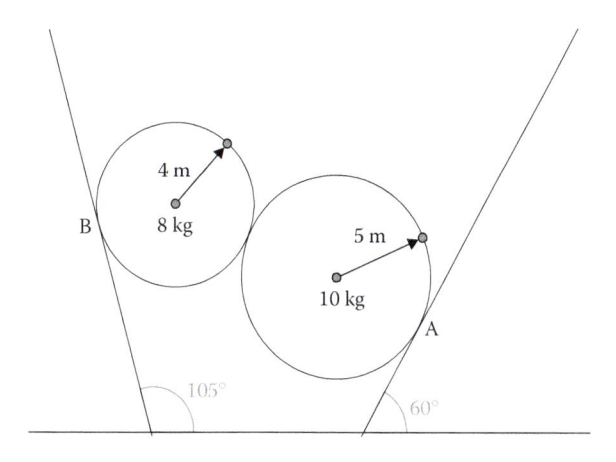

5. A bracket is subjected to the forces and couple shown. Determine an equivalent force–couple system to be applied at point 12 cm below the 300 N force.

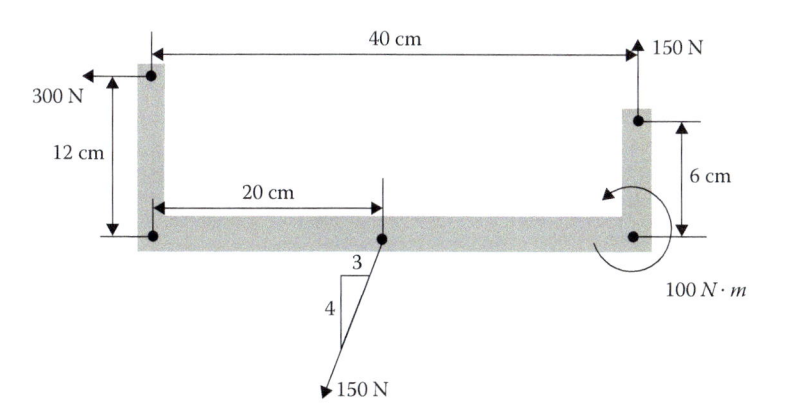

6. Four bolts connect support A to the ground. Determine the design load for each of the four bolts. (Free-body diagram is required.)

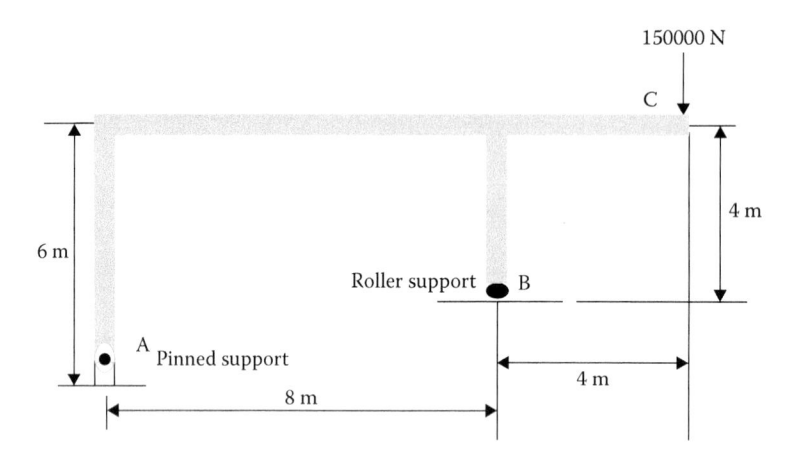

7. Determine the x- and y-coordinates of the centroid of the shaded area in the figure below.

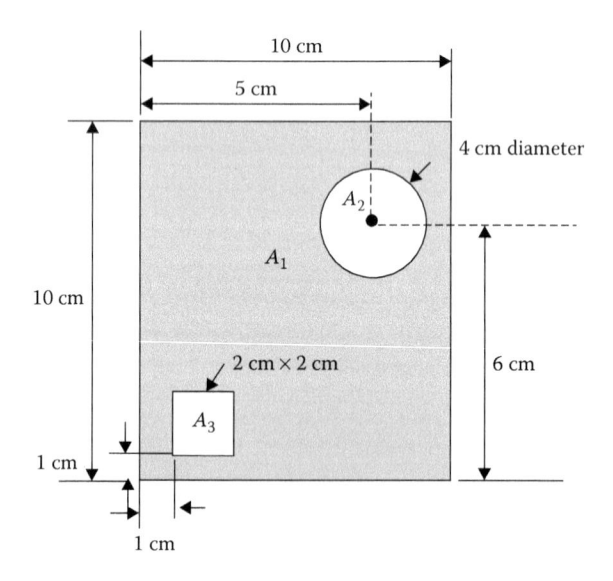

Statics

8. Determine the polar moment of inertia about the composite centroid in the figure shown below.

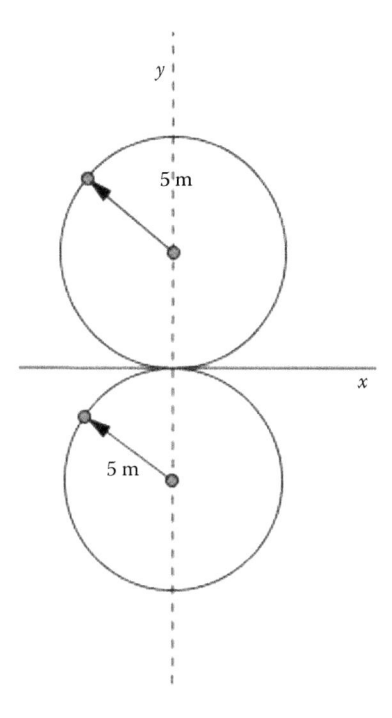

9. From the figure below, please answer the following questions:
   a. What are the x- and y-coordinates of the centroid?
   b. What is the moment of inertia about the x-axis?

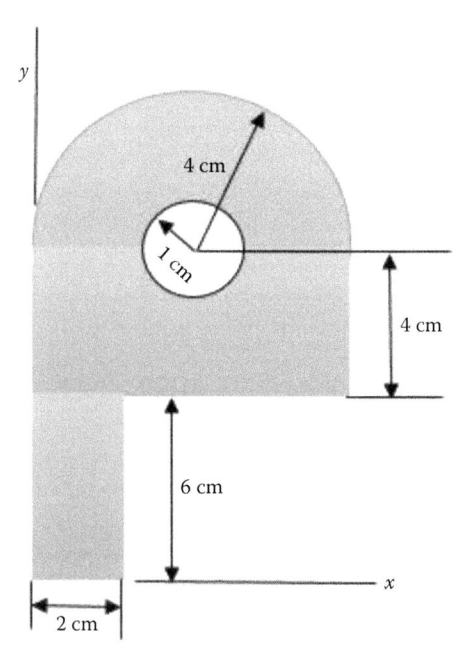

6
Materials Engineering

CHAPTER OBJECTIVE AND STUDENT LEARNING OUTCOMES

After completing this chapter, students should be able to

1. Describe metals, ceramics, and polymers based on their atomic bonding.
2. Relate how atomic bonding affects material properties.
3. Describe properties of metals, ceramics, and polymers.
4. Explain how processing affects material properties. [ABET outcome e, see Appendix C]
5. Describe composite materials and the rule of mixtures. [ABET outcome h, see Appendix C]

6.1 Introduction

Materials science is an enabling discipline. Advances in materials science have enabled advances in many other technologies, including silicon chips in computers, tool materials in manufacturing, aluminum in aircraft, aluminum oxide chips in mobile telephones, and a multitude of polymer applications from food service to packaging to automotive materials. As they design the products of tomorrow, engineers in all disciplines need to have a broad understanding of what materials can and cannot do.

6.2 Metals

Metal is a very important material in our society; it forms the structure of many of our buildings, the engines of our automobiles, and the tools, motors, and appliances that shape much of our lives. If asked, most people can point to metal objects such as a doorknob or a snowplow. But what is metal? How is it different from other materials such as plastics or ceramics, and what makes it different?

Metals are substances made up from metallic elements. They are, for the most part, shiny and opaque; that is, light does not pass through them but instead reflects off them. They are both strong, which means that they carry considerable stress, and fairly ductile, which means that they deform under sufficient stress. Metals are also highly conductive of both heat and electricity. Common elements such as copper, aluminum, and iron are metals. Other less common elements including titanium, vanadium, chromium, nickel, manganese, cobalt, zinc, gold, silver, platinum, and many others are also metals. Mixtures of metallic elements are called alloys. Brass, bronze, carbon steel, stainless steel, and pewter are examples of alloys.

The properties of metals can be traced to their atomic structure. Metallic atoms are not different from any other type of atom; they consist of nuclei around which electrons orbit. In metallic atoms, however, the outermost or valence electrons are somewhat loosely bound to the nucleus. In a solid piece of metal, the nuclei and the inner electrons hold well-defined places in a regular three-dimensional array. However, part of what defines metals is the arrangement of the outer electrons. These valence electrons are spread throughout the array. This type of bond can be visualized as a sea or cloud of electrons (Figure 6.1).

The nonlocalized nature of the electron cloud gives metals some of their unique properties. For example, because the electrons exist as a cloud, the bonds between metal atoms are the same in all directions. This is in contrast with the types of bonds that occur in polymers. Polymers have interatomic bonds that are highly directional, and this directionality affects their properties. The cloud-like nature of the valence electrons in metals also accounts for their reflective and opaque nature. Light rays are reflected off electrons. Because the electrons in metals are diffused throughout the entire array of nuclei, a ray of light will be reflected from anywhere it strikes a metal.

By contrast, polymers and ceramics tend to be transparent. This is in part because ceramic and polymer atoms have their electrons restricted to relatively few locations. Light can go between these localized electrons, thus making polymers and ceramics translucent. Note, however, that both polymers and ceramics can be made opaque with the addition of pigments or other coloring agents; crystallinity and grain boundaries can also interfere with light transmission.

Metals are highly conductive of electricity. This feature is another consequence of the metallic cloud of electrons. Because the cloud electrons are not tightly bound to the nuclei, it takes only a small voltage to make them begin moving. Electric current is defined by moving electrons; thus metals conduct electricity very well compared to ceramics and polymers. The free electrons that make metals good conductors of electricity also make them conduct heat very well compared to most polymers and ceramics. Heat can be thought of as vibration or kinetic energy. Electrons vibrate more energetically at high temperatures than at low temperatures. In a gas, heated atoms transfer their energy to other atoms through collisions. Similarly, because the outer electrons in a metal are in something like a cloud, they can transfer heat along the metal piece by colliding with other electrons. This makes most metals conduct heat rapidly.

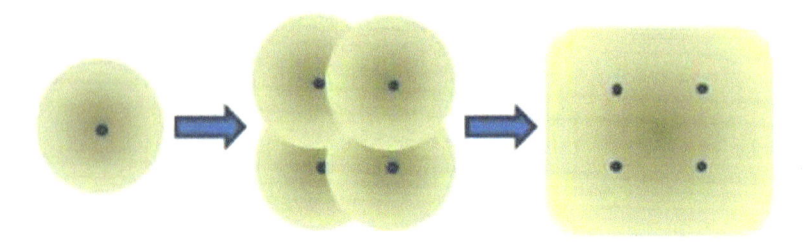

FIGURE 6.1
A single metal atom combines with others.

Because the electron cloud makes metallic bonding nondirectional, the nuclei and inner electrons of metals have no electron-related restrictions on the number and orientation of their nearest neighbors. This means that metal atoms can stack up in the three-dimensional arrangement that packs them together most efficiently. In other words, there is little empty space between atoms of metal. This contrasts with the atoms making up polymers; their bonding electrons are limited to relatively small areas, so that the number of nearest-neighbor atoms is limited. This is the reason why metals tend to have higher densities than polymers; their atoms can pack together more closely because their electrons exist as a nondirectional cloud rather than in a few specific locations. While the electrons in metals exist as a cloud, the nuclei and inner electrons (called the ion cores) are arranged in regular patterns under most circumstances. These regular patterns make metals crystalline for the most part; that is, the ion cores are arranged in a repeating array over long distances in terms of the size of the atoms. This regular arrangement accounts for some of the properties of metals.

There are many types of crystal structures; some are simple and some are complex. Most of the common metals have one of three crystal structures. These three crystal structures are face-centered cubic (FCC), body-centered cubic (BCC), and hexagonal close packed (HCP). Copper, aluminum, nickel, and silver have FCC structures. Chromium, tungsten, and iron at room temperature are BCC. Cobalt and zinc have HCP structures. The smallest repeating arrangement in a crystal is called a unit cell. As the name suggests, the unit cells for FCC and BCC are in the shape of cubes, while for HCP, the unit cell is a hexagon-shaped polyhedron.

The simple cubic is a specific type of unit cell. The atoms can be visualized as a sphere, simple cubic unit cell with one atom at each corner. BCC unit cells are similar to simple cubic, but with the addition of another atom at the center of the cube, like a baseball in a box. FCC unit cells have an atom at each corner and also an atom at each face of the cube. FCC unit cells can be pictured with the idea of a rigged gambling dice. The FCC dice would have single dots on all six faces, and also at the corners.

The BCC unit cell is a simple cubic cell with an additional atom at the center of the cube (red atom, center). The FCC unit cell (right) has atoms at the corners of the cube but also contains an atom at each cube face; these are the red atoms. While Figure 6.2 shows a total of 14 atoms per FCC unit cell, there are actually a total of 4 atoms per FCC unit cell. Each corner of the cube shares an atom with seven other unit cells, so each corner counts as 1/8 of an atom. Because there are eight corners to a cube, the corners account for one atom per FCC unit cell. FCC unit cells have one atom on each of its six faces. However, each face-centering atom is shared with another unit cell; this means that the face-centering atoms contribute a total of three atoms to the unit cell, for a total of four.

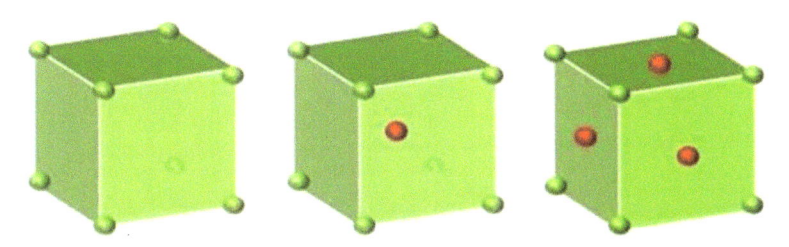

FIGURE 6.2
Simple cubic unit cell (left), BCC (center), and FCC (right).

In the FCC lattice, it should be noted that the atoms are treated as hard spheres and that these spheres only touch on the faces of the cube. Simple geometry calculations then show that the edge length, a, for an FCC unit cell should be

$$a = 2R\sqrt{2} \tag{6.1}$$

BCC unit cells have an atom at each corner as well as a single atom at the exact center of the cube. There are no atoms at the faces of the cube in BCC unit cells. This arrangement gives BCC unit cells two atoms. Again, each atom at a corner contributes 1/8 to the unit cell for a total of one atom, while the body-centering atom accounts for the second. The geometric calculations for the edge length of a BCC unit cell are slightly more complex because the atoms touch only on the body diagonal (the diagonal that runs, e.g., from the upper-left to lower-right corners of the cube). They yield the following relationship:

$$a = \frac{4R}{(\sqrt{3})} \tag{6.2}$$

The geometry of HCP unit cells is based on triangles rather than cubes and is more complex. The exterior of an HCP unit cell consists of two hexagon shapes connected by six rectangles. Each hexagonal face has atoms at the six corners as well as another at the center of the hexagon. In addition, there is a triangle containing three more atoms at the center of the unit cell. HCP unit cells contain a total of six atoms. The three at the center of the unit cell are contained as a whole. The two atoms at the centers of the hexagonal faces account for one more atom in the unit cell count. There are two halves, because each atom is shared with another unit cell. Finally, there is one atom for each hexagonal face as each corner of the hexagonal faces contributes 1/6 of an atom to the unit cell total.

Example 6.1

A theoretical metal atom has a diameter of 0.4 nanometers (nm). What is the volume of its unit cell if the atoms are arranged in a BCC lattice?

Solution:

$$a = \frac{4R}{(\sqrt{3})}$$
$$= 4 * 0.2 * 1.732$$
$$= 1.3856 \, nm$$

Volume of a cube is length cubed, so

$$V = (1.3856)^3$$
$$= 2.6604 \, nm$$

6.2.1 How Do We Obtain Metals?

Most of us are aware that metals come from ores mined from the earth's crust. But how do we transform mineral ores into metals?

Iron is a good example of an oxide-based ore. Much of the iron ore that is available for mining around the world is high in magnetite or hematite, both of which are oxides of iron. The ore must be dug up from mines or open pits. Typically the ore is found embedded in other types of rock such as quartz. This means that the ore must be beneficiated, or concentrated. The ore is crushed or milled into particles of smaller sizes, and then separated by gravity or some other method. Magnetite ore, if crushed finely enough, can be separated from quartz with the use of a magnet. Hematite ore is denser than magnetite; for hematite beneficiation typically involves gravity separation.

Once the iron ore is beneficiated, refining to metal can begin. Iron refining is done in blast furnaces, large tubular vessels lined with firebrick. Iron ore and coke (a form of carbon obtained from coal) are loaded at the top of the blast furnace and heated air is introduced at the bottom. Limestone is added to help remove the traces of quartz in iron ore. It does so by lowering the melting temperature of the quartz, just as salt lowers the melting temperature of ice on the roads in winter. The resulting material is known as slag. As these materials work their way down the blast furnace, the following chemical reactions take place:

$$2C + O_2 \rightarrow 2CO$$

$$Fe_2O_3 + 3CO \rightarrow 2Fe + 3CO_2$$

At the bottom of the blast furnace, liquid iron can be tapped just as a spigot lets water out of the bottom of a barrel. Slag also works its way to the bottom of the furnace, but it floats on top of the denser iron. It is tapped at a level above the iron tap in order to keep the two separate.

The iron coming from a blast furnace is known as pig iron because hundreds of years ago, blast furnace iron was fed into ingots whose shapes reminded the workers of piglets. Pig iron has high carbon levels, up to 4%; this makes pig iron too brittle for many uses.

Instead, pig iron is made into steel, which is considered to have 2% or less of carbon. This carbon reduction is accomplished by melting pig iron and blowing oxygen into the liquid. The oxygen oxidizes the carbon into carbon dioxide. This reduces the carbon in the iron to an acceptable level, while leaving the majority of the iron in the metallic state. Note that scrap iron can be, and typically is, added to pig iron as well; the tonnage of recycled steel is higher than any other recycled material such as aluminum or plastic.

Other metals are typically found in the form of sulfur-based ores. Copper is a good example of a sulfur-based ore; many types of copper ore are sulfides. Like iron ore, copper ore must be beneficiated before refining can begin, again by grinding, milling, and separating. The grinding step reduces the ore to fine particles. When the particles are fine enough, each individual particle is essentially made from a single compound—ore or waste mineral.

This fine particle size makes froth flotation separation possible. Froth flotation separation depends on the fact that ore versus mineral particles have different levels of affinity for water—that is, they are hydrophobic (water-fearing) or hydrophilic (water-loving) to varying degrees. The ground material is mixed with water and a surfactant chemical is added. The correct surfactant makes the ore particles hydrophobic but does not affect the mineral particles. When air is bubbled through the slurry, creating a froth, the now-hydrophobic ore particles attach themselves to the bubbles. This allows the ore-rich foam to be collected, separating it from the unwanted minerals.

In many cases, flash smelting takes place next. In this process, powdered ore is dispersed in a heated stream of air or oxygen, with the chemical reactions taking place in midair. When the particles settle, they form a liquid copper-rich layer covered by slag; as in a blast furnace, the slag floats on top of the metal because it is less dense than the metal. The liquid metal contains 30%–70% copper, mostly as a sulfide. Air blown through the liquid metal removes much of the sulfur as sulfur dioxide. This gas is captured and, when bubbled through water, turned into sulfuric acid for use in other chemical processes:

$$2CuS + 3O_2 \rightarrow 2CuO + 2SO_2$$

$$CuS + O_2 \rightarrow Cu + SO_2$$

In contrast to the above two thermal processes, aluminum refining relies primarily on chemical and electrical processes. Aluminum ore, known as bauxite, is found as a mixture of aluminum oxide and silicon dioxide (silica). The bauxite is crushed and milled to a fine particle size. Then it is immersed in hot sodium hydroxide, where it is converted to sodium aluminate in the Bayer process:

$$Al_2O_3 + 2NaOH \rightarrow 2NaAlO_2 + H_2O$$

This step also dissolves the silica. To remove the silica, lime (calcium carbonate) is added; this makes the silicon precipitate out as insoluble calcium silicate. Once this step is complete, carbon dioxide can be bubbled through the liquid solution of sodium aluminate, converting the dissolved aluminum oxide to aluminum hydroxide:

$$2NaAlO_2 + CO_2 \rightarrow Al(OH)_3 + Na_2CO_3 + H_2O$$

The aluminum hydroxide is then heated to a high temperature, where it decomposes into aluminum oxide (alumina) and water. At this point, aluminum metal can be obtained through the Hall–Héroult process. In this process, the alumina is dissolved in molten cryolite, Na_3AlF_6, or sodium aluminum fluoride. This molten salt bath behaves like a plating solution—when exposed to low voltage, the dissolved aluminum ions plate out at the cathode, while carbon dioxide plates out at the anode. Note that, although the voltage is low, the current is very high; aluminum refining consumes large amounts of electricity. The cell is run at over 1000°C, so the aluminum melts as its melting temperature is around 650°C. It settles at the bottom of the bath and spigot equipment is used to periodically pour off the liquid metal.

6.2.2 Properties of Metals

As mentioned above, metals cover a range of strengths and are usually at least somewhat ductile. In order to discuss the strength and ductility of metals, however, we must ask what is meant by these terms. Regarding strength, our common sense tells us that steel is stronger than, for example, nylon. How do we know this, and how do we measure it?

Suppose we are being asked which material should be used to build a bridge; our choices are steel and nylon. We would like to know which material is stronger because we want the bridge to be a strong, reliable structure. As a test, consider hanging a 100 lb load from a steel bar and the same weight from a similar nylon bar. We expect that the nylon might stretch and break while the steel would be affected little, if at all. However, suppose the nylon bar was 6 in. in diameter, while the steel bar was 1/4 in. in diameter. In that case, the 100 lb load would likely have little effect on the nylon while possibly breaking the steel. Which test gives us the better estimation of strength? Should bridges be built from nylon or from steel?

The obvious answer is that the strength of a bar depends both on its load-bearing cross section and the material from which it is made. In order to make a fair comparison between two materials, we need to normalize the results, or put them on the same basis. With this in mind, we can express the strength of materials in terms of load per area. In English units, this is pounds per square inch (psi) in the metric system, Newtons per square millimeter, or megapascal (MPa). In this way, we can remove the effects of size and make direct, valid comparisons between materials. The engineering term for strength is stress, symbolized by σ:

$$\sigma = \frac{F}{A_o} \quad (6.3)$$

where:
  σ is stress
  F is the force applied to the part
  A_o is the initial cross-sectional area that will carry the load

The change in shape of the part carrying the load is also an important factor. This factor, strain, is symbolized by ε:

$$\epsilon = \frac{(l_i - l_o)}{l_i} \quad (6.4)$$

where:
  l_i is the instantaneous length
  l_o is the original length

Strain, in other words, is change in length divided by initial length.

Testing of this type is carried out in a tensile tester. This machine applies a load to a tensile bar of a specified shape, pulling or compressing to failure while recording load and deformation. A typical tensile tester uses heavy threaded uprights between rigid cross members. The entire system uses relatively massive members to maintain a high degree of stiffness. When the system applies a load to a sample, there is essentially no flexing of the test machine; all the deformation occurs in the test sample.

The moving cross member is driven by threaded uprights, as in Figure 6.3. In this type of tester, the crosshead can apply tension or compression to a test sample. Instrumented fixtures called load cells measure load, while other equipment measures lengthening in the sample. Standard-sized bar samples are used. They have a narrow gage section between two wider sections; the wide sections are gripped by the tensile tester. The gage section acts to concentrate the load (force/area), and this generally causes the failure to occur in the gage section.

In a tensile tester, the grips hold the tensile test specimen or tensile bar. During tensile testing the tensile bar specimen is pulled from both ends using the threaded uprights drive connected to the moving crosshead. The axial forces in the bar are measured by the load cells in the grip. The axial force or tension is increased gradually until the bar fails or breaks into two pieces. As the force is applied, the bar goes to initial deformation as it stretches to yield point, ultimate point, and fracture of the specimen.

When metals are tested to failure in tension, a typical stress–strain curve results. There is an initial linear portion where stress is proportional to strain, typically at low load and

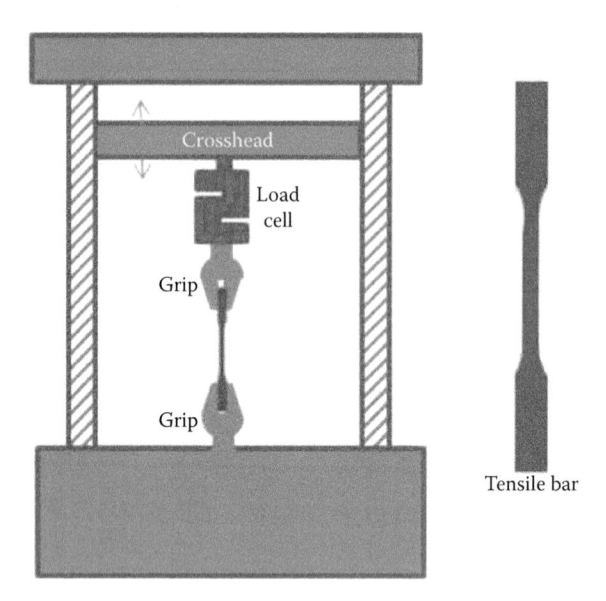

FIGURE 6.3
A tensile testing equipment and a tensile specimen (bar).

strain levels. In this region, parts are said to experience elastic loading. At an atomic level, during elastic loading, stretching of the atomic bonds accounts for the changes in the part's length. If the part is unloaded at this point, it will return to its original length. In this region,

$$\sigma = E\,\epsilon \qquad (6.5)$$

where:
  σ is stress
  E is a constant
  ϵ is strain

The above equation is known as Hooke's law. The constant E is Young's modulus or the modulus of elasticity; it is a material constant. Strong metals like tungsten have Young's moduli near 407 GPa (60 million psi); weaker materials like aluminum have Young's moduli around 69 GPa (10 million psi). Steel is midway at 207 GPa (30 million psi).

During a tensile test, if the load is continually increased, a yield point will be reached. That is, stress will cease being proportional to strain; permanent deformation will occur in the sample and it will not return to its initial shape if it is unloaded. On a stress–strain graph, the linear curve will begin to bend, or for some metals, will go through a sharp drop before rising again. This is called the yield point. In terms of atoms, the deformation has caused atomic bonds to stretch so much that they break, with an atom losing the bonds with its nearest neighbors and reforming them with the next neighbor.

As the sample continues to be pulled, load on the sample increases to a maximum and then decreases somewhat until the part fractures. The point of ultimate tensile stress, or UTS, is this maximum load divided by the original cross-sectional area. Up to the UTS point, the sample will have deformed uniformly in the gage section. However, as the load rises above the UTS, the deformation will become concentrated at a single point, causing a local thinning called necking. Eventually fracture will occur at the neck.

The drop in stress after the UTS point is reached is apparent, not real. It occurs because we calculate stress based on the initial cross-sectional area. This stress is called the engineering stress. Because the cross-sectional area drops during the test, the true stress, which is load divided by the true cross-sectional area, continues to rise.

The stress–strain curve in Figure 6.4 was developed for two different metals. The solid line shows a yield point drop for a metal, while the dashed line shows no drop in yield for another metal.

With these definitions in mind, we can give metals an approximate rank regarding their strength and ductility. Generally, the common metals can be ranked approximately as follows, with the understanding that processing can affect strength markedly:

Low strength—Lead, tin, lithium, calcium, magnesium, zinc

Moderate strength—Silver, copper, aluminum, gold

High strength—Iron, nickel, chromium, titanium, molybdenum

Ductility is another important property of metals. In a tensile test, ductility is the amount of strain (elongation) that occurs at the point of fracture. In a more general sense, ductility tells manufacturers how much plastic deformation (permanent deformation) is expected when forming a metal by cold rolling, forging, drawing, and so on. Ductility is calculated as the change in length divided by initial length, where l_f is the final length and l_o is the initial length. Note that the final length is calculated by fitting the broken part back together and measuring the total length at the time of the fracture:

$$\%\text{Elong} = \frac{100(l_f - l_o)}{l_o} \quad (6.6)$$

Similarly, ductility can be calculated as a reduction in area for a part pulled to fracture in tension. Because mass is conserved, a part that becomes longer under tension must also lose cross-sectional area; if not, matter would be created, and of course this is impossible

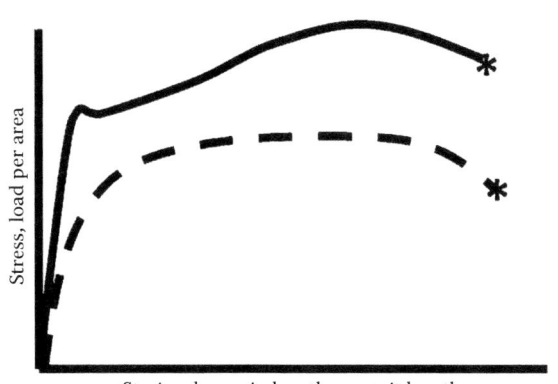

FIGURE 6.4
Stress–strain curve for metals.

under normal conditions. Like the elongation calculation, reduction in area (RA) is a percentage calculation:

$$\%RA = \frac{100(A_o - A_f)}{A_o} \tag{6.7}$$

where A_o and A_f are the initial and final cross-sectional areas, respectively. Like elongation, the final cross-sectional area must also be measured after fracture. With this understanding of ductility, some common metals and alloys can be given an approximate rank as shown below, again with the proviso that processing can markedly affect ductility:

High ductility—Brass, gold, silver

Lower ductility—Steel, titanium, molybdenum

Example 6.2

A tensile bar made from brass is pulled to fracture on a tensile tester. The initial dimensions of the gage section are 0.762 mm (0.030 in.) thick × 12.75 mm (0.502 in.) wide, and the gage section is initially 50.8 mm (2.000 in.) long. The bar is observed to withstand 2958 N (665 lb) maximum load before load drop begins. Fracture occurs at a load of 2068 N (465 lb). After fracture, the part is fitted back together and the gage section now measures 85.344 mm (3.360 in.). Calculate engineering stress values for fracture stress and UTS. What is the ductility of this brass?

Solution:

Fracture stress:
$$\sigma_f = \frac{F_f}{A}$$
$$= \frac{2068}{9.7}$$
$$= 213\,\text{MPa}\,(30{,}985\,\text{psi})$$

UTS:
$$\sigma_u = \frac{F_u}{A}$$
$$= \frac{2958}{9.7}$$
$$= 304\,\text{MPa}\,(44{,}100\,\text{psi})$$

Ductility, percent elongation:
$$\%EL = \frac{100\,(l_f - l_o)}{l_o}$$
$$= 68\%$$

Another important property of metals is their electrical conductivity. As mentioned above, when metals are exposed to a voltage, the electrons in the cloud are free to move, thus inducing current. Note that the resistivity of a metal is a material property not affected by the part's size, although the overall resistance is affected by size. Resistivity, ρ, is calculated as

$$\rho = \frac{RA}{\ell} \qquad (6.8)$$

where:
   R is the overall resistance as measured on an ohmmeter
   A is the cross-sectional area of the conductor
   ℓ is the length of the conductor

Resistivity is expressed in units of ohm-centimeters, ohm-inches, or similar terms. Most metals are good conductors, having low resistivity compared with ceramics or polymers. However, some metals are better conductors than others. One of the reasons for these differences involves impurity atoms. The concentration of impurity atoms in a metal affects resistivity. If a metal contains a single type of a secondary atom, and the secondary atom easily substitutes for the first atom in the unit cell (i.e., the secondary atom forms a solid solution with the first), resistivity is affected as shown in the following equation:

$$\rho_i = A c_i (1 - c_i) \qquad (6.9)$$

where:
   ρ_i is the resistivity at a certain impurity level
   A is a constant
   c_i is the atom fraction of the impurity

As this equation suggests, alloys will always have higher resistivities than the pure metals from which they are made. The resistivity of metals also rises with temperature, mostly because increasing temperature increases thermal vibrations in the atoms of the lattice. This resistivity change, except only a range below approximately –150°C, increases linearly with temperature as follows:

$$\rho_t = \rho_o + \alpha(T - T_0) \qquad (6.10)$$

That is, resistivity at temperature T is a constant, ρ_o, plus a slope term, α, times the difference between the test temperature and the reference temperature.

In addition to their high electrical conductivity, metals are also good conductors of heat. This is another property that is related to the electron cloud. In metals at relatively low temperatures, heat is conducted by electrons, so the best electrical conductors are also the best conductors of heat. Silver and copper, for example, are low in electrical resistivity and conduct heat well. By contrast, stainless steel does not conduct electricity as well as the two pure metals, and its ability to conduct heat is also lower than that of copper or silver.

6.2.3 Processing of Metals

The extractive metallurgical processes described in a previous section usually yield large, simple pieces of metal such as ingots or slabs. These large pieces of metal must be worked into smaller pieces and shaped into practical forms in order to be useful. In casting, the metal is remelted and poured into molds that are near to the final shape. In mechanical forming processes such as forging or rolling, external stress utilizes the ductility of the metal to deform it into its final shape.

Casting is most often used if the final shape desired is complicated and would be too expensive to form by other means. A familiar example is an automotive engine block. With multiple cylinders and cooling channels, an engine block would be quite expensive to machine or forge out of solid material. In other cases, the material to be cast is so difficult to form by other methods that casting is the only process that will work. The casting process is easy to picture, with liquid metal being poured into a mold, cooled, and removed. However, castings are often plagued with internal porosity. This will not be a problem if the casting is an art object. However, if the part is aimed at a load-bearing engineering application, porosity could act as a stress riser and cause catastrophic failure. To prevent porosity in castings, it is first necessary to understand how it forms.

Liquid metals cool from the outside inward; they lose heat through their surfaces, so naturally their surfaces cool first. Large castings form a *skin* of solid metal at the walls of the mold. At some point during the solidification process, this skin covers the entire surface of the metal while there is still liquid metal inside. The internal pool of liquid metal continues to cool. At the same time, most metals contract when they solidify—that is, metals are denser as solids than as liquids. So when the internal pool of liquid metal cools inside the skin formed at the outside of the casting, the last liquid to solidify will leave porosity behind. Shrinkage porosity of this type is often found toward the center of the casting, but varying solidification rates can cause different distributions of the pores.

To counteract this tendency to form pores, risers are designed along with the casting. Risers are reservoirs of liquid metal. As the metal freezes and shrinks inside the mold, the liquid metal in the riser continues to supply the casting. Risers also act as heat reservoirs; metallurgical engineers design them to solidify last, after the casting has solidified. This keeps liquid metal in the riser to supply the casting. If there is any porosity, it is contained in the riser rather than the casting. To design risers, an equation called Chvorinov's rule is used. It relates solidification time to the shape of the casting:

$$\text{TST} = C_m \left(\frac{V}{A} \right)^n \tag{6.11}$$

where:
  TST is total solidification time
  V is the volume of the casting
  A is the surface area of the casting
  n is an exponent around 2
  C_m is a constant based on the mold material, the relative pouring temperature versus the melting point, and the thermal properties of the metal being cast

When designing risers, engineers consider that the casting and the riser are separate castings. Then the values for C_m are the same for both riser and casting. This allows us to make a simple comparison: in order for the riser TST to be greater than the casting TST, we must make the volume/area ratio higher for the riser than for the casting. In theory, the best riser would be a large sphere because spheres have the highest ratio of volume to surface area. However, spheres are not easy to cast. Most foundries use a cylinder and find that this shape works well.

Also as per Chvorinov's rule, a riser that is shaped like a radiator fin would not work well. With its low ratio of volume to surface area, it would be highly likely to solidify before the casting. This shape has such a high surface area, and therefore such a tendency to lose heat through the surface, that it is almost guaranteed to produce porosity in the casting. Riser size must also be optimized. In the ideal case, large risers will make it very unlikely that the casting will contain porosity. However, risers represent scrap material that must be removed from the casting prior to use. Even if the riser material is recycled by remelting, there is a cost associated with machining the riser material off the casting. Metallurgical engineers use casting simulation software to calculate the optimum size and placement for risers. This software calculates heat flow, liquid flow, solidification, and so on to minimize the total cost of risers and other excess metal in the casting.

Mechanical forming processes involve deforming metals into their final shapes. Rolling, extrusion, drawing, and forging are the main branches of mechanical forming. In rolling, pieces of metal are passed through rolls set to a separation smaller than the thickness of the metal. The circular rolls pinch the metal and pull it in, forcing it to take on the shape of the rolls. Many rolls are flat, producing sheet stock; however, shaped rolls can also produce bar stock, V-channel, I-beams, diamond plate, and other special shapes.

Most rolling mills use tandem setups, where a single slab of metal will run through one rolling mill directly to the next, each reducing the cross section slightly more and also accelerating the speed of travel of the metal. This acceleration is caused by the thinning of the metal under the rolls. The mass of the metal is constant, and its width changes little during rolling while its thickness changes a great deal. Because mass is conserved, the thinner metal emerging from the rolls must travel slightly faster than that entering the rolls. If the material goes through several tandem mills (mills set up one after the other), the stock can exit the last mill at surprisingly high speeds, up to 600 surface feet per minute in some cases.

Rolling is commonly performed with the metal maintained at high temperatures. Because the metal is softer at the elevated temperatures of hot rolling, very large reductions in thickness can be achieved. However, they come at the expense of surface quality. At hot rolling temperatures, the metal forms surface oxides or scale that leaves the stock with a rough surface. Cold rolling, by contrast, is done at room temperature. Cold rolled material normally has excellent surface quality because it does not oxidize as hot worked material does. However, cold rolling reductions are limited to approximately 50%; after this, many metals become too hard and brittle to roll further unless they are first softened by annealing, a moderate-temperature furnace treatment.

A forging process goes back to the blacksmith's shop; it involves deforming metal (generally at high temperatures) via hammer blows or squeezing in a shaped die. Because casting often leaves coarse microstructural features in the metals, forging or hot rolling may be used to break up and refine these coarse features.

In an extrusion process, a bar stock is pushed into a die with a shaped opening smaller than the stock entering. High pressure is used to squeeze the metal out the die opening like toothpaste from a tube. Tubing, rods, and bars with complex cross sections can be extruded. Much extruded metal is aluminum-based as this soft alloy will deform under lower pressures more than, for example, steel.

Another important method of metal forming is wire drawing. In this process, bar stock is forced through a die that has a cone-shaped hole. The material coming out of the die is caught and pulled, reducing the cross section and increasing the strength through cold work. Successively smaller dies are used until the desired wire diameter is reached.

6.2.4 How Does Processing Affect the Properties of Metals?

In addition to the above processes, all of which mainly aim to change the shape of the part, thermal processes change the properties of the materials. Annealing, quenching, and precipitation hardening are all processes that affect the properties of metals. All of these processes are monitored by measuring hardness. Often the desired properties involve tensile strength, but hardness correlates well with tensile properties and is much quicker to test.

Hardness is the resistance of a material to penetration by an indenter. There are several types of hardness testers; one of the most widely used types is the Rockwell tester. Rockwell testers have hard indenters connected to a mechanical system that applied load to the indenter. A sample is put below the indenter, an initial low load is applied, and then a higher load is applied. The depth to which the indenter sinks is proportional to the hardness. With the extremely wide range of hardnesses in materials, from superhard ceramics to soft tin or polymers, Rockwell testers use a range of indenters. For hard materials such as ceramics and hardened steel, diamond indenters are used. For soft materials, round indenters of hardened steel are used, with larger diameters for softer materials.

6.2.4.1 Annealing

As mentioned in Section 6.23 on cold rolling, cold work hardens most metals because the deformation introduces irregularities into the regular lattice of the unit cells. Metals deform through the motion of these irregularities, which are called dislocations. As cold work begins, dislocations glide relatively easily through the matrix of atoms. However, as successively more cold work is introduced to the metal, large numbers of dislocations are formed. New dislocations are difficult to move through these tangles of existing dislocations. On a macroscopic scale, cold rolling increases hardness and strength of metals while at the same time decreasing their ductility.

If a highly polished and etched surface of a metal sample is examined under a microscope, a structure of grains is revealed. Grains are small volumes of metal in which the unit cells are aligned. The orientation of the unit cells is relatively random. The boundaries between grains are the irregular areas where these misaligned arrays of atoms meet.

Cold work flattens and stretches metal grains, similar to the way a rolling pin flattens and stretches pie dough. At the same time, high concentrations of dislocations are introduced in response to the cold work, as discussed above. When the cold-worked metal is heat treated above its recrystallization temperature, small equiaxed grains begin to grow within these flattened, deformed grains. The metal becomes softer and more ductile with a lower yield point due to this recrystallization.

Recrystallization takes place during a process called annealing. Because recrystallized grains are small, cycles of cold work and annealing can be used to refine the microstructure of a part. In general, smaller grain size correlates with higher strength. Material that is initially cast into ingot or slab forms usually has a coarse microstructure with large grains, so cycles of cold rolling and annealing can be used to refine it. Note that hot rolling takes place above the recrystallization temperature, so it can also refine the microstructure at the expense of a rough surface finish due to oxidation.

6.2.4.2 Quenching

Iron is BCC at room temperature and FCC at temperatures above approximately 920°C. The BCC and FCC forms of iron are phases, just as ice is a phase of water. The BCC phase of iron is called ferrite, while the FCC form is called austenite. These two phases are stable under equilibrium conditions—that is, when austenite is cooled infinitely slowly, it will form ferrite.

By contrast, when steel is cooled rapidly, as when red-hot parts are quenched into water, a third phase will form. This phase is called martensite. Instead of the cubic forms of lattice structure, this phase has unit cells that are body-centered tetragonal, or BCT. The BCT unit cell is similar to the BCC, except that one dimension of the cube has been stretched. A BCT unit cell has two square faces and four that are rectangular.

The FCC to BCT transformation takes place because carbon atoms, present in steels at low levels, fit into the FCC lattice better than they do in the BCC lattice. If the cooling process is slow, there is time for the carbon atoms to diffuse into favorable positions even in the tight-fitting BCC lattice. However, if the cooling process is rapid, the carbon atoms are trapped in lattice spaces that are too small to hold them. The resulting strain to the lattice causes it to distort into the BCT form. The lattice strain also makes martensite much harder and stronger than ferrite or austenite.

6.2.4.3 Precipitation Hardening

Some alloy systems can be hardened without the drastic quenching described above. The properties of alloys in which two phases form under equilibrium conditions are markedly affected by the size and distribution of the second phase particles. Under most conditions, fine, even dispersions of second phase particles increase strength. Precipitation hardening alloys allow the engineer to tailor the dispersion for maximum strength.

Precipitation hardening relies on three characteristics of an alloy. First and most important, the alloy must contain a second phase. Second, the hardness of this second phase must be greater than that of the base alloy. Third, the alloy must dissolve the second phase at an elevated temperature and then precipitate it at lower temperatures—that is, solubility for the second phase must decrease as temperature is decreased.

Several aluminum alloys meet these three criteria, so that they can be precipitation hardened. Aerospace applications in particular are well known for using precipitation hardening aluminum alloys. In these alloys, second phases that are rich in magnesium, silicon, or zinc as well as other elements can be dissolved at high temperatures, approximately 550°C or so. Then the part is rapidly cooled, not to change the phases present as in a steel quench but simply to speed the process along. The part is then reheated to an intermediate temperature, perhaps 150°C–250°C. At the intermediate temperature, fine precipitates form, hardening and strengthening the part. In a few alloys, precipitation can even take place at room temperature. These alloys are cooled below room temperature after the high-temperature treatment, then formed (into rivets, for example) and allowed to harden in place.

In general, the lower the temperature at which the precipitates form, the more favorable the properties will be, but the longer the process will take. Some aluminum alloys reach peak strength after months-long hold times at the intermediate temperature. However, hold times in the multiple months are not usually practical in a production environment, so shorter, higher-temperature treatments are usually chosen.

6.2.5 What Technologies Are Made Possible by Metals?

Different properties of metals allow engineers to develop products to improve the world and human life. Their electrical conductivity enables us to utilize electricity, freeing us from much physical labor and making the entire universe of electronics available. The hardenability of steel makes tools possible for all types of manufacturing, from automobiles to washing machines, from clothing to medicines, from agriculture to mining. The properties of aluminum make modern aircraft viable. The strength of steel is designed into steel-framed buildings and vehicles, and thousands of other applications.

Computers make use of many types of metals. Because the central processing units (CPUs) can generate excessive heat during operation, metal heat sinks are used to help

FIGURE 6.5
Aluminum heat sink with a fan in a computer circuit board.

manage temperatures. The best metals for heat sink applications have good thermal conductivity. Copper and aluminum are typical choices. Copper has better heat conductivity, and its high melting point makes it safe from the temperatures generated by the CPU. However, copper is denser than aluminum, so copper heat sinks add to the weight of the computer. Aluminum has reasonable heat conductivity and is lightweight; it is often used in portable devices such as laptop computers. In Figure 6.5, the fan-cooled heat sink is made from aluminum.

6.3 Ceramic Materials

Like metals, most people can identify ceramic objects such as a teapot or a brick. But what is a ceramic material? Again like metals, a good definition for ceramic materials relates to the types of bonds between atoms. Most ceramic materials are compounds made up of a combination of metallic and nonmetallic elements. These elements are connected with ionic bonds, or mostly ionic bonding with some degree of covalent bonding as well. Ionic bonds occur between elements that have large differences in their electronegativities. The large difference in electronegativity makes ionic bonds strong, so ionically bonded materials have high melting points.

Ceramic materials tend to be crystalline with well-defined unit cells, although there are more noncrystalline ceramics than metals. Glass, in particular, is a noncrystalline ceramic material used widely in different engineering application such as windshield of an automobile. Ceramic unit cell structures are more complex than those of metals because ceramic materials have two or more types of atoms, often of different sizes and in varying ratios. Consider that sodium chloride (NaCl), silicon dioxide (SiO_2), and barium titanate ($BaTiO_3$) are all ceramic materials; the unit cells become more complicated to visualize. However, many ceramic unit cells can be thought of as simple cubic or FCC arrangements of anions, with cations residing in the interstitial sites; this concept can help to simplify ceramic unit cells.

6.3.1 How Do We Obtain Ceramic Materials?

Ceramic materials are essentially minerals that are mined from the earth. Silica (silicon dioxide, one form of which is quartz) is one of the most abundant minerals in the earth's crust, for example. Most sand is silica, and silica is typically mined in the form of gravel or sand. Clay, another important ceramic mineral, also occurs naturally and is dug from the ground for use. As mentioned above, bauxite ore is the raw material from which aluminum is extracted. If the refining process is halted after the Bayer process, aluminum oxide is obtained. Similarly, periclase is another mineral that is mined from the earth; when refined, it yields magnesium oxide, or magnesia.

6.3.2 What Are the Properties of Ceramic Materials?

Ceramic materials are known for having high heat capacity and low heat conductance—that is, they are good thermal insulators. Similarly, they are typically good electrical insulators as well. Because ionic bonds tie electrons up so effectively, it would take a very large voltage to force most ceramic materials to become electrically conductive. Some ceramics, however, are semiconductors and a few exotic compounds are superconducting. Ceramics are also normally very hard and strong, but quite brittle as well. All of these properties are related to or affected by the atomic bonds in ceramic materials.

The mostly ionic bonds in ceramic materials give great strength to the material at a macroscopic level. In metal oxide ceramic materials, the metal typically loses an electron to the oxygen. This makes its electron configuration more stable and its overall charge positive. The inverse is true for the oxygen; it gains an electron, stabilizing its electron configuration and imparting a net positive charge. Ionic bonding is strong because the electrostatic force between the positive and negative ions is quite strong as well as nondirectional. The outermost electrons in ceramic materials are typically tightly bound to the oxygen (or the other nonmetallic element in the compound), so they are not available to conduct heat. Instead, heat is conducted through the atomic nuclei as increased vibration. As an atom heats up, it vibrates at a higher amplitude. It then imparts some of this increased energy to its nearest neighbors, analogous to spheres connected to springs. This transmits heat from atom to atom; the process is slower than electronic transmission, so ceramic materials are generally classed as insulators compared to metals.

Ceramic materials can be broadly divided into two categories based on whether they are crystalline or noncrystalline. Crystalline ceramic materials include products such as bricks, tiles, and porcelain objects. Noncrystalline ceramic materials are almost all glass of some type. Crystallinity affects other properties of ceramic materials, but its most obvious effect is on transparency. Ceramics tend to be totally transparent when they are either 100% noncrystalline or 100% crystalline. Hundred percent crystallinity occurs when the part is a single crystal, as in a crystal of quartz. Materials between 0% and 100% crystalline are translucent rather than transparent because the grain boundaries between crystals break up the transmission of light. The same is true for loose powder particles of ceramics. Titanium dioxide in the form of large single crystals is transparent, but in the form of submicron powders, it is so effective at blocking light that it is used as the active ingredient in some sunscreens.

Ceramic materials are also known for being strong but brittle, with little ductility. It is difficult to run standard tensile tests using tensile bars for ceramic materials. Because ceramics are brittle, the tensile tester grips are likely to break the samples even before tension is applied. The brittleness of ceramic materials also makes it necessary to align the samples perfectly, and as perfection is not possible, the results tend to be highly variable. Instead, three-point or four-point bending tests (flexural strength or modulus of rupture) are used for ceramic materials. In a ceramic sample with a rectangular cross section, modulus of rupture (MOR) is described by the following equations for rectangular bars:

$$\text{MOR} = \frac{3FL}{(2bd^2)} \text{ (three-point bending)} \tag{6.12}$$

$$\text{MOR} = \frac{3F(L-L_i)}{(2bd^2)} \text{ (four-point bending)} \tag{6.13}$$

where:
F is the fracture load
L is the length of the support span, or the outer span if four-point bending is considered
b is the width of the test bar
d is the thickness of the bar

In the case of four-point bending, L_i is the distance between the loading or inner span. Figure 6.6 shows a schematic of a four-point bending test.

Example 6.3

A test bar of aluminum oxide 12.7 mm (0.5 in.) wide and 3.81 mm (0.15 in.) thick is tested in a four-point bending test. It breaks at a load of 4337 N (975 lb). The support span (L) is 25.4 mm (1.0 in.) and the inner span (L_i) is 12.7 (0.5 in.). What is the MOR for this bar?

Solution:

$$\text{MOR} = \frac{3F(L-L_i)}{(2bd^2)}$$

$$= \frac{3*4337(25.4-12.7)}{(2*12.7*3.81^2)}$$

$$= \frac{165{,}239}{368.7}$$

$$= 448.1 \text{ MPa } (65{,}000 \text{ psi})$$

Material engineers generally prefer to test MOR using four-point bending based on the stress states in the bars being tested. In both cases, the bar is in compression at its upper surface and in tension at its lower surface. However, in three-point bending, this tension increases from the support points to the central load point. In four-point bending, the

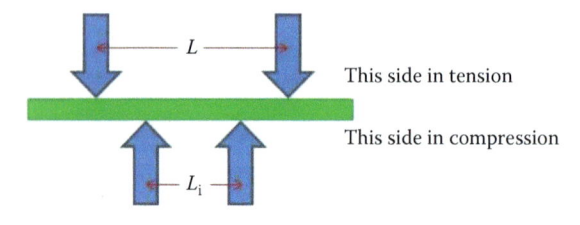

FIGURE 6.6
Four-point bending test of a bar.

stress increases between the outer support points and the two inner load points; however, there is constant stress between the two load points. This gives more repeatable results for the four-point test. The variabilities in MOR testing can be influenced by the flaw sensitivity of ceramic materials. When ceramic materials are placed in tension, they are highly sensitive to the size and placement of flaws. All ceramic materials can be thought of as containing a multitude of tiny flaws with a range of sizes, shapes, and orientations. These flaws act as stress concentrators for tensile stresses. If a flawed ceramic part is subjected to a macroscopic tension, the stress at the crack tip is given by

$$\sigma_t = 2\sigma_o \sqrt{\left(\frac{c}{\rho}\right)} \tag{6.14}$$

where:
σ_t is the stress at the crack tip
σ_o is the macroscopic stress
c is the length of the crack
ρ is the radius of the crack tip

because ρ can be extremely small, the stress at the crack tip can rise to the very high levels needed to break the ionic bonds in ceramic materials. In contrast, ceramic materials are quite strong in compression. Intuitively, compressive stresses can be pictured as holding the preexisting flaws closed while tension opens them. Prestressing takes this idea to its logical conclusion. Prestressed concrete is an example; steel bars are held under elastic tension while concrete is cast around them. When the concrete is cured, it bonds to the steel. The external tension on the steel is then released. The steel maintains the tension and this puts the concrete into compression. Any stress on the finished concrete part must exert a tension high enough to overcome this initial compressive load in order to crack the part.

Similar to electrical and thermal properties, the mechanical properties of ceramics can also be related to the bonds between atoms. Because ionic bonds are very strong, it takes high stresses to break them. And because ceramic materials are so brittle, the work involved in breaking interatomic bonds goes into creating new surfaces—that is, fracturing the part. This is in contrast to metals, where the atom-to-atom bonds can *stretch* and allow atoms to slip over their neighbors. On a macroscopic level, this leads metals to exhibit relatively high amounts of plastic deformation before fracturing as compared to brittle ceramic materials.

6.3.3 How Do We Process Ceramic Materials?

Once ceramic raw materials are dug from the ground, they are processed into usable materials through milling, batching, mixing, forming, drying, and firing. The first step, milling, reduces particles to finer sizes. This may involve anything from crushing coarse rocks to refining powders that are already at microscopic sizes. There are several types of mills that can be used to reduce the size of ceramic particles. One type is a ball mill, a cylindrical container with milling media inside (pellets or balls of hard material). The cylinder is rotated on revolving rolls. The rolling action causes energetic impacts between pieces of the media as they tumble and fall; these impacts break up the particles into finer sizes.

Batching is the preparation of the various ceramic materials from different ingredients. It involves precise weighing and other, more involved steps as needed. Once the batch materials are prepared, they are mixed in ribbon or pug mills. This mixing step may be performed wet or dry.

Once the batch is mixed, the ceramic parts are formed into the shapes needed. Forming may involve pressing, extrusion, or slip casting. In pressing processes, simple shapes are pressed in dies using ceramic powders with small amounts of binder materials. Spark plug insulators are often pressed into shape, for example. Extrusion involves forcing a semi-liquid ceramic material through a shaped die, somewhat like pressing toothpaste from a tube. Large tonnages of ceramic material are extruded in the shape of oval honeycombs for catalytic converters.

In slip casting, a liquid slurry of ceramic material takes on the internal shape of a mold. A slurry of clay particles suspended in water is poured into a plaster of Paris mold. Plaster of Paris, a hydrated form of calcium sulfate, has many tiny pores that pull water from the clay slurry by capillary action. As water is removed from the clay slurry, a semisolid layer of clay is caked onto the inside of the plaster of Paris mold. When this layer is thick enough, the mold is upended and the remaining liquid slurry is poured out. The plaster of Paris continues to remove moisture from the layer of wet clay, and eventually the clay is dry enough to remove from the mold.

During the time when the liquid slurry is inside the mold, the layer of clay thickens at a rate that is parabolic with time—that is, the rate is initially rapid, and then it slows down. This is because, once the layer of clay is formed, moisture must be pulled through this layer into the plaster of Paris. The thickness of the layer becomes the rate-limiting step

$$X = \sqrt{kt} \tag{6.15}$$

where:
X is the thickness of the layer
t is time
k is a constant

Drying removes excess liquid from the formed part. Drying processes may involve applying heat to speed the drying of formed objects such as wet slip cast parts. Care must be taken because drying processes that are too aggressive can cause cracks. The dried parts are also quite fragile and must be handled gently. Drying processes may also be used to dry a powder that is mixed while wet. In this case, a process called spray drying is often used. As the name suggests, spray drying involves atomizing a liquid slurry of ceramic powder through a spray nozzle. The powders dry rapidly because the fine particle sizes expose a very high surface area to the atmosphere.

Finally, ceramic parts are fired to increase density. The mainly ionic bonds between atoms in ceramic materials give them high melting points, as mentioned previously. Aluminum oxide, or alumina, melts at 2072°C (3762°F), for example. Silicon dioxide (silica) melts at 1600°C (2910°F). These extremely high melting points mean that melting is not a feasible way to form many ceramic materials, with the exception of glass. Instead, for non-glass ceramics, the formed and dried parts are sintered. Sintering is a high-temperature firing process that densifies materials as atomic diffusion drives the elimination of powder surface area. Surfaces (i.e., solid–vapor interfaces) have relatively high free energy compared to solid–solid interfaces. At high temperatures, this energy is available to densify the part. Because surface area drives sintering, the engineer can help the sintering process to proceed more rapidly by controlling the particle size; this is the reason why much of ceramic sintering is done with submicron particle sizes.

Example 6.4

The slip-casting rate constant for a porcelain component is measured to be 0.81 mm/min (0.032 in./min). How thick will the slip-cast wall be after a hold time of 15 min?

Solution:

$$X = \sqrt{kt}$$
$$= \sqrt{(0.81 * 15)}$$
$$= 3.48 \text{ mm} (0.134 \text{ in.})$$

6.3.4 How Does Processing Affect the Properties of Ceramic Materials?

The strength of ceramic materials is highly dependent on their processing. Factors that affect strength include the degree of porosity in the finished part and the number, size, orientation, and shape of flaws in the part. Porosity has a marked effect on the strength of ceramic materials. MOR decreases exponentially as porosity increases:

$$\text{MOR}_p = \text{MOR}_o e^{(-nP)} \tag{6.16}$$

where:
MOR_p is the modulus of rupture at a given porosity level
MOR_o is the modulus for the fully dense material
n is a constant
P is the volume fraction of porosity

Porosity is affected by sintering history as well as the average size and size distribution of the powder particles being used to manufacture the ceramic part. Sintering is a thermally activated process; the rate of sintering is typically exponential in temperature, so small changes in furnace temperature can greatly affect the degree of densification. Particle size also affects sintering rate, as mentioned above; sintering rate is approximately proportional to particle size. Fine particles drive faster sintering rates, while coarse particles slow the rate of sintering.

The number, size, orientation, and shape of flaws in ceramic materials also strongly affect their strength. Tiny surface cracks or scratches, damage from thermal shock or handling, pores, and even the points at which the corners of grain meet can all act as flaws. The plane strain fracture toughness of a ceramic material expresses its ability to resist fracture and is calculated in terms of the critical flaw size:

$$K_{IC} = Y\sigma\sqrt{\pi a} \tag{6.17}$$

where:
Y is a constant
σ is the applied stress
a is the length of a surface crack (or half the length of an internal crack)

Generally, the material will not crack until the right-hand side of the equation is greater than the fracture toughness, and this condition is strongly affected by the size of the flaw, as the equation shows.

6.3.5 What Technologies Are Made Possible by Ceramic Materials?

By amount of usage, traditional ceramic materials dominate over technical ceramics. Traditional ceramics include bricks, tiles, whiteware (bathroom fixtures), pottery, and glass. These support our way of life, with buildings and floors protected by bricks and tiles, while whiteware supports the sanitation that lengthens our lives and glass brings light into our homes. The technical applications of ceramics cast a long shadow as well. Specialty tools and abrasives made of ceramic materials enable exotic manufacturing processes; electrical insulating parts make long-distance power transmission possible; and piezoelectric ceramic materials make it possible for us to store data.

Another recent application of ceramic materials can be observed in millions of mobile devices. Corning® Gorilla® Glass is used for the screens of mobile devices such as smart phones and the tablet computer shown in Figure 6.7. This glass is designed to be resistant to breaking so that it can be used in thin layers for mobile devices. The strengthening mechanism is somewhat similar to reinforcing concrete with tensioned steel rods. In this case, the glass is immersed in a bath of molten potassium salts. Ion exchange occurs between the glass and the salt bath. Bulky ions of potassium are taken into the surface of the glass. Because the potassium ions are larger than the ions that they replace, the glass surface is in compression after the treatment; the process is informally called *ion stuffing*. This compressive stress at the glass surface must be overcome to break the glass, so that dropping a phone or a tablet computer no longer guarantees it will be ruined.

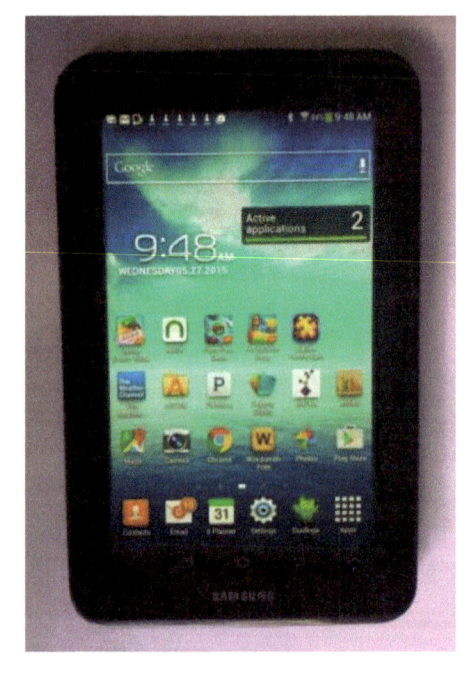

FIGURE 6.7
Gorilla® Glass used in the screen of a Samsung tablet computer.

6.4 Polymers

6.4.1 What Is a Polymer?

Plastic materials are normally all around us; the everyday items in our homes and workplaces usually contain many types of plastic items. Appliances, toys, packaging, food service items, computer housings, telephones, and many other items we take for granted partly owe their existence to the development of plastics.

Plastics, also known as polymers, are materials made from very large molecules that are often based on petrochemicals. Polymers are composed of many repeats of smaller molecules. Some polymers are formed in a long-chain shape; others form branching, brush, ladder, and other various shapes. The long chains, some of them having many thousands or even millions of repeat molecules, typically fold and tangle and bend. Chain-type polymers are often classified as thermoplastic materials, meaning that they can be molded or melted with the application of heat. Polyethylene, polypropylene, polystyrene, and polyvinyl chloride (PVC) are all examples of thermoplastic polymers (Figure 6.8).

Another polymer family structure consists of three-dimensional arrays. These arrays may be long-chain molecules with occasional links between the chains. The links between the chains are known as crosslinks. Other three-dimensionally arrayed polymers may have more than two bonds per unit, so their *chains* can bond in three or more spots. This gives molecules that bond like a net in three dimensions. This type of polymer is known as thermosetting. Once these three-dimensional bonds are formed, they may char at extreme temperatures, but they will not melt. Epoxies and phenolics are examples of thermosetting polymers. Within the molecules of polymers, atoms are bonded with covalent bonds. In covalent bonds, two atoms share one or more pairs of electrons. In many cases, because they can share a pair of electrons, the atoms are more stable with the bond than without it. The nitrogen and oxygen in our atmosphere are present in the form of covalent-bonded molecules, for example. In polyethylene molecules, the links in the chain are covalent bonds. So are the three-dimensional bonds in epoxies.

For thermoplastic polymer molecules, other forces such as van der Waals also affect material properties. Van der Waals forces are based on the separation of charge within a molecule. This charge separation means there is a slightly positive and a slightly negative side to the molecule, although the overall net charge on the molecule is zero. Water is a good example of a molecule with a slight charge separation. The two hydrogens line up on one side of the oxygen and give that side a slight positive charge, while the oxygen charge is negative. In the same way, other molecules can also have varying degrees of charge separation, leading to van der Waals forces between molecules. Van der Waals forces are weaker than the covalent bonds within the molecules. Thermoplastic polymers may be crystalline or noncrystalline (amorphous). In some thermoplastic polymers, the long-chain molecules

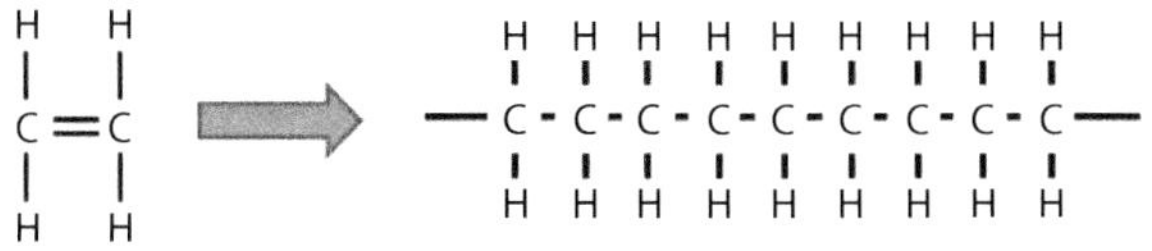

FIGURE 6.8
Ethylene gas (left) is the raw material for polyethylene polymer.

may have islands of regular arrangement. During cooling from the melt, portions of the chains fold into an ordered array like fan-fold printer paper. If many of these layers of folded molecules come together, a crystallite will form. Several thermoplastic polymers are partially crystalline, including polyethylene and polypropylene.

The degree of crystallinity in a polymer is an important parameter because crystallinity affects material properties markedly. Degree of crystallinity (by weight) is calculated using accurate measurements of density. Density measurements are sensitive to the degree of crystallinity because crystalline portions of the polymer are denser than amorphous regions; this density difference comes from the closer packing of the crystalline molecules:

$$\% \text{Crystalline} = 100 \frac{\{\rho_c(\rho_s - \rho_a)\}}{\{\rho_s(\rho_c - \rho_a)\}} \tag{6.18}$$

where:
ρ_c is the density of the completely crystalline polymer
ρ_a is the density of the completely amorphous polymer
ρ_s is the density of the sample polymer whose degree of crystallinity is to be determined

Amorphous polymers, by contrast, have no regular arrangement of the chains. These are often polymers with highly branching chains. The complex shapes of these molecules make alignment into regular crystals very difficult. Examples of amorphous polymers include polysulfone and polyetherimide (Figure 6.9).

An important characteristic of thermoplastic polymers is their molecular weight. The average molecular weight affects thermal and mechanical properties markedly. The molecular weight of a polymer must be calculated because a range of weights is normally found in each material batch. One useful equation is the number-average molecular weight, which is expressed as

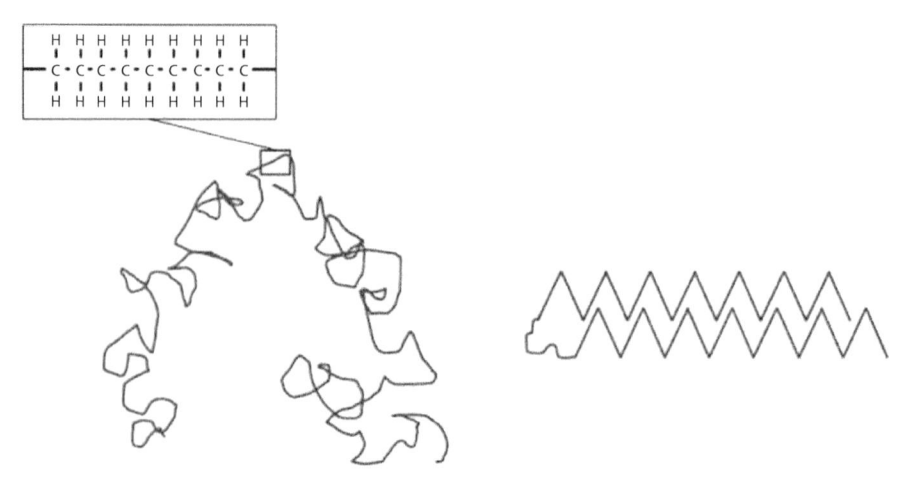

$$M_n = \sum x_i M_i \tag{6.19}$$

FIGURE 6.9
Amorphous (left) and crystallizing (right) polyethylene.

where:
M_n is the overall number-average molecular weight
x_i is the fraction of the number of molecules in a size range i
M_i is the average molecular weight in the same size range i

Example 6.5

A sample of polyethylene is measured to have a density of 0.92 grams per cubic centimeter (g/cc). 100% crystalline polyethylene has a density of 1.004 g/cc, and 100% amorphous polyethylene's density is 0.853 g/cc. How crystalline is this sample of polyethylene?

Solution:

$$\% \text{Crystalline} = 100 \frac{\{\rho_c(\rho_s - \rho_a)\}}{\{\rho_s(\rho_c - \rho_a)\}}$$

$$= 100 \frac{\{1.004(0.92 - 0.853)\}}{\{0.92(1.004 - 0.853)\}}$$

$$= 48.4\% \text{ crystalline}$$

This polyethylene sample's molecular weight fractions have been measured, and the ranges are recorded below. What is the number average molecular weight for this sample?

Molecular Weight Range (grams/mole)	Fraction of Chains at This MW (x_i)
35,000–45,000	0.02
45,000–55,000	0.13
55,000–65,000	0.18
65,000–75,000	0.24
75,000–85,000	0.22
85,000–95,000	0.17
95,000–105,000	0.04

$M_n = \Sigma x_i M_i$

$= (0.02 * 40,000) + (0.13 * 50,000) + \cdots + (0.04 * 100,000)$

$= 71,800 \text{ g/mol}$

6.4.2 What Are the Properties of Polymers?

The combination of covalent and van der Waals forces in polymers means that, for thermoplastic polymers, the forces keeping their molecules intact are strong, while the forces between one molecule and another are weaker. Because of this, thermoplastic polymers tend to be relatively weak. If tested in a tensile tester, for example, polyethylene might have an ultimate tensile strength of 1000–4000 psi (approximately 7–27 MPa). For comparison, the ultimate tensile strength of low-carbon steel is approximately 55,000 psi (approximately 380 MPa). Microscopically, during a tensile test an amorphous thermoplastic polymer experiences different phases of deformation. At the early stage, while stress is relatively low, the randomly arranged chain molecules are simply pulled into straighter alignment.

If the stress were released at this early stage, the polymer tensile bar would go back to its original length.

At later stages of the tensile test, after the molecules have been pulled straight, they begin to accommodate the stress by sliding over each other. The weak van der Waals forces allow this sliding to occur at comparatively low stresses. At this point, the polymer tensile bar has undergone permanent deformation; if the tensile test was halted, the part would not return to its original length. Finally, as the molecules continue sliding over each other, voids form and fracture occurs. Molecular weight affects the strength of thermoplastic polymers; higher molecular weights give stronger polymers on average. This is because the longer chains of high molecular weight materials have many more van der Waals bonds per molecule than short chain materials. Ultimate tensile strength can be expressed as

$$\text{UTS} = \text{UTS}_{\text{inf}} - \frac{A}{M_n} \tag{6.20}$$

where:
UTS is ultimate tensile strength of the sample
UTS_{inf} is the tensile strength at infinite molecular weight
M_n is the number average molecular weight of the sample
A is a constant

When a semicrystalline polymer is pulled in a tensile test, its behavior is more complex than that of an amorphous polymer. Initially the semicrystalline part deforms through the straightening of molecules in the amorphous regions of its microstructure. Once this avenue of deformation is exhausted, the crystalline portions of the material begin to experience rotation, deformation, and breakage as the stress continues to rise. This causes semicrystalline polymers to be both stronger and more brittle than amorphous polymers. Because the deformation of the crystalline regions is complex, semicrystalline polymers may experience yield drop during a tensile test, unlike amorphous polymers.

The thermal properties of polymers are also related to the van der Waals forces between molecules. Because van der Waals forces are much weaker than covalent bonds, thermoplastic polymers have relatively low melting points, most of them well below 300°C. Thermoplastics also have a glass transition temperature (T_g). Below T_g, polymers behave in a more brittle manner; above T_g, they behave more like a rubber. Thermally and electrically, polymers are typically very good insulators. Their electrons are tied up in covalent bonds and are therefore not available to conduct heat or electric current.

6.4.3 How Do We Obtain Polymers?

Most of the engineering polymers are hydrocarbons, obtained from petroleum or natural gas products. When petroleum is pumped from underground deposits, it is a mixed material, with high, medium, and low molecular weight fractions included, from tars through vapors. This natural mixture is separated into usable fractions by evaporation of the light portions and cracking of the heavier portions. Cracking uses heat, catalysts, or both to break large molecules into smaller fractions. In this process, the carbon–carbon bonds in the long-chain molecules are broken to give the desired smaller molecules.

Once the desired smaller molecules are formed, they act as building blocks to build up polymers. To make polyethylene, for example, ethylene gas (C_2H_4) is passed over a titanium (III) chloride catalyst on a silicon dioxide substrate. As the ethylene combines into polyethylene molecules on the catalyst bed, the reaction gives off heat; the heat in turn helps to drive the reaction. The result is a long chain of carbon atoms, each one bonded to two hydrogen atoms and two other carbon atoms. A different type of process is used to make PVC. In this process, liquified vinyl chloride monomer (C_2H_3Cl) is mixed with water in a pressure vessel. The water typically contains suspending agents, and high-speed mixing is also employed; both factors help to form small droplets out of the vinyl chloride. Peroxide initiators are used to begin the reaction from the monomer to the polymer. These initiators break bonds within the monomer, allowing the monomers to rejoin into long-chain molecules. The reaction is highly exothermic, so cooling must be employed. The process yields a suspension of PVC particles as the result of the reaction. This slurry is dried to give a powder that can be processed further.

Polypropylene is chemically similar to polyethylene, but some of the hydrogen atoms in polyethylene are replaced by methane groups in polypropylene. The placement of the relatively bulky methane groups is very important in determining the properties of the polypropylene. If the methane groups are all on one side or regularly alternating, the resulting polymer will be much more prone to crystallize than with randomly placed methane groups. Like polyethylene, polypropylene production depends on the use of catalysts. Special Ziegler-Natta or metallocene catalysts are often used. The type of catalyst used dictates the placement of the methane groups and therefore many of the properties of the polymer. Overall, the reaction can take place in a gas phase process. The raw material is propylene gas, C_3H_6. Gas phase processing may take place either in a fluidized bed reactor or a stirred reactor. In both cases, the reaction takes place when gases are passed over a finely divided catalyst. As with all chemical processes, temperature and pressure are important variables and must be precisely controlled.

6.4.4 How Do We Process Polymers into Usable Forms?

The polymers coming out of the processes described above are still raw materials. The PVC and polypropylene are often in the form of powders, and the polyethylene may emerge in the form of small pellets. Secondary manufacturing must be done to make usable products from these raw materials. As part of this secondary manufacturing, many polymers have chemicals added to them. These additives can include plasticizers, fillers, colorants, and stabilizers. Plasticizing additions improve the ductility and reduce the brittleness of plastics. Plasticizers are normally small molecules that take positions between the long-chain polymer molecules. They act somewhat like lubricants, reducing the van der Waals forces between the polymer molecules so that the final polymer is softer and weaker as well as more ductile.

Fillers may improve the properties of polymers, such as their strength and abrasion resistance, or sometimes they may simply be used to replace part of the more expensive polymer. Filler materials are usually small particles such as wood flour, silica, or clay particles. Colorants are dyes or pigments used to give color to a polymer. Dyes are complex chemical formulations that must be carefully chosen to dissolve in the polymer without reacting with it. Pigments are nondissolving materials such as carbon (black) or titanium dioxide (white) that are mixed with the polymer but may settle out if not carefully processed. Stabilizers are added to polymers to protect them from deteriorating under environmental exposure such as ozone and ultraviolet (UV). The use of carbon black in automobile tires is a good example of this type of stabilizer; the carbon black protects tires against both ozone and UV.

Once these additives are mixed with the polymers, forming processes take place. For both thermoplastic and thermosetting polymers, much of this processing takes place at elevated temperatures. For crystalline thermoplastic polymers, the processing temperature is normally above the melting temperature. Amorphous thermoplastics can be processed at lower temperatures between the glass transition and the melting temperature. In both cases, pressure is usually applied to force the part into the desired shape; to retain the shape, the pressure is held until the part cools.

These molding techniques used with thermoplastic polymers include injection molding, blow molding, and extrusion. Injection molding uses a cylinder to force pelletized or powdered raw material through a heated chamber toward a mold. The heat from the walls of the heating chamber melts the thermoplastic material. Then the cylinder forces the viscous liquid polymer into the mold chamber. Pressure is maintained until the parts have cooled. Blow molding is widely used for making plastic bottles. In blow molding, a length of polymer tube is joined at the bottom to form something like a heavy-walled bag. This bag, called a parison, is moved while still hot to a mold. The mold is formed in two pieces around the parison. Then air or steam is used to pressurize the inside of the parison; this forms the exterior of the parison to the interior of the mold. When the bottle is cooled, the mold is opened.

In the extrusion process, pressure is used to force a raw material through a shaped die. In the case of plastic extrusion, a screw is used to push polymer powder or pellets into and through a heating section; this melts the polymer. The screw also generates pressure to force the viscous liquid polymer through the die. Tubes, pipes, sheets, and various shapes (L-channel, etc.) can be formed by extrusion. Films may be formed directly by extrusion or by blowing. In blowing, a tube is extruded and its inside is pressurized with a gas. This forms a bubble that is drawn to thin the film further. The continuous bubble may be slit to form a single layer of film, or it may be laminated and cut for uses such as garbage bags.

When thermosetting polymers are to be processed into usable form, there are normally two steps, similar to the preparation of a two-part epoxy. In the first step a resin is made. This resin is called the prepolymer, and it normally has a low molecular weight. In the second stage, the prepolymer is formed into the final shape in a mold. Polymerization, or curing, takes place when heat and/or pressure and/or catalysts are added to the prepolymer.

6.4.5 How Does Processing Affect Polymer Properties?

At the raw material stage, the conditions under which polyethylene is formed (temperature, pressure, and various catalyst factors) dictate the molecular weight. High pressure favors low-density polyethylene (LDPE). This polymer has a low density because it is highly branched and also because there are a wide range of molecular weights in the product. Low pressure processing using an aluminum-based catalyst yields high-density polyethylene (HDPE). HDPE has a much more linear structure with less branching. This low-pressure process also yields a narrow range of molecular weights that are much higher, on average, than LDPE. In general, contaminants going into this process, such as hydrogen or acetylene, interfere with polymerization and thus yield undesirably low molecular weights.

As with polyethylene production, the characteristics of PVC raw material are markedly affected by the manufacturing conditions. Temperature and pressure must be tightly controlled. The peroxide initiators used must also be given careful consideration. Some initiators are initially very active but decay rapidly. Others begin their activity slowly but last longer. If only the first type of initiator were used, the polymerization would proceed rapidly but the resulting PVC would have quite a low average molecular weight. If only the second type of initiator were used, molecular weights would be higher, but overall process productivity would be lower. Normally a mix of fast and slow initiators is used, with the mixture tailored to give desirable molecular weights with reasonable productivity.

Manufacturing polypropylene is very dependent on the properties of the catalyst used, as mentioned above. The ideal catalyst will allow chain formation, but will restrict the monomers in the chain to a single orientation. This will yield polypropylene molecules with all the methyl groups on one side of the chain (*isotactic* polypropylene) or possibly with the methyl groups attached on opposite sides in a regular order (*syndiotactic* polypropylene). Atactic and syndiotactic polypropylene are desirable because these regularly arranged molecules can solidify with a high degree of crystallinity. If the methyl groups are arranged randomly along the chain (*atactic* polypropylene), the resulting polypropylene will be amorphous. The random orientation of the methyl groups will make crystallization almost impossible, and the resulting polymer will be softer, weaker, and more rubbery than crystalline versions.

When partially crystalline thermoplastic polymers flow under the heat and pressure of processing equipment, variations in crystallization can occur locally in the finished parts. Just as pulling a tensile bar aligns the crystalline portions of the polymer to some extent, so does the flowing motion in injection molding, blow molding, and extrusion. Because crystallinity makes polymers stiffer, stronger, and less ductile, the process can affect the material properties of the part. Many objects made from crystalline polymers are observed to fracture at the corners of the part, where stretching and thus crystallization are most pronounced. Thermoplastic polymer processing can also affect the material properties of the finished part through the use of regrind material. Thermoplastic materials can in theory be recycled indefinitely. Thrifty manufacturers normally regrind and reuse scrapped parts, excess material, and so on. A typical target is to add approximately 20%–25% regrind to the fresh material. However, the heat needed to process the polymer degrades the stabilizing additives and antioxidants. If material is reground and reprocessed several times, lower molecular weights can result, and as mentioned above, this will result in lower strengths if all other factors are held equal.

Regrinding can also draw contaminants into the parts if care is not taken. Many contaminants are simply cosmetic defects, such as colored particles in a clear polymer. However, one harmful contaminant is water. If the reground material tends to absorb moisture, as nylon, Lexan, and other polymers do, then adding regrind material to fresh material can affect the properties of the finished parts. As the polymer is heated during processing, the moisture-laden regrind can form bubbles and the water can hydrolyze the molecules. The bubbles result in visible defects, and the hydrolysis is another process that reduces the average molecular weight, again yielding weaker parts.

Whether thermoplastic or thermosetting, the temperature of the process has a marked effect on the properties. In particular, overheating the materials can break the polymers down. In thermoplastics, this can result in lowered molecular weight, as the long chains oxidize because of the heat. Overly high process temperatures can also drive off some of the desirable additives in the raw material, such as plasticizers and stabilizers. In thermosetting polymers, excessive heat will tend to char the molecules. In all cases, the net result is to degrade the properties of the material.

6.4.6 What Technologies Are Made Possible by Polymers?

Polyethylenes, polyvinyl chlorides, and polypropylenes are among the highest tonnage plastics. Uses for polyethylene include packaging and containers of all sorts. PVC accounts for millions of tons in the form of building materials—flooring, windows, siding, and composite decking. Polypropylene is also used in packaging as well as rope, carpeting, and automotive components.

Polycarbonate is another polymer with widespread uses, although it is not made at volumes as applications (*bullet-proof* glass, dome lights, sky lights, automotive headlamps, safety glasses, etc.); other uses include electronics and electrical hardware, data storage as CD disks, cases for electronics, and so on.

FIGURE 6.10
Polycarbonate protector for a smart phone.

When a smart phone is dropped, the toughened glass mentioned above may protect the display; however, the remainder of the phone is still vulnerable to the fall. Another use for polycarbonate is as molded protectors for smart phones, as shown in Figure 6.10. The same impact resistance that makes polycarbonate a good choice for safety glasses helps to preserve the fragile electronics of cell phones.

6.5 Composite Materials

6.5.1 What Are Composite Polymer Materials?

Metals, ceramics, and polymers all have strong and weak points. Metals are generally strong, ductile, and electrically conductive, but they are also dense and vulnerable to corrosion. Ceramics are insulating, strong, and generally resistant to corrosion, but they are also brittle. Polymers have low density values, resist corrosion, and are good insulators, but they are relatively weak.

To obtain materials with the best of all the above properties, materials scientists devise composite materials. Composites are materials that combine two classes of materials in order to optimize material properties. There are metal matrix composites, typically designed to have metals being reinforced with ceramic fibers such as silica whiskers. Ceramic composites normally pair a ceramic fiber with a ceramic matrix, although concrete can be classed as a composite material as well, with the gravel acting to reinforce the cement matrix.

However, if concrete is excluded, fiber-reinforced polymers probably make up the highest volume of composite materials. Fiberglass is a very common example, but polymers may be reinforced with carbon fibers and aramid fibers like Kevlar as well. Polymers may also be reinforced with particles. Carbon black particles reinforce automobile tires, for example, and fumed (very small particle size) silica is also used to reinforce both auto tires and the elastomers used for the soles of shoes.

Materials Engineering

6.5.2 How Do We Obtain/Manufacture Composite Materials?

Unlike metals, ceramics, or polymers, the raw materials for composite materials are not directly obtained from the earth. With few exceptions (wood is considered by some to be a composite material), composites are manufactured from existing materials. Manufacturing of composite materials depends on the design of the material. Fiber-based composites demand different techniques than particulate composites. Manufacturing techniques for fiber-based composites include the use of prepreg material, in which mats or woven fibers are coated with partially reacted polymer resin. These prepreg sheets can then be cured using heated, shaped dies. In other cases, parts are built up using alternating layers of resin that are painted into a mold, followed by sheets of fibers. Preformed reinforcements may be infused with resin, possibly using pressure, vacuum assistance, and so on; the resin is then cured. Manufacturing for tubular products such as pipes and gas cylinders may use various winding techniques (helical, circumferential, etc.) to apply layers of fibers and resin. Particulate composites are manufactured using blending techniques. Reinforcing materials such as chopped fibers or powders are blended into thermoplastic melts or thermosetting resins.

6.5.3 What Are the Properties of Composite Materials?

Many factors affect the degree to which the strength of the matrix is improved by the reinforcing phase, including the strength of the reinforcing phase and the degree of bonding between the matrix and the reinforcement. If fibers are used for reinforcement, factors affecting the strength improvement of the matrix include the length of the fibers, the shape of their cross sections, their orientation with respect to the stress on the part, and their distribution.

The largest effect on the strength of the composite material is simply the concentration of the reinforcing phase. To a first approximation, the properties of the composite will be a weighted average of the properties of the two materials used to make it. For example, the value of Young's modulus can be roughly estimated by

$$E_c = E_m V_m + E_p V_p \tag{6.21}$$

where V is volume fraction and the subscripts c, m, and p refer to composite, matrix, and particulate, respectively. The same equation can be used to estimate the density of the composite as well as strength in the longitudinal direction for fiber composites. When density is estimated, the equation becomes

$$\rho_c = \rho_m V_m + \rho_p V_p \tag{6.22}$$

Example 6.6

Polyester resin has a density of approximately 1.35 g/cc. Glass reinforcing fiber has a density of approximately 2.2 g/cc. What density will a glass-reinforced polyester composite have if it contains 28% glass by volume?

Solution:

$$\rho_c = \rho_m V_m + \rho_p V_p$$
$$= (1.35 * 0.72) = (2.20 * 0.28)$$
$$= 1.588 \, g/cc$$

6.5.4 How Does Manufacturing Affect the Properties of Composites?

The strength of composite materials depends on the presence of excellent adhesion between the matrix and the reinforcing material. For many of the techniques described above, the inclusion of air bubbles—areas of the composite where the resin does not contact the reinforcement—is an Achilles heel, resulting in weak spots. Moisture or other contamination may also affect the strength of the resin-reinforcement bond. In some cases, distribution of the reinforcement may also be an issue, even with good resin coverage and bonding. This may be the case when a dense reinforcing phase settles to the bottom of a liquid resin, for example, or when layers of resin and reinforcement are unevenly applied by hand.

6.5.5 What Technologies Are Made Possible by Composites?

The area that has benefitted most from composites is the aerospace industry. The possibility of combining light weight with high stiffness and strength makes resin–fiber composite materials widely used in aerospace applications; they often replace metal components for weight savings and corrosion resistance. Sports equipment such as racing bicycles and automotive applications are also becoming more frequent.

In an application closer to home, particulate composites are being developed for use in the metal heat sinks used for computers (see Section 6.1). Aluminum is often chosen for heat sinks because of its good heat conductivity and low density. However, aluminum melts at around 660°C; even with this relatively high melting temperature, today's computer chips can push aluminum to the edge of its safe use temperature. For this reason, a composite of silicon carbide powder and aluminum metal is under investigation for use as heat sink material. The silicon carbide, being a ceramic material, is hard and strong at high temperatures; its inclusion does not greatly reduce thermal conductivity or increase density. The net result is a composite heat sink that can stand up to higher temperatures.

6.6 Conclusion

Advances in materials science have enabled many new inventions and developments. The engineer seeking to design new products needs to be aware of the basic properties of materials and the possible effects that processing may have on material properties.

Problems

1. A metal atom has a diameter of 0.36 nm; it is arranged in an FCC lattice. How long is a side of the unit cell for this metal?

2. A tensile bar fractures at a load of 2001.7 N (450 lb). Its original gage dimensions are 6.35 mm wide × 1.52 mm thick (0.25 in. wide × 0.06 in. thick). What is the fracture stress?

3. The tensile bar from problem 2 is measured after fracture. Its width is now 6.17 mm (0.243 in.) and its thickness is now 0.864 mm (0.034 in.). What is the ductility of this test sample?

4. A wire is 0.02 cm in diameter and 15 cm in length; it has a resistance of 0.2 Ω. What is its resistivity?

5. A nickel resistor has a resistance of 10.5 Ω at room temperature (25°C). It has a temperature coefficient of resistance of 0.005866/°C. What will its resistance be at 100°C?

6. Two castings are proposed for a study. One casting is a cube 10.16 cm (4 in.) on a side; the other is a rectangle 10.64 cm (16 in.) tall × 10.16 cm (4 in.) wide × 2.45 cm (1 in.) thick. Both are cast from the same metal at the same temperature, and the mold material is the same in both cases. If the cube-shaped casting solidifies in 14.5 min, how rapidly do we expect the rectangular casting to solidify?

7. A zirconium oxide bar 12.7 mm wide and 7.6 mm thick (0.5 in. wide and 0.3 in. thick) is fractured in three-point bending. The support span is 25.4 mm (1 in.) and the fracture load is 3002 N (675 lb). What is the modulus of rupture?

8. An aluminum oxide bar with the same dimensions as the zirconium bar above is to be fractured in four-point bending, with the support span at 50.8 mm (2 in.) and the inner span at 25.4 mm (1 in.). We expect its modulus of rupture to be approximately 413 MPa or 60,000 psi. At approximately what load will this bar fracture?

9. A crack in a ceramic material is 100 μm long and its radius is 2 μm. How much greater is the stress at the crack tip than the nominal stress applied to the part?

10. The modulus of rupture for aluminum oxide is around 413 MPa (60,000 psi) when it is fully dense. Assuming that the constant in Equation (6.16) is 2, what modulus of rupture would we expect to see if the aluminum oxide contained 10% porosity?

11. A batch of polyester has the molecular weights shown below. What is the weight-average molecular weight for this batch?

Molecular Weight Range (g/mol)	Fraction of Chains at This MW (x_i)
40,000–50,000	0.12
50,000–60,000	0.13
60,000–70,000	0.11
70,000–80,000	0.12
80,000–90,000	0.22
90,000–100,000	0.23
100,000–110,000	0.07

12. A fiberglass composite is made from polyester resin with 10% by volume glass fibers. The resin has a Young's modulus of 3.4 GPa (0.5×10^6 psi) and the glass fibers have a Young's modulus of 69 GPa (10×10^6 psi). What is the Young's modulus for the composite material?

7
Design and Analysis

CHAPTER OBJECTIVE AND STUDENT LEARNING OUTCOMES

After completing this chapter, students will be able to

1. Design components by calculating stress, strain, and materials' selections. [ABET outcome c, see Appendix C]
2. Analyze beam deflection and critical load in column. [ABET outcome e, see Appendix C]
3. Solve problems using Bernoulli's equation and the continuity equation. [ABET outcome e, see Appendix C]
4. Perform analysis using conduction, convection, and radiation equations. [ABET outcome e, see Appendix C]

7.1 Introduction

Design and analysis is a multifaceted task that requires knowledge, skills, and abilities in different subjects related to the design. The complexities associated with engineering analysis can vary significantly from simple stress analysis of a part to complex analysis of a system where multiple components and their interfaces require analysis. Design is an iterative process with many interactive phases until all specifications and requirements are met. Many resources exist to support the designer, including different sources of information and an abundance of computational design tools. The design engineer needs to not only develop competencies in their field but must also cultivate a strong sense of responsibility and professional work ethic. There are roles to be played by codes and standards, ever-present economics, safety, and considerations of product liability. The survival of a mechanical component is often related through stress and strength. Matters of uncertainty are ever present in engineering design and are typically addressed by the design factor and the factor of safety, either in the form of a deterministic (absolute) or statistical sense.

The fundamental principle of design involves performing the required analysis to meet the specifications and design intent. For example, for designing a bracket to support an object, the load or force on the object is used to perform stress analysis. Stress analysis also requires the geometry of the initial bracket design as stress is a function of both force and geometry. The second step in the design process is to select a material that will satisfy the stress developed because of the applied force. The relationship between the stress and strength of the material is determined by the safety factor, where the strength must be equal to or greater than stress in the part. For example, a design with the 1.5 safety factor requires material with 1.5 times more strength than stress in the part. Most engineering analysis requires knowledge in engineering mechanics, fluid mechanics, and heat transfer. Therefore, the fundamental principles and analysis procedures used in the above three subjects are provided in this chapter.

7.2 Stress and Strain

Elastic behavior of the material and the equilibrium of members are the fundamental principles in engineering analysis. Stress and strain can be analyzed in axially loaded members, thin walled pressure vessels, beams, eccentrically loaded members, and columns.

The selection of appropriate material with strength that exceeds the calculated values from stress analysis is important for design. Commonly used materials in engineering application are steel, aluminum, cast iron, and wood. Composite, polymer, and other materials are increasingly used in engineering design. The material properties of these engineering materials are presented in Table 7.1 in both US units (Mpsi) and metric units (GPa).

7.2.1 Stress

The term *stress* is used to express the internal reaction in a member when forces are applied to the cross-sectional area of an object. From the perspective of load, stress is the result of applied force or a system of forces that tends to deform a body. From the perspective of what is happening within a material, stress is the internal distribution of forces within a body that balances and reacts to the loads applied to it. Stress is the force per unit area. Typical units of stress are lbf/in^2 or psi, ksi, and MPa. There are two primary types of stresses: *normal stress* (σ) and *shear stress* (τ). In *normal stress*, force is normal to the surface area, and in *shear stress*, force is parallel to the surface area:

$$\sigma = \frac{P_{normal}}{A}$$

$$\tau = \frac{P_{parallel}}{A}$$

TABLE 7.1

Materials Properties

		Materials			
	Units	Steel	Aluminum	Cast Iron	Wood
Modulus of elasticity, E	Mpsi	29	10	14.5	1.6
	GPa	200	69	100	11
Modulus of rigidity, G	Mpsi	11.5	3.8	6.0	0.6
	GPa	80.0	26.0	41.4	4.1
Poisson's ratio, v		0.30	0.33	0.21	0.33

Mpsi, millions of pounds per square inch.

7.2.2 Strain

Strain or linear strain (ϵ) is a change of length per unit length. Linear strain can be described as in/in, mm/mm, and so on, making it dimensionless. Shear strain (γ) is the strain due to angular deformation resulting from torsion or twist of a member. Shear strain may be presented as percent deformation or as angles in radians or

$$\epsilon = \frac{\delta}{L}$$

$$\gamma = \frac{\delta_{\text{Parallel to area}}}{\text{Height}} = \theta$$

7.2.3 Hooke's Law

The linear relationship between elastic stress and strain is defined as Hooke's law: *Stress is proportional to strain*. For normal stress, the constant of proportionality is the *modulus of elasticity (Young's modulus), E*. When stress (y-axis) and strain (x-axis) are plotted during tensile testing to characterize the material property, the slope of the stress–strain curve in the elastic zone is the modulus of elasticity. The modulus of elasticity of steel is approximately 30×10^6 psi or 210 GPa. The higher value of modulus of elasticity is due to smaller value of strain in steel due to applied forces. Hooke's law can be expressed by the following relationship:

$$\sigma = E\epsilon$$

7.2.4 Poisson's Ratio (v)

Poisson's ratio is a constant that relates the lateral strain to the axial (longitudinal) strain for axially loaded members:

$$v = -\frac{\epsilon_{\text{Lateral}}}{\epsilon_{\text{Axial}}}$$

The typical value of Poisson's ratio is between 0.3 and 0.35 for steel and aluminum. For an element with torsional loading, the shear stress is proportional to the shear strain with a proportionality constant. This proportionality constant is called shear modulus or *modulus of elasticity, G*:

$$\tau = G\gamma$$

For an elastic, isotropic material, the modulus of elasticity, the relationship between modulus of elasticity (E) and shear modulus (G), can be expressed as

$$G = \frac{E}{2(1+v)}$$
$$E = 2G(1+v)$$

In general, the modulus of rigidity or shear modulus (G) is approximately 2.6 times less than the modulus of elasticity (E) for steel and aluminum. Therefore, the modulus of rigidity can be estimated if the modulus of elasticity is known. Poisson's ratio can be approximated as 0.3 if the value is not known.

7.3 Uniaxial Loading and Deformation

The deformation or deflection of a member with axial load can be derived from Hooke's law. If the original length of the member is L, with axial load P, the deformation can be calculated using the following relationship (Figure 7.1):

$$\delta = L\epsilon = L\left(\frac{\sigma}{E}\right) = \frac{PL}{AE}$$

An axial member with different cross-sectional areas and lengths requires the use of superposition to determine the total deformation. The deformations of different cross-sectional areas and lengths are added to calculate the total deformation of the member:

$$\delta = \sum \frac{PL}{AE} = P \sum \frac{L}{AE}$$

$$\delta = \int \frac{PdL}{AE} = P \int \frac{dL}{AE}$$

The final length after deformation, L_f, is the original length, L, plus the total deformation of different sections of the member:

$$L_f = L + \delta$$

7.4 Maximum Normal Stress Theory

The maximum normal stress theory states that the failure of a component occurs when the principal tensile or compressive stress is larger than the ultimate strength of the material. This theory is primarily used for brittle materials under static or biaxial loading conditions. Although brittle materials have higher compressive strength, both tensile and compressive stress analysis are required during design.

Factor of safety is an important design consideration as the load or force used in calculating stress may vary because of unknowns in determining loads, and the component may experience sudden, intermittent higher load in the future. There are uncertainties associated with material strength due to variations in materials' processing and degradation of material properties over time. The factor of safety can account for the above

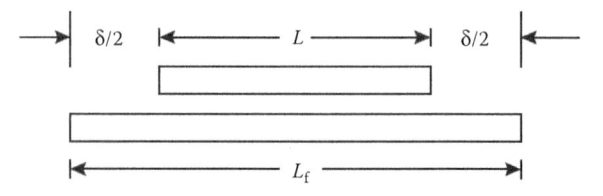

FIGURE 7.1
Deformation of a bar with length.

Design and Analysis

variations or uncertainties and should be incorporated in the design. If the uncertainties are higher, a higher value of the safety factor should be used. Typical values of the safety factor range from 1.5 to 3.0. The factor of safety, FS, is the ratio of ultimate strength of the material, S_u, divided by the calculated stress in the part, σ. If the factor of safety is known in advance, the allowable stress, σ_a, can be calculated by dividing the ultimate strength by FS,

$$FS = \frac{S_u}{\sigma}$$

$$\sigma_a = \frac{S_u}{FS}$$

The failure criterion is

$$\sigma_a > \frac{S_u}{FS}$$

Example 7.1

A solid aluminum bar is compressed with an axial load (P) of 200 kN. The modulus of elasticity (E) of the aluminum is provided in Table 7.1. Calculate the axial deformation (δ) and axial compressive stress.

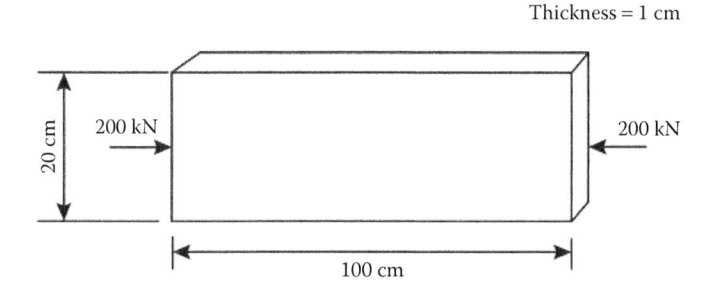

Solution:

Compressive normal stresses are negative.

The cross-sectional area, $A = 0.2\,\text{m} \times 0.01\,\text{m} = 0.002\,\text{m}^2$. From Table 7.1, the modulus of elasticity of aluminum, $E = 69\,\text{GPa} = 69 \times 10^9\,\text{Pa} = 69 \times 10^9\,\text{N/m}^2$ (1 Pa = 1 N/m²)

$$\text{Axial deformation,}\ \delta_{axial} = \frac{PL}{AE}$$

$$\delta_{axial} = \frac{(-200 \times 10^3) \times 1}{0.002 \times 69 \times 10^9} = -0.00145\,\text{m}$$

$$= -1.45\,\text{mm}\ (\text{negative sign represents decrease of length})$$

$$\text{Compressive stress,} = \frac{P}{A} = \frac{-200 \times 10^3}{0.002}$$

$$= -10^8\,\text{Pa}\ (\text{negative sign represents compressive stress})$$

Example 7.2

Calculate the final thickness of the bar of Example 7.1.

Solution:

Calculation of final thickness requires

1. Calculation of axial strain
2. Calculation of lateral strain using Poisson's ratio for aluminum
3. Change in thickness = lateral strain times original thickness
4. Final thickness = original thickness − change in thickness

Poisson's ratio for aluminum can be obtained from Table 7.1 as 0.33,

$$v_{axial} = \frac{\delta}{L} = -\frac{0.00145}{1} = -0.00145 \, m/m$$

$$\text{Poisson's ratio, } v = -\frac{\epsilon_{lateral}}{\epsilon_{axial}}$$

$$\epsilon_{lateral} = -(v \times \epsilon_{axial}) = -(0.33 \times 0.00145) = 0.0004785 \, m/m$$

$$\Delta t = \epsilon_{lateral} \times t = -0.0004785 \times 0.01 = 0.00004785 \, m$$

$$t_{final} = 0.01 + 0.00004785 = 0.01004785 \, m = 1.004785 \, cm$$

When a bar is compressed, the length decreases and the thickness increases.

7.5 Pressure Vessel

Pressure vessels are important components used in applications such as propane tanks; oxygen cylinders used in hospitals; fire extinguishing cylinders; and oil, water, and gas pipelines. Tanks under internal pressure experience circumferential, longitudinal, and radial stresses. If the wall thickness is small, the radial stress component is negligible and can be disregarded (Figure 7.2).

A cylindrical tank with wall thickness greater than 10% of its inside radius is classified as a thick cylinder and less than 10% thickness is classified as a thin cylinder. A thin- or thick-wall cylinder can develop radial, longitudinal, and circumferential stresses. The radial stress in thin-wall cylinders should be considered during design in addition to the circumferential stresses. Because of the severity of potential damage from failure of pressurized tanks, pressure vessel design uses special standards. American Society of Mechanical Engineers Boiler and Pressure Vessel code is widely used for safe design of pressure vessels (ASME BPV code, Section VIII Division I).

7.5.1 Hoop Stress

If the wall thickness of a cylindrical pressure vessel is very small compared to its radius, the resulting radial stress may be small compared to the tangential stress. If an internal pressure P is exerted on the wall of a cylinder of thickness t and inside diameter d, then

Design and Analysis

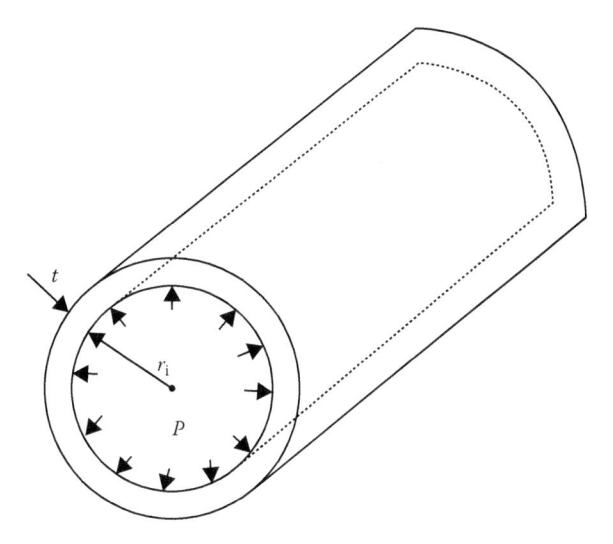

FIGURE 7.2
Stresses in the pressure vessel.

force Pd will tend to separate two halves of a unit length of the cylinder. This force is resisted by the tangential stress, also called the *hoop stress* σ_t, acting uniformly over the stressed area. The tangential or hoop stress can be calculated as

$$\sigma_t = \frac{Pd}{2t} = \frac{Pr}{t}$$

7.5.2 Axial Stress

In a cylindrical pressure vessel with enclosed ends such as a propane tank, the axial force at the ends produces a stress directed along the longitudinal axis known as the *longitudinal* or *axial stress*, σ_a. It can be observed that the longitudinal stress is half of the hoop stress and therefore, pressure vessel design calculation is performed using hoop stress:

$$\sigma_a = \frac{Pd}{4t} = \frac{Pr}{2t} = \frac{\sigma_t}{2}$$

Example 7.3

A compressed propane gas cylinder has internal gage pressure 5 MPa. The outside diameter of the cylinder is 50 cm. If the material has an allowable stress of 90 MPa, calculate the required wall thickness. Also check if it is a thin wall or not.

Solution:

Assume a thin-walled tank

$$\sigma_t = \frac{Pd}{2t}$$

$$t = \frac{Pd}{2\sigma_t} = \frac{5 \times 50}{2 \times (90+5)} = 1.3158\,\text{cm}$$

Thin wall assumption

$$\frac{t}{r_i} = \frac{t}{(d-2t)/2} = \frac{1.3158}{\{[50-(2)(1.3158)]/2\}} = 0.0556 < 0.1 \quad [\text{thin wall}]$$

7.6 Shear Stress

Shear stress occurs when a part is loaded with torque, moment, or torsion. The shear stress at the outer surface of a bar of radius r, which is loaded by a torque, T, is

$$\tau = \frac{Tr}{J}$$

The polar moment of inertia, J, of a solid round shaft is

$$J = \frac{\pi r^4}{2} = \frac{\pi D^4}{32}$$

For a hollow round shaft,

$$J = \frac{\pi}{2}\left(r_o^4 - r_i^4\right) = \frac{\pi}{32}\left(D_o^4 - D_i^4\right)$$

where:
r_i is the inner radius
r_o is the outer radius of the hollow shaft

For a solid shaft, the shear stress is zero at the center and maximum at the surface.
If a shaft of length L carries a torque T, the angle of twist can be calculated as (Figure 7.3).

$$\phi = \frac{TL}{GJ}$$

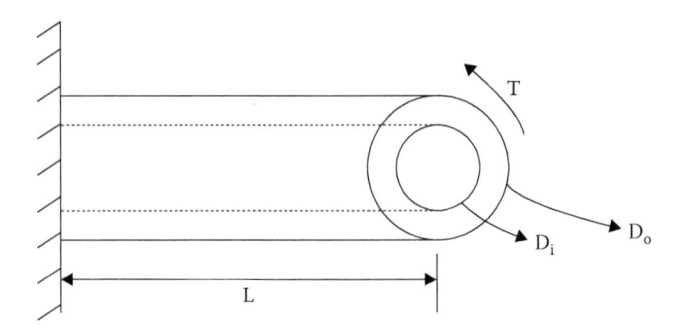

FIGURE 7.3
Hollow shaft with torsional load.

Design and Analysis

Example 7.4

Calculate the maximum shear stress in a 0.22-m-diameter shaft when 15,000 N m torque is applied to the shaft.

Solution:

$$J = \frac{\pi r^4}{2} = \frac{\pi}{2}\left(\frac{0.22}{2}\right)^4 = 2.3 \times 10^{-4} \, m^4$$

$$\tau = \frac{Tr}{J}$$

$$\text{Maximum shear stress} = \frac{15{,}000 \times 0.11}{2.3 \times 10^{-4}} = 7.2 \text{ MPa}$$

7.7 Stress in Beams

The shear stress, V, in a beam is the sum of all vertical forces between the section and one of end. The bending stress in a beam occurs when a moment is applied or results from forces applied to the beam. For a positive bending moment, the lower surface of the beam is under tensile stress and the upper surface is under compressive stress (Figure 7.4). The bending stress distribution passes through the centroid or *neutral axis* of the cross section. The centroid axis is also called the neutral axis where there is no stress as the stress changes from compressive to tensile stress at that axis. Bending stress varies with location; it is zero at the neutral axis and increases linearly with distance from the neutral axis:

$$\sigma_B = -\frac{My}{I}$$

where:
  y is the distance from the neutral axis
  M is moment
  I is the moment of inertia

Because the maximum stress is used in the design, the distance from the neutral axis y can be replaced with c:

$$\sigma_{B,max} = \frac{Mc}{I}, \quad \text{for compression } (-) \text{sign can be omitted}$$

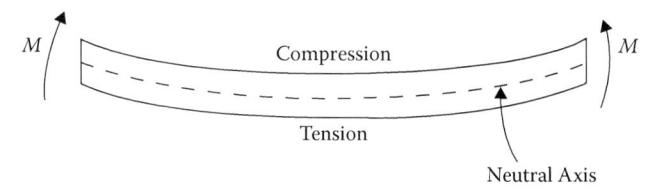

FIGURE 7.4
Stresses in a beam with positive load.

For a rectangular beam with width b and height h, the centroidal moment of inertia can be calculated as

$$I = \frac{bh^3}{12}$$

Example 7.5

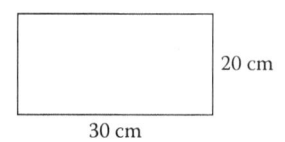

30 cm × 20 cm

Find the moment of inertia and maximum compressive stress of the above geometry if the maximum bending moment is 25 kN m.

Solution:

Given, $b = 0.3$ cm, $h = 0.2$ cm, $M = 25{,}000$ N m, $c = 0.1$ m

$$I = \frac{bh^3}{12} = \frac{0.3 \times 0.1^3}{12} = 2.5 \times 10^{-5}\ m^4$$

$$\sigma_{B,max} = \frac{25{,}000 \times 0.1}{2.5 \times 10^{-5}} = 100\ MPa$$

7.8 Buckling in Column

If a vertical load is applied through the centroid of a tension or compression member's cross section, the loading is said to be *axial loading* or *concentric loading*. *Eccentric loading* occurs when the load is not applied through the centroid. *Long columns* will buckle in the transverse direction that has the smallest radius of gyration.

The critical load is defined as the load at which a long column will fail (Figure 7.5). The critical load is the theoretical maximum load that a column can support without buckling. For columns without friction, the load is given by *Euler's formula*

$$P_{cr} = \frac{\pi^2 EI}{l^2}$$

Example 7.6

A rectangular steel bar 40 mm wide and 60 mm thick is pinned at each end and subjected to axial compression. The bar has a length of 1.80 m. The modulus of elasticity is 200 GN/m². What is the critical buckling load?

Solution:

$$P_{cr} = \frac{\pi^2 EI}{l^2}$$

$$= \frac{\pi^2 \times (200 \times 10^9) \times \left[(0.06 \times 0.04^3)/12\right]}{1.80^2} = 195\ kN$$

Design and Analysis

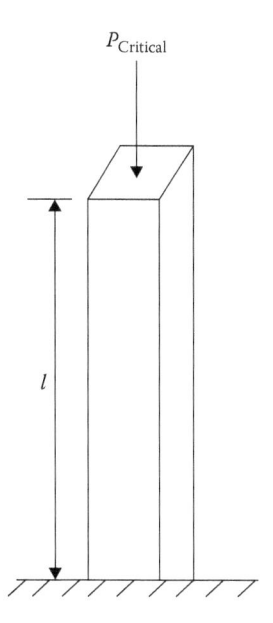

FIGURE 7.5
Critical load on column.

7.9 Thermodynamics

The subjects of heat transfer and thermodynamics are highly complementary and interrelated, but they also have fundamental differences. The transfer of heat and energy plays an important role in the first and second laws of thermodynamics because of the primary mechanisms for energy transfer between a system and its surroundings. While thermodynamics may be used to determine the amount of energy required in the form of heat for a system to pass from one state to another, it considers neither the mechanisms that provide for heat exchange nor the methods that exist for computing the rate of heat exchange.

7.9.1 First Law of Thermodynamics

The first law also addresses the ways in which energy can cross the boundaries of a system. For a closed system (a region of fixed mass), there are only two ways: heat transfer through the boundaries and work done on or by the system:

$$Q = \Delta U + W$$

where:
  ΔU is the change in the total energy stored in the system
  Q is the net heat transferred to the system
  W is the net work done by the system

The first law can also be applied to an open system (a control volume), a region of space bounded by a control surface through which mass may pass. Mass entering and leaving the control volume carries energy with it; this process, termed energy advection, adds a third way in which energy can cross the boundaries of a control volume. To summarize, the first law of thermodynamics can be stated for both a control volume and a closed system.

7.9.2 Second Law of Thermodynamics

The second law of thermodynamics is a directional law in that, in nature, processes only happen in one direction. Gravity, for instance, always *pulls* objects *toward* the earth. As it relates to thermodynamics, the transfer of heat will only occur naturally from a warmer object to a cooler object. Heat can be transferred to a warmer object but not without the aid of external energy. That is, work must be done on the system and losses will occur. Heat transfer ceases once temperatures reach an equilibrium. From the first law,

$$\text{Net work done} = \text{Net heat transfer}$$

$$\text{or, } \sum W = \sum Q$$

But in practice,

$$\text{Net work} < \text{Net heat transfer}$$

$$\text{or } \sum W < \sum Q$$

Because the work transfer is less than the heat transfer,

$$\sum Q - \sum W > 0 \quad \text{and has positive value}$$

This means that some heat transfer must be rejected and is lost as a result of inefficiency.

7.9.3 Third Law of Thermodynamics

The third law of thermodynamics is essentially a statement about the ability to create an *absolute* temperature scale, for which absolute zero is the point at which the internal energy of a solid is precisely 0.

Various sources show the following three potential formulations of the third law of thermodynamics:

1. It is impossible to reduce any system to absolute zero in a finite series of operations.
2. The entropy of a perfect crystal of an element in its most stable form tends to zero as the temperature approaches absolute zero.
3. As temperature approaches absolute zero, the entropy of a system approaches a constant.

7.10 Heat Transfer

Heat transfer can occur in a system or object simultaneously by three different phenomena known as *conduction, convection,* and *radiation*. A brief overview of these three phenomena is described in the following.

7.10.1 Conduction

Heat transfer through conduction occurs through a material. The transfer of thermal energy in the form of heat occurs because of atomic or molecular vibration in solids, or movement in liquids or gases. There may be free electron drift showing an energy flux in the direction of reducing temperature. Therefore, metals have higher thermal conductivity than liquids and gases. Example of heat transfer by conduction is the transfer of heat from steam in a pipe to the outside surface. The heat must pass through material of the pipe or be *conducted* through the pipe. The steady-state heat transfer by conduction through a flat plate is described by *Fourier's law of conduction*

$$Q = -kA\frac{dT}{dx} = -kA\frac{(T_2 - T_1)}{L}$$

where:
k is the coefficient of thermal conductivity, W/m K
A is the area of transfer, m²
T_1 is the inlet temperature, °C
T_2 is the outlet temperature, °C
L is the thickness of the plate through which heat flows, m

Example 7.7

The wall of a furnace is constructed from 0.15-m-thick fireclay brick with thermal conductivity of 1.8 W/m K (Figure 7.6). The inner and outer surface temperatures are 1500 and 1200, respectively. What is the rate of heat flow through a wall that is 0.7 m × 1.3 m on a side?

Solution:

$$Q = kA\frac{(T_1 - T_2)}{L}$$

$$= 1.8 \times (1.3 \times 0.7)\frac{(1500 - 1200)}{0.15}$$

$$= 3276 \text{ W}$$

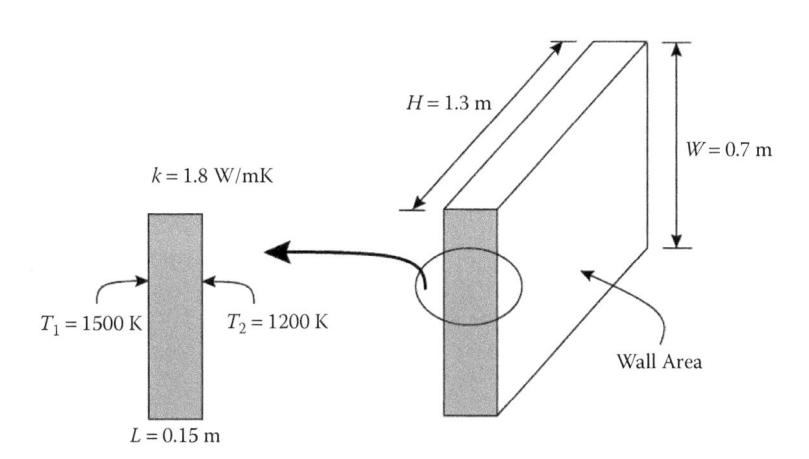

FIGURE 7.6
Example 7.8, heat transfer through a wall.

7.10.2 Convection

Heat transfer through convection occurs because of the movement of a fluid. Heat will transfer by conduction through the walls of the fluid container. From the container walls, the heat will transfer by conduction and radiation into the fluid causing convection to occur as the fluid moves.

Newton's law of convection calculates the forced convection heat transfer. Forced convection results when a fan, pump, or relative vehicle motion moves the fluid. If a single film dominates the thermal resistance, the heat transfer coefficient, h, is expressed as

$$Q = hA(T_w - T_\infty)$$

where:
T_w is the wall temperature
T_∞ is the fluid temperature

7.10.3 Radiation

Thermal radiation is the energy emitted by matter where no solid or liquid medium is present. The emission may be due to changes in the electron configurations of the constituent atoms or molecules. The energy of the radiation field is transported by electromagnetic waves (or alternatively, photons). Radiation transfer occurs most efficiently in a vacuum environment with no liquid or solid medium.

Radiation can be absorbed, reflected, or transmitted through a body. If α is the fraction of energy being absorbed (*absorptivity*), ρ is the fraction being reflected (*reflectivity*), and τ is the fraction being transmitted through (*transmissivity*), then the radiation conservation law is

$$\alpha + \rho + \tau = 1$$

7.11 Fluid Mechanics

A fluid is a substance in either the liquid or gas phase. Fluids cannot support shear stress and deform continuously to minimize applied shear forces. Liquids are considered incompressible while gases are considered to be compressible fluids.

7.11.1 Properties of Fluids

Understanding of different fluid properties is important for the analysis of fluid flow in engineering systems. General properties of fluids are described in the following section. The symbol usually used to represent the property is specified together with some typical values in SI units for common fluids.

> *Density*: The density of a substance is the quantity of matter contained in a unit volume of the substance. It can be expressed in three different ways.

Mass density: Mass density, ρ, is defined as the mass of substance per unit volume. Units: kilograms per cubic meter, kg/m^3.

Dimensions: ML^{-3}. typical values: water = 1000 kg/m^3, mercury = 13,546 kg/m^3, air = 1.23 kg/m^3, paraffin oil = 800 kg/m^3.

Specific weight: ω (sometimes γ, and sometimes known as specific gravity) is defined as the weight per unit volume. It could also be thought of as the force exerted by gravity, g, on a unit volume of the substance. The relationship between g and ω can be determined by Newton's second law, because weight per unit volume = mass per unit volume × g ω = ρg.

Units: Newton per cubic meter, N m^{-3}.

Typical values: Water = 9814 N m^{-3}, mercury = 132,943 N m^{-3}, air = 12.07 N m^{-3}, paraffin oil = 7851 N m^{-3}.

Viscosity: The viscosity of a fluid is a measure of that fluid's resistance to flow when acted on by an external force, such as a pressure gradient or gravity. Viscosity of fluids can be determined with a sliding-plate viscometer test. Consider two plates of area, A, separated by a fluid with thickness, δ. The bottom plate is fixed and the top plate is in motion at a constant velocity, v, by a force, F (Figure 7.7).

Experiments with many fluids have shown that the force, F, required to maintain the velocity, v, is proportional to the velocity and the area is inversely proportional to the separation of the plates:

$$F \propto \frac{vA}{\delta}$$

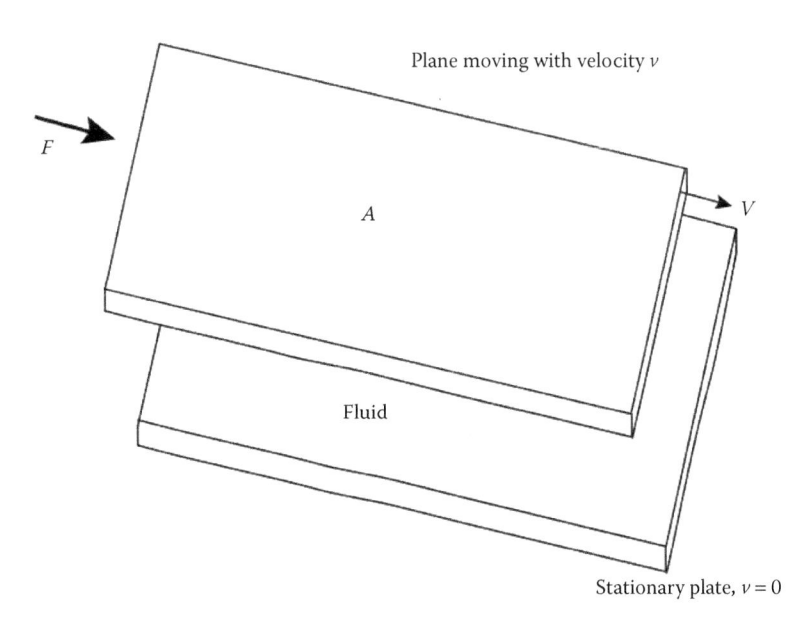

FIGURE 7.7
Sliding-plate viscometer.

The constant of proportionality is the *absolute viscosity*, μ, also known as the *absolute dynamic viscosity*. For a linear velocity profile,

$$\frac{v}{\delta} = \frac{dv}{dy}$$

Fluid shear stress,

$$\tau_t = \mu \frac{dv}{dy}$$

Another quantity with the name viscosity is the ratio of absolute viscosity to mass density. This combination of variables is known as *kinematic viscosity*, v. The primary dimensions of kinematic viscosity are $L2/\theta$. Typical units are ft^2/s and m^2/s.

7.11.2 Barometers

The *barometer* is a common device for measuring atmospheric pressure (Figure 7.8). It is constructed by filling a long tube open at one end with mercury and inverting the tube such that the open end is below the level of mercury in a filled container. If the vapor pressure of the mercury in the tube is neglected, the fluid column is supported only by the atmospheric pressure transmitted through the container fluid at the lower, open end. The atmospheric pressure is given by

$$p_a = \rho g h = \gamma h$$

If the vapor pressure of the barometer liquid is significant (as it would be with alcohol or water), the vapor pressure effectively reduces the height of the fluid column:

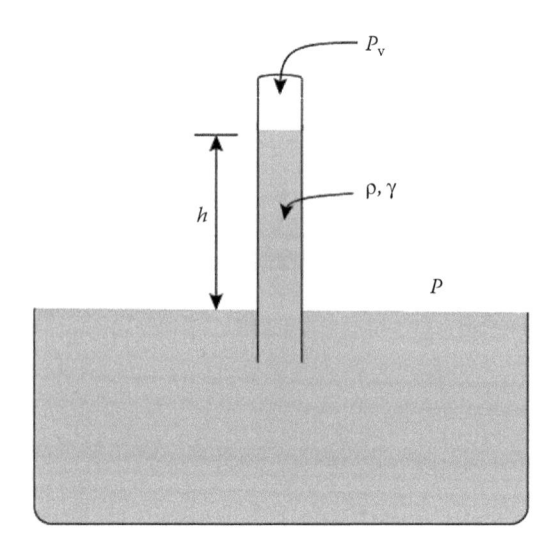

$$p_a = \rho g h + p_v$$

FIGURE 7.8
Barometer.

7.11.3 Forces on Submerged Plane Surfaces

The pressure on a horizontal plane surface is uniform over the surface because the depth of the fluid is uniform. The resultant of the pressure distribution acts through the center of the pressure of the surface, which corresponds to the centroid of the surface. The total vertical force for a horizontal plane of area A is given by

$$R = pA$$

The pressure on a vertical rectangular plane surface increases linearly with depth: the more the depth, the higher the pressure. The pressure distribution can be expressed in a triangular form as shown in Figure 7.9a, if the plane surface extends to the surface; otherwise, the distribution will be trapezoidal, as shown in Figure 7.9b.

7.11.3.1 Buoyancy

If an object is partially or completely submerged in fluid, the object will experience an upward force named *buoyancy force*. Buoyant force exists in all submerged objects and always acts upward to cancel the object's weight. *Archimedes' principle* states that the buoyant force on a submerged or floating object is equal to the weight of the displaced force. An equivalent statement of Archimedes' principle is that a floating object displaces liquid equal in weight to its own weight. The two forces acting on a stationary floating object are the *buoyant force and object's weight*.

Example 7.8

Calculate the height of a mercury column that is equivalent to 500 kPa pressure. The density of mercury is 13,500 kg/m³.

Solution:

$$p_a = \rho g h$$

$$h = \frac{p}{\rho g} = \frac{500 \times 1000}{13,500 \times 9.81} = 3.775 \text{ m}$$

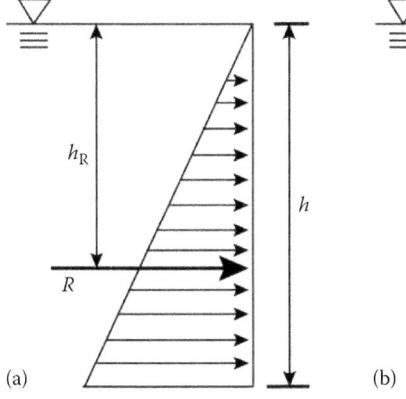

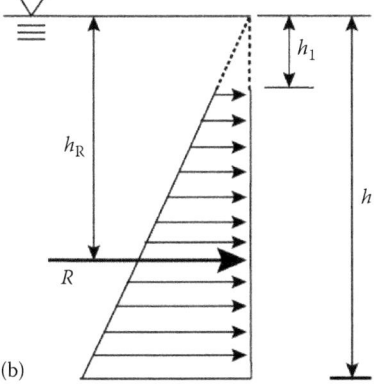

FIGURE 7.9
Hydrostatic pressure on a vertical plane surface.

Example 7.9

A fluid has a vapor pressure of 0.4 Pa and specific gravity of 18,000 kg/m³. If the fluid's column height is 2 m, what is the atmospheric pressure?

Solution:

$$p_a = \rho g h + p_v = (18{,}000 \times 9.8 \times 1000 \times 2) + 0.4$$

$$= 352{,}800.4 \text{ Pa}$$

7.11.4 Conservation of Mass

The conservation of mass states that mass cannot be added or removed from a system, regardless of the pipeline complexity, orientation of the flow, or type of fluid flowing. This single concept is often sufficient for solving simple fluid problems:

$$m_1 = m_2, \quad \text{where } m \text{ is the mass flow rate}$$

The conservation of the mass flow in fluids is known as the *continuity equation*. The continuity equation states that flow passing any two points of the system is the same:

$$m = \rho A v = \rho Q$$

$$\rho_1 A_1 v_1 = \rho_2 A_2 v_2$$

$$Q = A_1 V_1 = A_2 V_2$$

where:
ρ is the density of the fluid
A is the cross-sectional area of the system
v is the velocity of the fluid (Figure 7.10)

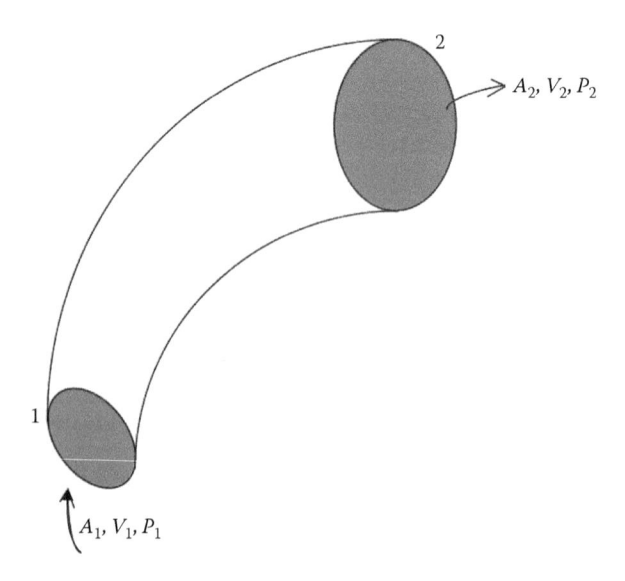

FIGURE 7.10
Generalized flow conservation.

7.11.5 Conservation of Energy

Bernoulli's equation, also known as the *field equation,* is an energy conservation equation that is valid for incompressible, frictionless flow. Bernoulli's equation states that the total energy of a fluid flowing without friction losses in the pipe is constant. *The total energy possessed by the fluid is the sum of its pressure, kinetic, and potential energies.* In other words, Bernoulli's equation states that the total head of any two points is the same. Head is a measure of pressure energy in terms of the height of the fluid:

$$\frac{p_1}{\rho_1 g} + \frac{v_1^2}{2g} + z_1 = \frac{p_2}{\rho_2 g} + \frac{v_2^2}{2g} + z_2$$

The *head loss due to friction* is denoted by the symbol h_f. This loss is added to the original Bernoulli equation to restore the equality. The *extended Bernoulli equation* accounting for friction is

$$\frac{p_1}{\rho_1 g} + \frac{v_1^2}{2g} + z_1 = \frac{p_2}{\rho_2 g} + \frac{v_2^2}{2g} + z_2 + h_f$$

7.11.6 Reynolds Number

The *Reynolds number, Re,* is the dimensionless number interpreted as the ratio of the internal forces to viscous forces in the fluid. The internal forces are proportional to the flow diameter, velocity, and fluid density. The viscous force is represented by the fluid's absolute viscosity, μ.

$$Re = \frac{vD\rho}{\mu}, \text{ where } v = \text{velocity}, D = \text{diameter}$$

Because μ/ρ is the *kinematic viscosity,* ν,

$$Re = \frac{vD}{\nu}$$

7.11.6.1 Laminar Flow

If all the fluid particles are moving in paths parallel to the overall flow direction, the flow is called *laminar flow.* This occurs when the Reynolds number is less than approximately 2100. Laminar flow is typical when the flow channel is small, the velocity is low, and the fluid is viscous. Laminar flow is illustrated by pouring viscous fluid such as honey, shampoo, or engine oil.

7.11.6.2 Turbulent Flow

Turbulent flow is characterized by a three-dimensional movement of the fluid particles superimposed on the overall direction of the motion. A fluid is said to be turbulent if the Reynolds number is greater than approximately 4000. The fluid is said to be in the *critical zone or transient region* when the Reynolds number is between 2100 and 4000:

Laminar flow, $Re < 2100$

Turbulent flow, $Re > 4000$

Transient, $2100 \leq Re \leq 4000$

7.11.7 Flow Distribution

With laminar flow in a circular pipe or between two parallel plates, viscosity makes some fluid particles adhere to the wall. The closer to the wall, the greater the tendency will be for the fluid to adhere. In general, fluid viscosity will be zero at the wall and will follow a parabolic distribution away from the wall. The *average flow velocity* is found from the flow rate and cross-sectional area:

$$V = \frac{Q}{A}$$

Because of the parabolic distribution, velocity will be maximum at the centerline, midway between the two walls. The maximum velocity for the laminar flow is

$$V_{max} = 2v \; [\text{flow in a circular pipe}]$$

$$V_{max} = 1.5v \; [\text{flow between plates}]$$

Example 7.10

Figure 7.11 shows a picture of a steel piping system with an inlet diameter of 200 mm and water flowing at 10 m/s. The outlet of the pipe is branched as shown in the figure. The diameter of the outlet 1 is 80 mm and outlet 2 is 120 mm. The velocity of the fluid at outlet 1 is 3 m/s. Find the velocity of the water at outlet 2 and the flow rate of the outlets (Table 7.2).

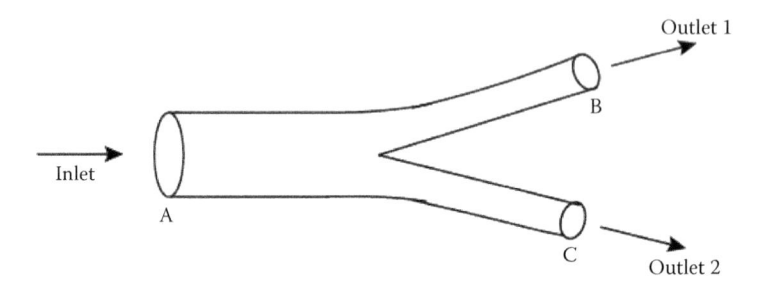

FIGURE 7.11
Flow through branched piping system.

TABLE 7.2

Specific Roughness of Typical Materials

Materials	ft. (ϵ)	mm (ϵ)
Riveted steel	0.003–0.0.	0.9–9.0
Concrete	0.001–0.01	0.3–3.0
Galvanized steel	0.00085	0.15
Commercial steel or wrought iron	0.00015	0.046
Drawn tubing	0.000005	0.0015

Solution:

$$\text{Given } Q_1 = 5000 \text{ gpm} = \frac{5000 \times 0.00378}{60} = 0.315 \frac{\text{m}^3}{\text{s}}; Q_2 = ?; Q_3 = ?$$

$$A_1 = \frac{\pi}{4}(0.2^2) = 0.0314 \text{ m}^2; \quad V_1 = 10 \text{ m/s};$$

$$A_2 = \frac{\pi}{4}(0.08^2) = 5.03 \times 10^{-3} \text{ m}^2; \quad V_2 = 3 \text{ m/s}$$

$$A_3 = \frac{\pi}{4}(0.12^2) = 0.0113 \text{ m}^2; \quad V_3 = ?$$

$$Q_1 = Q_2 + Q_3$$

$$\text{or} \quad A_1 V_1 = A_2 V_2 + A_2 V_2$$

$$\text{or} \quad (0.031416 \times 10) = (5.03 \times 10^{-3} \times 3) + (0.0113 V_3)$$

$$\text{or} \quad V_3 = 26.5 \text{ m/s}$$

$$Q_2 = 5.03 \times 10^{-3} \times 3 = 0.0151 \frac{\text{m}^3}{\text{s}}$$

$$Q_3 = 0.0113 \times 26.5 = 0.29945 \frac{\text{m}^3}{\text{s}}$$

7.11.8 Drag

Drag is the frictional force that acts parallel but opposite to the direction of motion. The total drag force (F_D) can be calculated from a dimensionless drag coefficient, C_D, that depends only on the Reynolds number:

$$F_D = \frac{C_D A \rho v^2}{2}$$

7.11.9 Lift

Lift is an upward force that is exerted on an object (flat plate, airfoil, rotating cylinder) as the object passes through a fluid. Lift combines with drag to form the resultant force on the object, lift force,

$$F_L = \frac{C_L \rho v^2 A_P}{2}$$

where:
C_L is the coefficient of lift

Problems

1. A solid round steel rod of 10 mm diameter and 500 mm length is rigidly connected to the end of a solid square brass rod 50 mm on a side and 400 mm long. The geometric axes of the bars are along the same line. An axial tensile force of 10 kN is applied at the ends of the assembly. For steel $E = 200$ GPa and for brass $E = 90$ GPa. Determine the total elongation of the assembly.

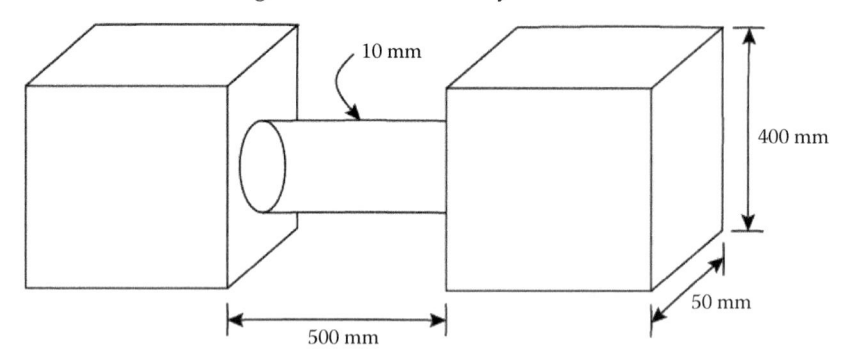

2. The pressure gage in an air cylinder reads 1000 kPa. The cylinder is constructed of 8 mm rolled plate steel with an internal diameter of 500 mm. What is the tangential stress in the tank?

3. A square steel column with a solid cross section is pinned at both its base and top. The column is 7 m in height and supports 4 MN. The modulus of elasticity of the steel is 200 GPa. What is the minimum cross-sectional size of the column to avoid buckling?

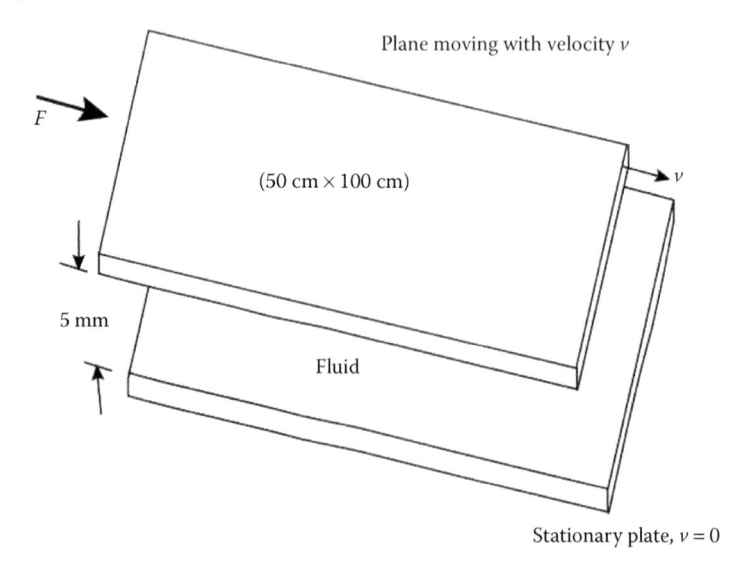

4. A sliding-plate viscometer is used to measure the viscosity of Newtonian fluid. A force of 30 N is required to keep the top plate moving at a constant velocity of 7 m/s. What is the viscosity of the fluid?

5. A liquid with a density of 900 kg/m³ is stored in a pressurized, closed tank. The tank is cylindrical with 12 m diameter. The absolute pressure in the tank above the

liquid is 300 kPa. What is the initial velocity of a fluid jet when a 0.5-m-diameter orifice is opened at point A?

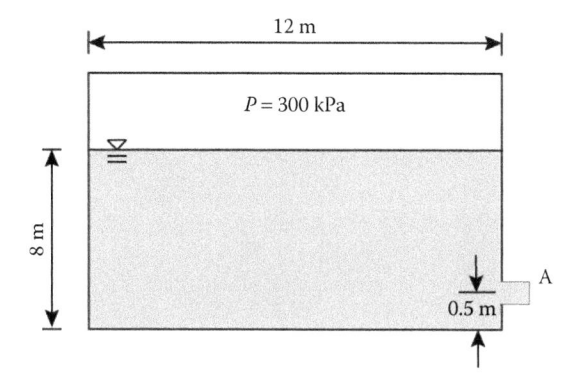

6. A 400-mm-long rigid metal cylindrical container with a diameter of 250 mm is closed at one end. The container is held vertically, nearly submerged with the open end down as shown in the figure. The atmospheric pressure is 101.3 kPa. The water rises 80 mm inside the container under these conditions.

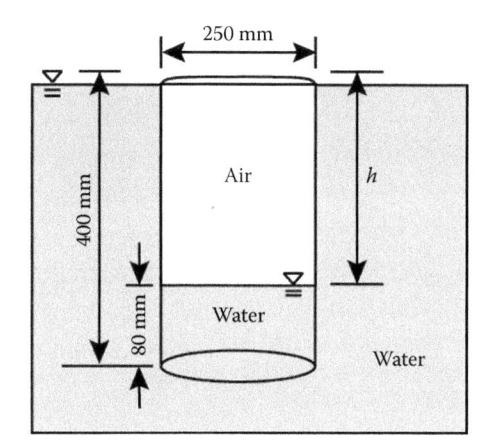

   a. What is the approximate total pressure of the air inside the container?
   b. The container is slowly moved vertically downward until the pressure in the container is 110 kPa. What will be the depth of the water surface measured from the free-water surface (i.e., what is the vertical distance between the free-water surface and the water surface in the container)?

7. Consider water flowing through a converging channel as shown and discharging to the atmosphere at the exit. What is the gage pressure at the inlet?

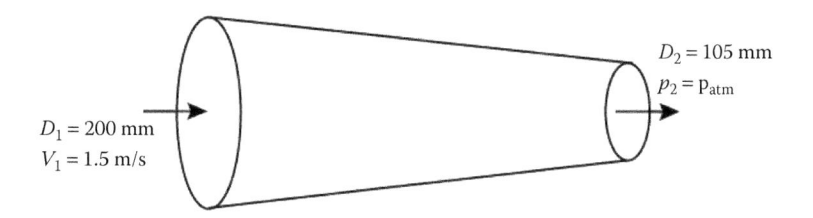

8

Electric Circuits and Components

CHAPTER OBJECTIVE AND STUDENT LEARNING OUTCOMES

After completing this chapter, students will be able to

1. Define terms such as electric charge, voltage, current, and amps.
2. Perform circuit analysis using Ohm's law and by combining resistors, capacitors, and inductors. [ABET outcome e, see Appendix C]
3. Apply diodes and transistors in analog and digital circuits.
4. Describe integrated circuit manufacturing processes and common IC devices. [ABET outcome k, see Appendix C]

8.1 Introduction

The advancement in the realm of the digital world has provided numerous breakthroughs in many aspects of modern life. Today we are surrounded by equipment such as cell phones, laptops, tablets, calculators, and pacemakers, which are innovations of the information age. One such key invention that has positively impacted many human lives is the advent of *robotic-arm surgery*. With computer-controlled precision cuts, this form of surgery provides extremely high success rates and promises speedy recovery, less pain, and minimal chances of infection. Robotic-arm-assisted techniques provide efficient operating procedures, such as when a patient comes in for a bladder or kidney removal he only has to stay overnight. Standing only a short distance away from the operating table, doctors can maneuver a robot more reliably to perform specific tasks. Back in the day, doctors would often have to practice their surgical skills on human or animal cadavers to stay on top of their game. Now, they can hold multiple practice sessions, like a pilot runs in a flight simulator, to increase their accuracy in a field where the slightest lack of precision could easily claim someone's life. Robotic arm surgery could be termed as a revelation in the healthcare sector and that has been possible because of the 1947 invention of transistors, one of the fundamental components in electric circuits. In this chapter, we will learn about all the electronic components that come together to form a circuit.

8.2 Understanding Electricity

Before getting into the details of circuit design and analysis, it would be a good idea to refresh some terms that you have been introduced to in physics. Basic knowledge of charge, electric field, current, and potential difference will prove to be extremely valuable as we begin our journey through these sections.

8.2.1 Electric Charge

When there is a mismatch in the total number of electrons and protons within an object, there will be a build-up of electrostatic charge. If there are more electrons than protons, we say the object is *negatively charged*, and if there are fewer electrons than protons, the object is *positively charged*. If the number of protons and electrons are the same, then we can say the object has no charge and is thus, *neutral*. The proof that electrostatic charges exist can be observed when you slide across the carpet on a day with low humidity, and you get zapped the moment you try to touch a doorknob. This is because you have a build-up of charge, meaning you were not a neutral object before touching the doorknob. When you make contact with the metal doorknob, electrons transfer into your body that feels like getting zapped.

8.2.2 Conduction and Insulation

A conductor is a material that lets electrons flow through it very easily, while an insulator is made up of materials that oppose the flow of electrons through it. Metals are usually the best conductors because their valence electrons have weak bonds, and these electrons are free to move around the solid object. The structure of a metal can be best described as some positive ions located in a sea of free flowing electrons. This is why metals like copper are used to manufacture electric wires. On the other hand, insulators have no free electrons and thus no flow of electrons. Because most elements in the periodic table are metals and likely to be good conductors, the best insulator tends to be plastic. The insulating material of wires is made of plastic that ensures that the conductive metal inside is not exposed.

8.2.3 Coulomb's Law

Charges employ either an attractive or a repulsive force on each other. Like charges will always repel and opposite charges will attract. Knowing magnitude of the forces that two charged objects apply on each other proves vital in understanding the fundamental concepts of electricity. Thanks to the work of Charles Coulomb, we now have an easy formula to find the magnitude of the force applied:

$$F = \frac{K|q_1||q_2|}{r^2} \tag{8.1}$$

The force applied is the same on both objects regardless of the difference in charge between them. In the above equation, K is Coulomb's constant = 8.99×10^9 N m^2 C^{-2} and r stands for the distance between two objects. The SI unit of charge is the *coulomb* (C). The number of coulombs in one fundamental unit of charge equates to 1.6×10^{-19} C.

8.2.4 Electric Field

An electric field is a point in space where a charged particle will feel an electric force. All charged particles create an electric field and this field will exert a force on other charged particles to interact with them. A positively charged particle has the electric field vectors pointing out from it in every direction. A negatively charged particle has electric field vectors pointing inward from all directions. If the positive and negative charges are placed nearby, the field pattern will have field lines going from positive to negative charge, as shown in Figure 8.1a.

8.2.5 Charge Distribution

In a charged wire or plate, there are many electrons each inducing an electric field, but we cannot possibly tell them apart because they are densely packed. This is why we assume in cases of shapes like plates, the charge is evenly spread out, meaning it is uniformly charged. Because it is not possible to track the electric field from every single electron, we have a general model to represent the electric field in its entirety. Any charged flat surface is referred to as an *electrode*, which is modeled as an infinite plane of charge. While placing two oppositely charged electrodes face to face, you are making a *capacitor* and thus creating a uniform electric field between them. It is a uniform electric field because the electric field strength is the same at every point within the field, as opposed to electric fields created from point charges that decrease in strength as the distance from the point increases. The field vectors point from the plate with the higher charge to the plate with the lower charge. They do not necessarily have to be positive and negative plates. Notice that the term *capacitor* has been mentioned in this section. We will discuss more about this electronic component later in the chapter, but for now hold onto the thought.

8.2.6 Charged Particles

As stated previously, a charged particle always feels a force in an electric field. Like charges feel a repulsive force between them and opposite charges feel an attractive force between them. It is the same in a capacitor. If a positively charged particle is placed in between the plates, it will accelerate toward the negative plate. If a negatively charged particle is placed in between, it will accelerate toward the positive plate. However, the plates do not necessarily have to be of opposite charges. And because the plates are assumed to have a uniform electric field, there is constant force acting on a charged particle placed between them. Because of this force exerted on the particles, the plates will experience acceleration that can be determined with the following equation:

$$a = \frac{qE}{m} \qquad (8.2)$$

where:
  E is the electric field strength between metal plates
  m is the mass of an electron = 9.11×10^{-31} kg

8.2.7 Potential Difference

Alongside the electric field, electric potential is also a property of charges. Making a charge separation can create a potential difference, as in the case of a capacitor. There is a uniform electric field within the plates, but there are also equipotential lines that are

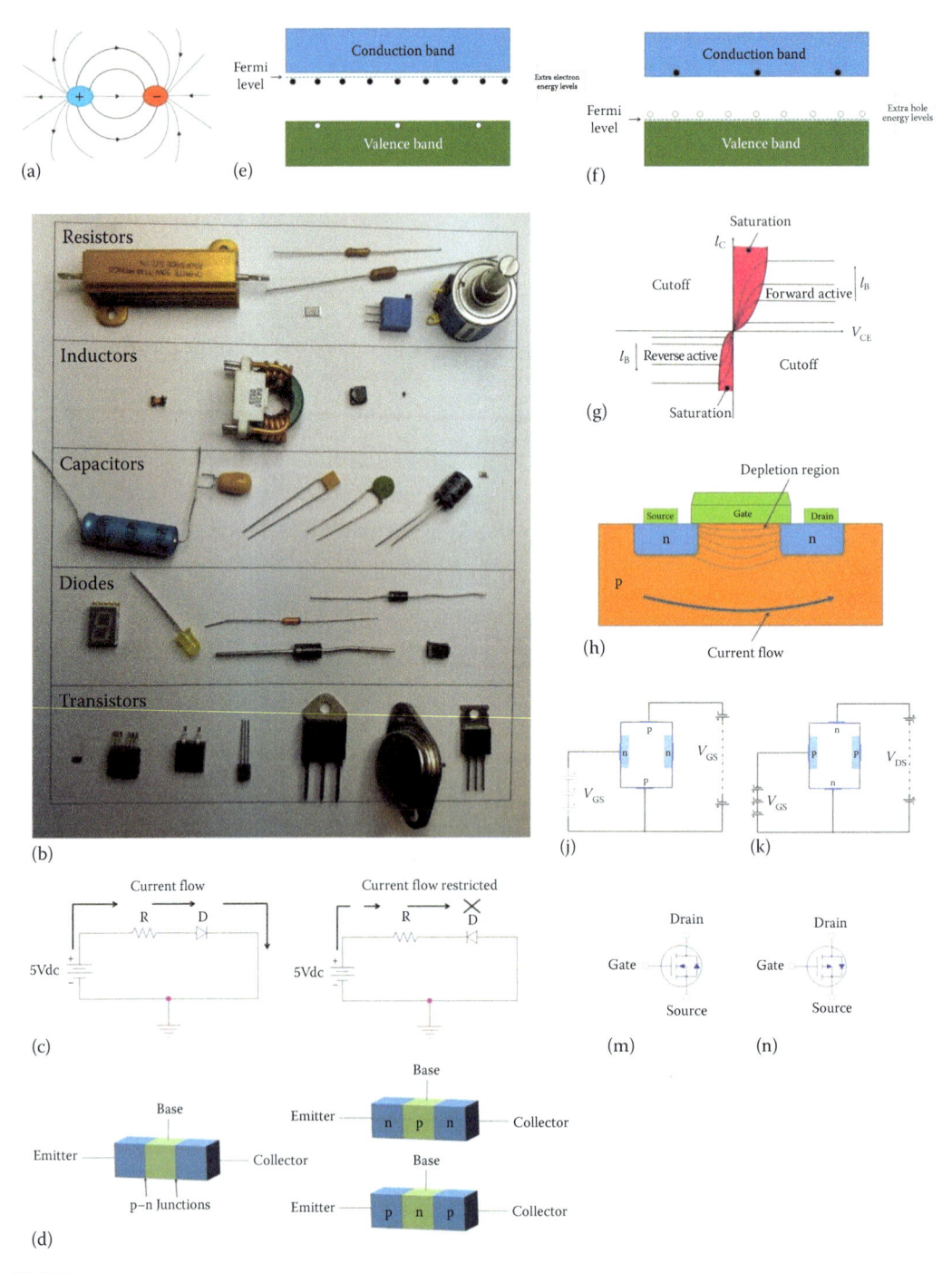

FIGURE 8.1
(a) Electric field lines. (b) Different types of resistors, capacitors, inductors, diodes, and transistors used in circuits. (c) Diode allowing current to flow in one direction. (d) n–p–n and p–n–p transistors. (e) An n-type semiconductor. (f) A p-type semiconductor. (g) The regions of operation for a diode. (h) A field effect transistor. (j) A p-channel JFET. (k) An n-channel JFET. (m) Circuit symbols for an n-channel FET. (n) Circuit symbol for a p-channel FET.

Electric Circuits and Components

perpendicular to the electric field lines. Electric field vectors go from the plate with higher electric potential to the plate with lower electric potential. So, in the case of a capacitor, the plate with a positive charge has more electric potential. This means that negative charges travel to the plate with higher electric potential within the electric field. This flow of negative charge, or electrons, is termed as electric *current*.

8.2.8 Current

Current is defined as the flow of charge. But a charge needs a conductor to flow through. This is evident from the fact that when you leave a charged-up capacitor idle, the capacitor will remain charged for some amount of time. This is because when the electrodes are left alone, air is present between them, and air being a good insulator keeps the electrodes in their charged state. However, if the metal plates were to touch directly or be connected with the help of a metal wire, both the plates would then become neutral, or in other words, get discharged. This is possible because metals are good conductors of electricity. The structure of metals, with a bunch of positive ions in a sea of electrons, allows for charges to move freely. The electrons, which are charge carriers, can transport the charges from one end to the other with minimal resistance. In other materials that conduct electricity, ions are the charge carriers.

If a wire connects the two electrodes of a capacitor, the charges flow from one end to the other with the help of charge carriers. But as mentioned before, because there is a potential difference between the electrodes, an electric field is created within the wire. The electrons flow from the zone with low electric potential (negative plate) to the zone with a higher electric potential (positive plate), thus creating a continuous flow of charge within the wire. This flow of charge is referred to as a *current*. However, traditionally current is the flow of charges from a positive to a negative terminal, as charge carriers were initially thought to be positive. Because the situation was found to be otherwise, we say that electrons travel from the negative to the positive terminal, and the current flow is in the reverse direction (positive to negative terminal). Another way of saying this is that electrons flow from lower to higher potential in a given circuit.

8.3 Circuit Components

In this section, we will introduce the basic electrical components that make up a circuit. These elements are present in all analog/digital applications we see around us. The majority of these components can be purchased online through distributors such as Digikey and Mouser. Passive components—resistors, capacitors, diodes, and some transistors—are inexpensive, readily available, and perform simple tasks, as we will discuss later in this chapter. Custom-designed components, commonly known as integrated circuits (ICs), perform application specific tasks that may require some level of software integration and are not always available off-the-shelf. Such types of devices are microprocessors in your computers, FPGAs, programmable memory, and so on.

8.3.1 Resistors

A *resistor* is a passive component that is used to restrict the flow of current in a closed circuit. Sometimes referred to as a load, multiple resistors are often used in combination to lower the voltage level of a given circuit. It is usually a two-terminal component. A resistor with three or more terminals is known as the *potentiometer* and provides variable resistance. Resistors are one of the two most fundamental components in a circuit, the other being capacitors (see the next section). You will hardly find a circuit that does not include a resistor. Resistors, measured in the unit of ohms (Ω), are available in a variety of resistance values to meet design requirements. Figure 8.1b shows some commonly used resistors.

8.3.2 Capacitors

A *capacitor* is a two-terminal device that stores electric charge temporarily. It is made of two metal plates (electrical conductors) separated by an insulator (dielectric). It comes in different flavors—film, ceramic, tantalum, and aluminum—each having unique properties that must be taken into consideration when choosing a specific type. Next to resistors, it is the most widely used component in a circuit. Capacitors come in a variety of ranges and ratings. Capacitance is usually measured in microfarads (μF). Figure 8.1b shows some different sizes of capacitors.

8.3.3 Diodes

A *diode* is a passive element used in electric circuits to make current flow in a given direction. Because of their nature, diodes are often used in circuits with an alternating current source to block the direction of current. Figure 8.2b shows a diode symbol. In a DC circuit, where the voltage source is a battery, the anode of the diode is connected to the higher potential side, and the cathode is connected to the lower potential side of the circuit, which allows current to flow from positive to the negative terminal of the battery. If the polarity of the diode is reversed (Figure 8.1c), current is blocked. Common types of diodes seen in circuits include LED, Laser, Schottky, Zener, and Tunnel. Advanced applications of diodes are discussed in more detail in the next section.

8.3.4 Transistors

A *transistor* is a semiconductor device with three terminals—base, collector, and emitter—that act as switches to control the flow of current across two of its terminals (collector and emitter) when the appropriate voltage is applied to the base terminal. Transistors are the key building blocks to electronic devices and can be found in numbers embedded in ICs. Their invention brought about a revolution to the electronics domain, and their inventors were awarded the 1956 Nobel Prize in Physics. Transistors come in two flavors—BJT and FET—as described below. Figure 8.1b shows the n–p–n type transistor commonly used in digital circuits.

8.3.5 Bipolar Junction Transistor

A *bipolar junction transistor* (BJT) is a semiconductor device that is formed with the combination of two diode junctions, commonly referred to as the *p–n junctions* (see Figure 8.1d). The combination of these junctions is used to classify two types of transistors: n–p–n and p–n–p. The n–p–n type transistor is formed with a thin layer of p-type semiconductor in between two n-type semiconductor layers (see Figure 8.1d). The p–n–p

Electric Circuits and Components

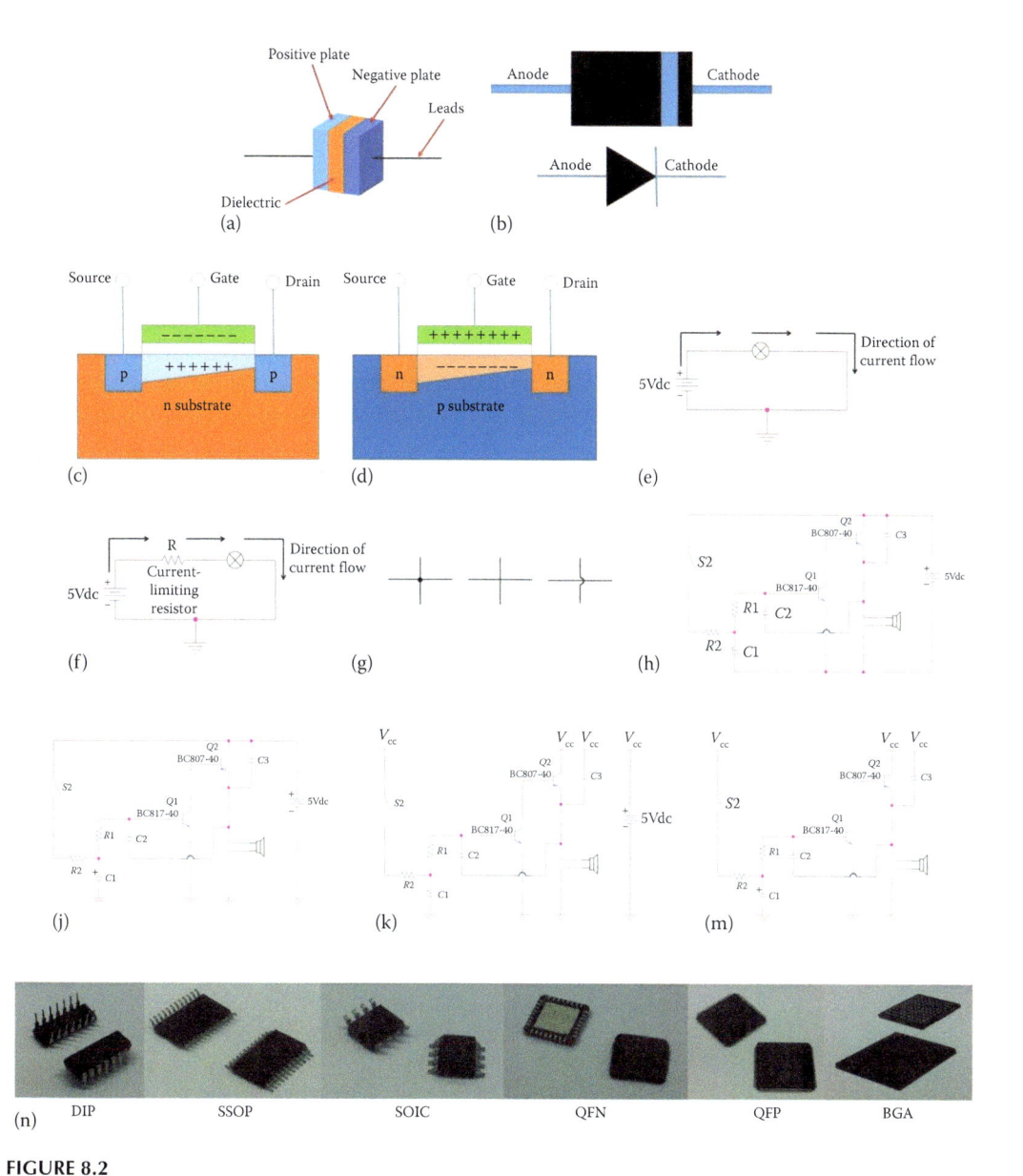

FIGURE 8.2
(a) Block diagram of a capacitor. (b) Circuit symbol of a diode. (c) p-channel MOSFET. (d) n-channel MOSFET. (e) Current traveling through a circuit. (f) Current-limiting resistor in a circuit. (g) Circuit connection symbols. (h) Example of a circuit design. (j) Common ground connections. (k) Common power supply connections. (m) Simplified circuit. (n) Integrated circuits.

type of transistor is formed with a thin layer of n-type semiconductor surrounded by the p-type semiconductor (see Figure 8.1d). All BJTs have three terminals—base, collector, and emitter. The two p–n junctions formed become the base–collector junction and the base–emitter junction, and their function is amplification of an electric current. This is because a small amount of base current can control the currents in the collector and emitter regions.

8.3.6 n-Type Semiconductor

This type of semiconductor has electron concentration in the conduction band that exceeds hole concentration in the valence band, as shown in Figure 8.1e. As a result, electrons become the majority charge carriers and holes, the minority charge carriers. In order to induce n-type behavior, donor impurities such as phosphorous or arsenic are added to the crystal structure of silicon.

8.3.7 p-Type Semiconductor

This type of semiconductor is just the opposite of n-type semiconductors. Here, the holes are the major charge carriers, and electrons are the minority charge carriers. The hole density in the valence bond leads to higher electron density in the conduction band (see Figure 8.1f). The addition of acceptor impurities to the silicon crystal structure induces p-type semiconductors. A common p-type dopant for silicon is boron.

8.3.8 BJT Operating Regions

There are essentially five distinct regions of operation for a transistor: forward active, reverse active, saturation, cutoff, and avalanche (breakdown). Figure 8.1g shows all these regions for a BJT. A more detailed description of each of these regions is provided later in the chapter.

8.3.9 Field Effect Transistors

A field effect transistor (FET) is a type of transistor that is used in applications for amplifying or switching analog or digital signals. In an FET, current flows along the semiconductor path known as the *channel*. Depending on the current carriers, either holes or electrons, the FET can be classified as *p-channel* (holes) or *n-channel* (electrons). There are electrodes at either end of the channel known as *source* and *drain* (see Figure 8.1h). Applying a voltage at the control electrode, called the *gate*, controls the current between the source and the drain. A small variation in gate voltage (V_{gs}), which signifies a FET's capability to amplify signals, can lead to large source-and-drain current (I_{ds}) variation. An FET has four terminals: *gate*, *source*, *drain*, and *body*. In most configurations, the body is connected to the source inside the package. FETs are divided into two major classifications—junction FET (JFET) and metal oxide semiconductor FET (MOSFET).

8.3.10 Junction FET

This type of device is the simplest form of FET that is often used as voltage-controlled resistance and electronically controlled switch. A p-type JFET consists of a semiconductor channel that contains a large portion of positive charge carriers (holes), whereas an n-type JFET has a semiconductor channel containing a large proportion of negative charge carriers (electrons). At either end of the JFET, there are two ohmic electric connections that form the drain and source. Electric charge flows between the source and drain through a high-resistivity semiconducting channel. If this channel is doped with donor impurities, primary charge carriers become electrons and the device is called the n-channel JFET, as shown in Figure 8.1k. Similarly, when the channel is doped with acceptor impurities, the flow of current through the channel becomes positive because of high density of holes, and the device is called the p-channel JFET, as shown in Figure 8.1j. The n-channel JFET is a more efficient conductor compared to a p-channel JFET because electrons have higher mobility through a conductor than holes do.

8.3.11 Metal Oxide Semiconductor FET

This type of transistor is a voltage-controlled FET that is used in circuits for switching or amplifying electronic signals. It differs from the JFET in that it has a metal oxide gate electrode that has a thin layer of insulating material, usually silicon dioxide, to electrically insulate it from the n-channel or p-channel semiconductor. The main advantage a MOSFET provides over other types of transistors is that it requires a small amount of current to switch ON while sourcing a large current downstream in the circuit. Most MOSFETs have a high gate voltage threshold in the range of 2.5–4 V. In a digital signal switching application (inverter circuit explained in Section 8.3) MOSFET is a benefit because the oxide layer between the gate electrode and the channel prevents current from flowing through the gate, thereby providing high input impedance and reducing power consumption. MOSFETs are usually three-terminal devices—gate, source, and drain—and come in two flavors: p-channel (see Figure 8.2c) and n-channel (see Figure 8.2d). Some high-power MOSFETs will have additional terminals, attached to source or drain, for dissipating heat from the package during normal operation. Thermal dissipation is key for proper operation of a MOSFET, especially in fast switching circuits, to prevent damage to the device. The MOSFET circuit symbol for an n-FET is shown in Figure 8.1m and for a p-FET in Figure 8.1n.

8.3.12 Integrated Circuits

An *IC* can be viewed as a component that contains multiple active and passive components, forming an electronic circuit within a package to perform more complicated and application specific tasks. These components are the building blocks of modern electronics, and can be found in many electrical applications such as television, mobile phones, computers, and automotive vehicles. Some examples of ICs include operational amplifiers (op-amps), logic gates, timers, sensors, power supplies, memory chips, and microcontrollers. They come in different package sizes and shapes—DIP (dual in-line package), SOIC (small outline IC), SSOP (shrink small outline package), QFP (quad flat package), QFN (quad flat no-leads), BGA (ball grid array), and so on, as shown in Figure 8.2n.

8.4 Circuit Symbols and Schematic Design

Imagine you are in a car driving from Detroit to Chicago without any maps, GPS, or any form of navigator. How would you know which highways to travel, and which exits to take? The chances are highly likely that you will get lost, or worst get stranded in the middle of nowhere, without the aid of a navigation system. Just like that, if you do not have a schematic diagram for the circuit you are designing, you will easily get lost trying to following the path in which the current will be flowing. *Schematics* provide you with a map of the current, from higher potential to lower potential, in a circuit. We all know that maps have symbols to designate specific landmarks such as bridges, railroads, tourist spots, and

TABLE 8.1

Circuit Symbols

Symbols	Description
—⋀⋀⋀—	Resistor
—⋀⋀— (with arrow)	Variable resistor
—⊢⊣—	Capacitor
—▷⊢—	Diode
—▶⊢— (with arrows)	Light emitting diode (LED)
—⌒⌒⌒—	Inductor
(symbol)	Transistor (n-p-n)
(symbol)	Transistor (p-n-p)
(symbol)	n-channel FET
(symbol)	p-channel FET
—/ —	Switch
(symbol)	Speaker
(symbol)	Transformer

water bodies. Similarly, circuit schematics have symbols for all active and passive components that help you identify the proper devices to be connected in an appropriate manner. Table 8.1 shows the commonly used circuit symbols. Note that there could be variants for circuit component symbols that are not shown here such as for Zener diodes or different types of switches, fuses, relay, and so on. Details about fuses and relays have not been discussed in this chapter.

8.4.1 Getting Started with Schematics

We all have some source of artificial lighting in our houses. The simplest circuit for that system is shown in Figure 8.2e with a two-terminal lamp connected to a DC source (5 V battery). One terminal of the lamp is connected to the positive terminal of the battery, while the other terminal is connected to the negative terminal. This circuit is commonly referred to as the ground reference and is always considered to be at zero volts. According to the laws of physics, current always flows from higher potential (positive battery terminal) to lower potential (negative terminal or ground reference). This path allows current through the lamp when the switch is turned ON. Even though the circuit is functionally appropriate, the lamp will experience a big inrush of current right at the moment the circuit is completed, which could potentially damage the lamp. We do not want that to happen, especially if the lamp is an expensive one. Therefore, one easy way to fix this situation is to resist the current using a resistor (as shown in Figure 8.2f). Now you have a more robust design that is safe to use. Note that if you change the value of the resistor, the brightness of the lamp changes. Do you know why this happens? Simply because lowering the resistor value allows more current through the lamp, making it brighter. The opposite happens when you lower the resistance in the circuit. *Always remember to connect all loose ends of a circuit or your design will not work properly. Current only flows through a closed-loop circuit. Refer to Figure 8.2g for circuit connection symbols.*

8.4.2 Reference Designators in a Circuit

Depending on the functionality of a system, the accompanying circuit maybe very complicated in design. There could be hundreds or thousands of components that you, as the designer, need to implement. The device symbols will give you a good idea of what type of component you are placing, but there is no way of identifying a particular component in a given portion of the circuit. In order to avoid confusion, we designate a symbol accompanied by a number to all the components. The convention for naming components in this manner is shown in Table 8.2. For example, if you have 10 resistors in your circuit, you can name them *R*1 through *R*10 in sequence, or pick 10 randomly selected numbers accompanying the letter *R* to designate that the components are resistors. The same rule applies to all other circuit components.

8.4.3 Circuit Design

When designing a circuit, you first need to create a table known as the *bill of materials* based on the components you need. For the circuit shown in Figure 8.2h, the bill of materials is given in Table 8.3. For now, it is not crucial to understand the functionality of the circuit. The goal is to give you an idea of what a circuit schematic looks like and how to read it.

As you can see, the terminals of the capacitors marked positive are aligned with the positive terminal of the battery that is the higher potential end of the circuit. The plus sign on the capacitor symbol indicates positive polarity of the terminal and should always be connected to the voltage source. The other terminal, commonly referred to as negative polarity, gets connected to the lower potential end of the circuit, usually the ground node. In the case of resistors, there is no need to identify polarity and hence can be connected

TABLE 8.2

Reference Letters

Letter	Components
R	Resistor
C	Conductor
D	Diode
L	Inductor
Q	BJT
M	FET
S	Switch
U	Integrated circuit
J	Connector
LED	Light emitting diode

TABLE 8.3

Bill of Materials

Component	Reference Designator	Value	Quantity
Resistor	R1, R2	42.2 K	2
	C1	100 µF	1
Capacitor	C2	0.022 µF	1
	C3	2.2 µF	1
Transistor	Q1	BC817-40	1
	Q2	BC807-40	1
Speaker	K1	—	1
Power source	BATT	9 V	1

as desired. For capacitors with small value, the polarity is not crucial. For a diode, not shown in this circuit, the connection must be made as described in Section 8.2.3 in order to have the proper current. For other components, such as transistors (BJT and FET) and ICs, connections need to be made per the design requirements, and the recommendations the manufacturer mentioned in the devices' datasheets.

If you are curious about how the circuit shown in Figure 8.2h works, here is an overview of the functional operation. There are terms here that you may not understand now, which is fine.

> The purpose of the circuit is to sound an alarm of variable frequency through the speaker when switch, S1, is closed. The transistors Q1 and Q2 form an oscillator whose frequency increases and decreases with the opening and closing of S1, respectively. The base of Q1 is biased from an RC circuit (formed with R1 and C1). When S2 is closed, C1 is charged through R2. As the voltage across C1 increases, the RC time constant decreases. This results in an increase in the frequency of the signal. When S1 is opened, C3 discharges through the speaker to the ground (negative terminal of the battery that has the lowest potential in the circuit) and the frequency of the tone decreases. The sound coming out of the speaker will almost be like a siren.

8.4.4 Simplified Ground Connections

The circuit in Figure 8.2h contains only a few components to perform the desired functionality. Imagine how the circuit will look for the motherboard of your computer. Or, for example, the circuits present inside your cell phones and tablets. How complicated do you think those would be? I am sure by now you have figured out that the electronic circuit in your cell phone, tablet, or computers must have numerous ground connections, and showing them with a single wire connected to the negative terminal of the power source (battery), as shown for the circuit in Figure 8.2h, would be impossible. Hence, the normal practice is to distribute the ground connection, using the symbols shown in Table 8.1, throughout the schematic using the ground symbol, as shown in Figure 8.2j. The ground connection is the common return path for current and is generally the lowest potential in the circuit.

8.4.5 Simplified Power Connections

In the schematic shown in Figure 8.2h, components C3, Q2, and S1 are all connected to the same voltage node—positive battery terminal. They can be replaced with a symbol, as shown in Figure 8.2k, denoting a common voltage node in order to simplify circuit representation. The circuit in Figure 8.2k is functionally the exact same as in Figure 8.2h, only

Electric Circuits and Components

represented in a simplified manner. In fact, due to such simplifications, you no longer need to show the battery connections with the rest of the circuit, because V_{cc} and ground symbols represent the positive and negative terminals of the battery, respectively, as shown in Figure 8.2m. You will, however, need to specify separately what the supply voltage (V_{cc}) is because you are no longer explicitly showing your source.

8.5 Understanding Circuits

In its simplest form, a circuit is a closed-loop path for current to flow. And from your high school physics class, you should remember that current is the flow of electric charge. To have a functional circuit, you need to have three basic elements present in your design: *power supply, load, and conductive wires*. The power supply acts as the voltage source providing enough potential difference across the entire circuit to generate current. This flow of charge needs to be consumed by a load to perform some sort of functionality. If there is no load present, there is no point of having a circuit because it is better to store all the charge in the power supply source. A load can be resistive, inductive, and/or capacitive. A circuit without any form of load is called a *short circuit*. Assuming you have a source and load, the last thing you need is to be able to connect them with wires that conduct electricity. The wires need to form a closed loop from the positive terminal of the source to the negative terminal, in addition to connecting all the intermediate electronic components. In Figure 8.2e, the battery is the power supply, the lamp is the load, and the connecting paths are conductive wires. Such a design is known as a *closed circuit*. If, for some reason, the current is interrupted between the power source terminals, you have designed an *open circuit*. In Figure 8.2f, where there is a resistive load along with the lamp, if the resistor goes bad, then the current to the lamp is interrupted and the lamp will never glow due to the open circuit phenomenon. In reality, a closed-loop circuit without the presence of any component will still offer minimal resistive load due to the small amount of resistance from the conductive wires. The wires will now have high power dissipation because all the charge stored in the power supply will flow through them, resulting in excessive heating, which can ultimately melt them, depending on the gage.

8.5.1 Series Circuits

Components in a circuit can be wired in two possible ways: *series* and/or *parallel* connections. In series connections, the components are wired in such a way that there is no discontinuity between the connecting terminals of the components and each component experiences the same amount of current. As shown in Figure 8.2f, the resistor and lamp are in series with each other because there is no discontinuity in connection between the components. Current travels through the resistor first, because it shares a terminal with the battery, and then through the lamp because of a solid linkage between the lamp and resistor. Let us add another lamp in series to modify this circuit and make it a little interesting (see Figure 8.3a). We now have three components all sharing power from the 5 V source. The direction of current remains the same, except that current is now flowing through the additional lamp as well.

How will the brightness of the lamp in Figure 8.2f be compared to the brightness of the lamps in Figure 8.3a? If you are thinking that the brightness of the lamps in Figure 8.3a will be

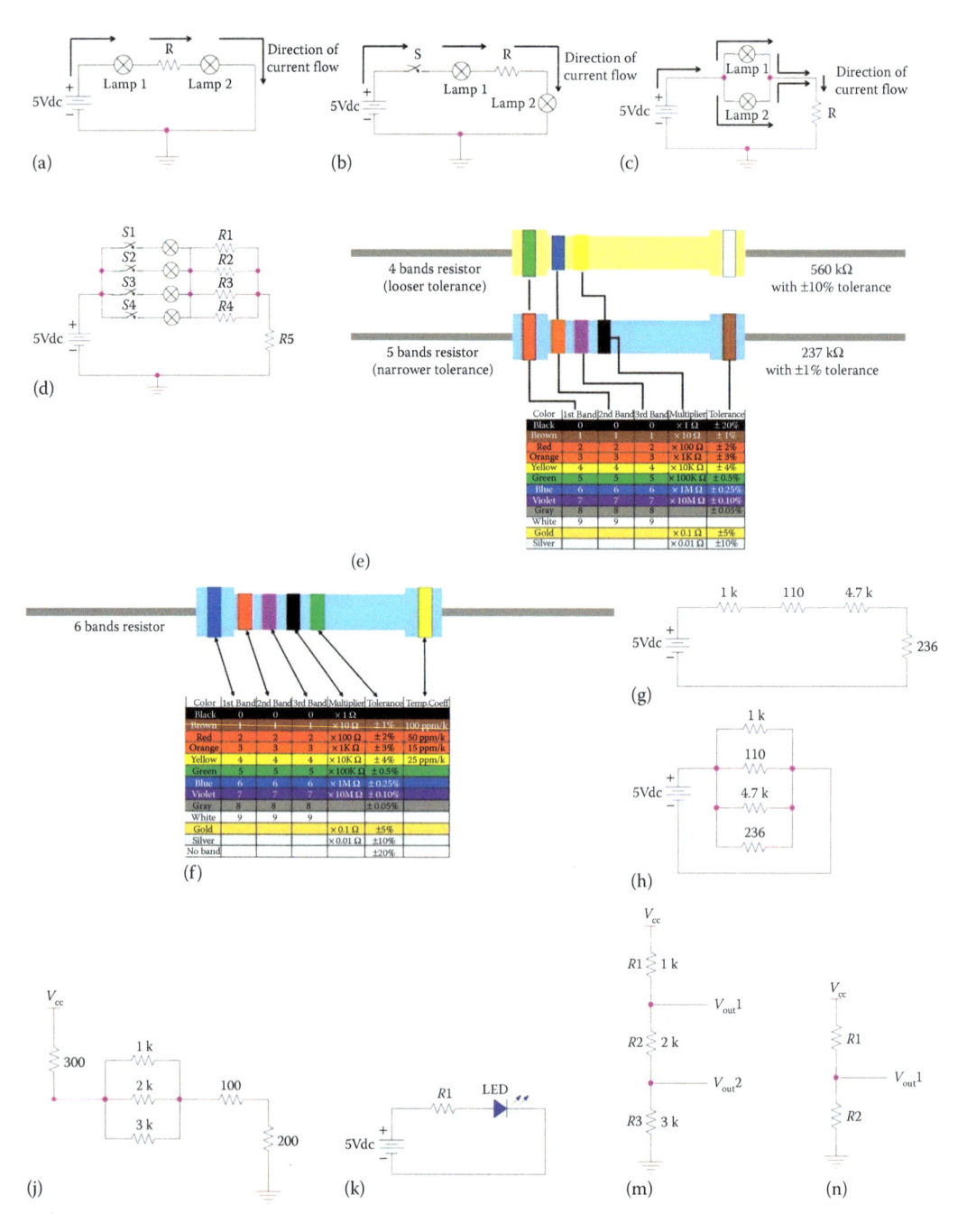

FIGURE 8.3
(a) Circuit with two lamps and a resistor. (b) Two lamps in series with a current-limiting resistor and a switch. (c) Two lamps in parallel. (d) Simplified household lighting system. (e) Four- and five-band resistor color codes. (f) Six-band resistor color codes. (g) Resistors in series. (h) Resistors in parallel. (j) Series and parallel combination for resistors. (k) Resistors used as a current-limiting device. (m) Three-resistor network for a voltage divider circuit. (n) Voltage divider circuit with two resistors.

lower, then you are on the right track. Because the power source remains the same, the same amount of charge will now be shared by the two lamps in Figure 8.3a, which is going to cut down the brightness. Another way of looking at this is in terms of resistive load in the circuit. The addition of a lamp adds more resistance to the circuit that makes it harder for the power source to push current through the lamps. As a result, total current in the circuit decreases due to *Ohm's law*, which will be discussed in more detail later in this chapter. One important thing to note here is that *the current through components connected in series remains the same*. Even though the total current was reduced in the circuit shown in Figure 8.3a, the same current will be flowing through the lamps and the resistor. One drawback to such a configuration is that if either lamp gets damaged, it opens the circuit, taking out the other lamp as well. Similarly, if the resistor fails and becomes an open circuit, both the lamps will be shut down because the current in a closed loop will be interrupted. This phenomenon is better observed if we add a switch in series with the rest of the components, as shown in Figure 8.3b. When you close the switch, both lamps glow and when you open the switch, both lamps go out.

8.5.2 Parallel Circuits

Components in parallel have the terminals directly connected to the main node in a circuit. This form of configuration allows multiple pathways for charge to flow. In Figure 8.3c current will split up at node A to take the route through each lamp, and reconcile at node B. The proportion of this split, at node A, will be dependent on the resistive load each lamp asserts. If both the lamps are identical, current will be divided equally among the lamps. If one lamp has a bigger load than the other, more current will flow through the smaller lamp. More current always travels through the path of least resistance. However, if either lamp gets damaged, current will flow through the other (working) lamp. This provides an advantage to components connected in parallel; if one device in the parallel network goes bad, the other devices do not have to suffer the consequences as they will keep functioning properly. This is how all household lights are connected to the main power supply unit. Each lighting device has a switch in series for you to be able to turn it ON or OFF, as shown in Figure 8.3d. The resistors, connected in series with each lamp, are provided for current-limiting protection for the circuit and $R5$ denotes any external load. Looking at one leg of the parallel circuit combination, we find that opening or closing $S1$ will only impact lamp 1 and have no impact on the other lamps. This is analogous to a household setting where usually one switch controls one lamp. You can form series and parallel combinations with switches to obtain different circuit operations. However, one thing to keep in mind is that adding components in parallel puts pressure on the power supply to source adequate amount of current. As a result, it is highly advisable to know the limitation of your circuit's power supply when adding resistive loads in parallel.

8.6 Resistors

As discussed previously a *resistor* is a passive electronic component that resists the flow of current in a circuit. Resistors come in different values and sizes for a wide range of applications. They can either be fixed or variable in value. There are four major types of fixed resistors—*thick film, thin film, wire wound, and carbon composite*. For the variable type you can get—*rheostat, trimmer, and potentiometer*.

Combining the right proportion of conductive and insulating materials helps manufacture a resistor of a specific value. Copper and aluminum are known to be good conductors of electricity, which is why most cables you buy from the store will be made of copper. Plastic and glass are examples of good insulators. Resistors, however, are generally made with a mix of carbon and ceramic. A resistor with higher carbon content is more conductive thereby reducing resistance value. A resistor with a higher mix of ceramic yields higher resistance. For the remainder of this section, we will focus on fixed resistors, which are widely used in circuits. Fixed resistors are two-terminal devices that can either be *surface mounted* or *through-hole* components, as shown in Figure 8.1b. Every object has some resistance even if it is a good conductor. The color code on the body of a through-hole thin-film and carbon-composite resistor, or a digital multimeter, helps measure resistance. Have you ever measured the resistance of your body? You can easily do this with the help of a meter by holding its probes in each hand. Within seconds the meter will read out your body resistance.

Resistance is measured in ohms. 1 Ω can be translated as the amount of resistance needed to make 1 A of current in a circuit that has a potential difference of 1 V. From a circuit perspective, if you connect a resistive load of 1 Ω across a 1 V battery, the current through the circuit will equate to 1 A. This is derived from Ohm's law, which is explained in the next section.

8.6.1 Ohm's Law

Ohm's law states that the current through a conductor is directly proportional to the potential difference across the conductor, assuming constant resistance. Mathematical relationship is given as follows:

$$V = IR, \quad \text{or} \quad I = \frac{V}{R}, \quad \text{or} \quad R = \frac{V}{I} \tag{8.3}$$

where:
V is the potential difference (voltage) across the conductor
I is the current through the conductor
R is the resistance value of the conductor

Given below are some examples on how to determine an unknown parameter, using Ohm's law, if the other two terms are known.

Example 8.1

You are given a circuit that has a 9 V battery connected in series with a 100 Ω resistor. Determine the current through the circuit.

Solution:
Current = I = V/R = (9 V)/(100 Ω) = 0.09 A or 90 mA

Example 8.2

A lamp with resistive load of 1 kΩ is connected to a power supply that is sourcing 5 mA of current. What is the potential of this power supply?

Solution:
Potential = V = IR = (5 mA) × (1 kΩ) = (0.005 A) × (1000 Ω) = 5 V

Electric Circuits and Components 197

Example 8.3

If you replace the power supply, from the previous example, with a 12 V battery and change the lamp such that 10 mA current is now flowing, what is the resistive load of the lamp?

Solution:
$R = V/I = (12\text{ V})/(10\text{ mA}) = (12\text{ V})/(0.01\text{ A}) = 1200\ \Omega$ or $1.2\text{ k}\Omega$

It is crucial for you to remember to convert units of voltage, current, and resistance to volts, amperes, and ohms, respectively. Otherwise your computational values will be inaccurate. Most times large resistance values are expressed in kilo-ohms (kΩ) and small current values in milliamperes (mA). Using Ohm's law to quantify the meaning of 1 Ω in resistance, we get: $R = V/I = (1\text{ V})/(1\text{ A}) = 1\ \Omega$.

8.6.2 Resistor Color Codes

To determine the resistance value of through-hole resistors with color bands on its surface, Figure 8.3e can be used for four- or five-band resistors and Figure 8.3f for six-band resistors. The stripes in four- or five-band resistors indicate two important parameters about the resistor—*resistance* and *tolerance*. In a six-band resistor, the sixth band represents the *temperature coefficient*. For the purposes of this chapter, we will mainly focus on four- or five-band resistors. Because fixed resistors do not have polarity indicators on the device, it may get confusing to determine how to read the bands. You will often find a few stripes bundled together on one end, and one colored stripe on the other end of the resistor. Read the bundled side of the resistor first, starting from the leftmost band and moving right. This bundle band will give you the numeric value of the resistance. The band on the other end of the resistor will provide you with the tolerance percentage.

8.6.3 Reading Color Code of a Four-Band Resistor

As indicated in Figure 8.3e the first two bands represent the two leftmost digits of the numeric resistance value. The third band denotes the multiplying factor and the fourth band, which is at the other end of the resistor, will give you the tolerance of the part. For example, if you read the bands as yellow, green, red, and silver, you will read a resistance value of 4500 Ω or 4.5 kΩ with ±10% tolerance:

Yellow	Green	Red	Silver
4	5	×100 = 4500 Ω or 4.5 kΩ	±10%

Similarly, if you have a resistor with bands brown, black, orange, and gold, the corresponding resistance will be 10 kΩ with ±5% tolerance. The derivation is shown as follows:

Brown	Black	Orange	Gold
1	0	×1 k = 10 kΩ or 10,000 Ω	±5%

Let us look at some example problems with four-band resistors below.

Example 8.4

A resistor has color bands—orange, yellow, brown, and silver. What are the ratings?

Solution:
Orange ≡ 3, yellow ≡ 4, blue ≡ 10 Ω. Resistance = 34 × 10 Ω = 340 Ω, ±10% tolerance.

Example 8.5

You are given a 5.7 kΩ, 5% resistor. What would the color bands look like?

Solution:
5.7 kΩ ≡ 5700 Ω and 5 ≡ green, 7 ≡ violet, ×100 Ω ≡ red, ±5% ≡ gold.

8.6.4 Reading Color Code of a Five-Band Resistor

Similar to the previous section, five-band resistor color codes can be identified from Figure 8.3e. The difference here is that the fourth band is the multiplying factor, and the fifth band is the tolerance rating. Let us look at examples that will help you understand how to read the color codes from the surface of the component and identify the resistor parameters.

Example 8.6

A resistor has color bands—red, blue, green, brown, and brown. What are the ratings?

Solution:
Red ≡ 2, blue ≡ 6, green ≡ 5, brown ≡ 10 Ω, and brown ≡ ±1%
  Resistance = 265 × 10 Ω = 2650 Ω or 2.65 kΩ, ±1% tolerance.

Example 8.7

You have a resistor with 5% tolerance that you measure to be 470 kΩ with a multimeter. What color bands would you see on the surface of the component?

Solution:
470 kΩ = 470,000 Ω, 4 ≡ yellow, 7 ≡ violet, 0 ≡ black, × 1 kΩ ≡ orange, 5% ≡ gold.

8.6.5 Resistor Tolerance

Even though you can read the color codes to identify the specific value of a resistor, the actual resistance measured with a multimeter will be slightly different. This is because of variations induced during the manufacturing process, which is why the last band indicates the tolerance of that particular resistor. The process can be controlled to have tighter tolerances, such as 0.5% or 0.25%, but these components will be more expensive compared to a 5% or 10% resistor. You might be asking: Does this tolerance impact my circuit, and if so what is the magnitude of the impact? To answer the first part of this question: yes, there will be some impact to your circuit. The magnitude of the impact can be determined after doing some analysis of your design.

For example, you have a circuit with a 5 V power source that is connected in series with a 100 Ω, ±10% resistor and a lamp that can handle 100 mA of current. According to Ohm's law, current through the lamp will be (5 V/100 Ω) = 50 mA, which is well within

Electric Circuits and Components 199

its rated current. With the tolerance specified, the resistor can have values in the range of 90–110 Ω impacting the current in the circuit. Therefore, current will be in the range of (5 V/110 Ω) = 45.5 mA and (5 V/90 Ω) = 55.6 mA. These values are within the rated current of the lamp, and hence the circuit will function properly. However, if your lamp was rated to 55 mA of current, then having a 10% resistor will be cutting close to the limit. In this case, you should opt for a more expensive solution and choose a tighter tolerance resistor, 1% or 5%.

8.6.6 Power Dissipation Ratings

Most electronic devices mention *power rating* in their datasheets that must be taken into consideration when designing a circuit. You do not want to electrically overstress your part and cause a thermal event. The power rating of a resistor gives you an idea of how much power it can dissipate in the circuit before it overheats enough to cause permanent damage. It is calculated in units of watts (W). Power dissipation can be calculated using the following mathematical formula, where V is the potential difference across the resistor and I is the current through the resistor:

$$P = V \times I = I^2 \times R = \frac{V^2}{R} \qquad (8.4)$$

Most through-hole resistors found in electronic circuits are rated for ⅛ or ¼ W. Most surface mount resistors are rated for 0.1 or 0.125 W depending on the package size. Higher power rated resistors come in larger packages, for both through-hole and surface mount resistors, and are generally more expensive.

Example 8.8

You have designed a circuit that has a 10 V power source connected in series with a 1 kΩ, ¼ W resistor. Will this resistor survive the circuit conditions?

Solution:
Power = $V \times I = V^2/R$ = (10)²/(1000 Ω) = 0.1 W, which is within ¼ W, and therefore the resistor should work fine in this circuit.

8.6.7 Resistors in Series

Any electronic circuit you look at will have multiple resistors of different sizes and ratings. They are either connected with each other or with other components in a similar manner, in series or in parallel. Let us first look at how to work with resistors that are connected in series, as shown in Figure 8.3 g. Resistors in series can be arithmetically added to give you the equivalent resistance, and in this case, the equivalent resistance (R_{eq}) will be 1 kΩ + 110 Ω + 4.7 kΩ + 236 Ω = 6046 Ω or 6.046 kΩ. Therefore, in generic terms, we can express equivalent resistance through the following equation for resistors connected in series, where n is an integer:

$$R_{eq} = R_1 + R_2 + R_3 + \cdots + R_n \qquad (8.5)$$

8.6.8 Resistors in Parallel

For resistors connected in parallel, equivalent resistance can be computed via the following equation, where n is an integer:

$$\frac{1}{R_{eq}} = \frac{1}{R_1} + \frac{1}{R_2} + \frac{1}{R_3} + \cdots + \frac{1}{R_n}$$

$$R_{eq} = \left(\frac{1}{R_1} + \frac{1}{R_2} + \frac{1}{R_3} + \cdots + \frac{1}{R_n}\right)^{-1}$$

(8.6)

Figure 8.3h shows a network of resistors connected in parallel whose equivalent resistance will be

$$R_{eq} = \left(\frac{1}{1000} + \frac{1}{110} + \frac{1}{4700} + \frac{1}{236}\right)^{-1} = 68.77\,\Omega$$

One thing you should note here is that when resistors are connected in parallel, their equivalent resistance will always be less than the smallest resistor value in the network. As shown above, R_{eq} is 68.77 Ω, which is less than the smallest resistor, 236 Ω, in the network.

8.6.9 Series and Parallel Combination

In a large circuit, it is highly likely to find resistors connected in both series and parallel combination. When that happens, it is good practice to identify the resistors in parallel and calculate their equivalent resistance first, and then combine the remaining series resistors. To solve the circuit shown in Figure 8.3j, let us work with the parallel resistors first—1, 2, and 3 kΩ. Their equivalent resistance is

$$R_{eq} = \left(\frac{1}{1000} + \frac{1}{2000} + \frac{1}{3000}\right)^{-1} = 545.45\,\Omega$$

Note that the equivalent resistance, in the parallel resistor network, is less than the smallest resistor in the network (1 kΩ). The next step is to add the resistors in series connection, shown as follows:

$$R_{total} = R_1 + R_2 + R_3 + R_{eq} = 300\,\Omega + 100\,\Omega + 200\,\Omega + 545.45\,\Omega = 1.145\,k\Omega$$

Even though the 300 Ω resistor is to the left side of the parallel resistor network, it is still in series with all the other resistors, and must be taken into consideration when computing total circuit resistance. Let V_{cc} be 10 V supply, then the current through the circuit will be (10 V/1.145 kΩ) = 8.73 mA according to Ohm's law.

8.6.10 Current Limiting with Resistors

By nature, resistors resist the flow of current in a circuit, and therefore are often used to limit the flow of current from the power supply in order to prevent damage to other components. A simple demonstration is shown in Figure 8.3k where a resistor is used to limit the current through the LED. The value of the resistor is yet to be determined, but it will heavily depend on the current rating of the LED. Most commercial LED datasheets specify

Electric Circuits and Components

current ratings in the range of 30 mA. Just to be safe, we will design the circuit for 25 mA and prevent damage to the LED. From Ohm's law, we can calculate R_1 to be

$$R_1 = \frac{V}{I} = \frac{5\,\text{V}}{0.025\,\text{A}} = 200\,\Omega$$

Therefore, the resistor can be any value greater than or equal to 200 Ω. What do you think will happen to the LED if you connect it directly to the battery? As you may have guessed, the LED will glow brightly for a small duration and then blow out. This happens because the LED offers very little resistance in the circuit that allows for a large inrush of current through the LED, damaging it permanently. The use of resistors to limit current is always advised for current sensitive components.

8.6.11 Voltage Divider Circuit with Resistors

When resistors are connected in series, as shown in Figure 8.3m, the potential difference across each resistor will be different and dependent on the size of the resistors. If you pick a node in between two resistors in series, the voltage at that node will be a fraction of the total voltage the power source supplies. This phenomenon is often used in electronic circuits to divide voltage using resistor networks. You may want to divide voltage to bring it down for components that cannot handle a large, continuous voltage. The following formula gives the voltage present at the node in between two series resistors:

$$V_{out} = V_{in} \times \left(\frac{R_2}{R_1 + R_2} \right) \tag{8.7}$$

V_{in} will be V_{cc} in the case of the circuit shown in Figure 8.3m. If the resistor network has more than two resistors, as shown in Figure 8.3n, then you need to compute V_{out} for each node in between the two series resistors separately, shown as follows:

$$V_{out}^1 = V_{cc} \times \left(\frac{R_2 + R_3}{R_1 + R_2 + R_3} \right) = 5\,\text{V} \times \left(\frac{2\,\text{k}\Omega + 3\,\text{k}\Omega}{1\,\text{k}\Omega + 2\,\text{k}\Omega + 3\,\text{k}\Omega} \right) = 4.17\,\text{V}$$

$$V_{out}^2 = V_{cc} \times \left(\frac{R_3}{R_1 + R_2 + R_3} \right) = 5\,\text{V} \times \left(\frac{3\,\text{k}\Omega}{1\,\text{k}\Omega + 2\,\text{k}\Omega + 3\,\text{k}\Omega} \right) = 2.50\,\text{V}$$

One thing to note is if R_1 and R_2 are of the same value, then your voltage will be divided in half, as shown in the case of V_{out}^2. Voltage from a divider circuit is always measured with respect to ground or lowest potential node in the circuit.

8.7 Capacitors

When a voltage is applied using a battery across a conductive wire, electrons flow from the negative terminal to the positive terminal of the battery due to the phenomenon that *opposite charges attract*. This flow of electrons is termed as *current*. Note that, even though electrons flow from the negative to positive polarity of the battery, direction of current in the circuit is always shown from higher potential to lower potential. The current-carrying conductor creates an electric field that, when applied across a capacitor, reaches right across the insulating material of the capacitor and traps charge. As a quick refresher, a capacitor is an electronic component that is used to store charge when connected to a voltage source. It is made of two metal plates separated by a *dielectric* material, as shown in Figure 8.2a. Four different types of capacitors have been mentioned in Section 8.2, and they are categorized using the type of dielectric material used in manufacturing. The ability to store electric charge is known as *capacitance*.

Even though capacitors are generally identified as charge storing devices, they have other useful applications in a circuit depending on the type of capacitor chosen. They can be used in circuits, in conjunction with resistors, and for providing time intervals; an example of such a circuit is shown in Section 8.11 with a 555 timer. They are used for filtering unwanted signal frequencies and suppressing high-frequency voltage transients—explained in more details in Section 8.7.7 for designing a low-pass filter. You can also design other kinds of filtering circuits by combining certain capacitance with appropriate components that is outside the scope of this chapter. If you are interested in learning more about filter design, I would recommend you to read about digital signal processing. In addition, capacitors are used to block unwanted DC voltage in AC systems. Analysis of how a capacitor does this is shown in Section 8.7.6 with some level of calculus.

Revisiting some familiar concepts from physics class, capacitance between two parallel plates, separated by a distance x, of area A can be calculated using the following equation:

$$C = \frac{\varepsilon_0 A}{x} \tag{8.8}$$

where ε_0 is the permittivity of free space = 8.854×10^{-12} F/m. In Figure 8.1b the biggest size capacitor is 360 µF. Suppose you connect this capacitor to a power supply of 12 V. Positive charge will accumulate in one plate and negative charge in the other. The charge stored in the capacitor will be directly proportional to the potential difference applied across the terminals of the capacitor, shown as follows:

$$Q = CV \tag{8.9}$$

Energy stored in the capacitor, beginning from being empty to having charges moving from one plate to the other until one plate has +ve charges and the other has –ve charges, is given by

$$W = \frac{1}{2}CV^2 \tag{8.10}$$

Remember to keep the polarities of the power source and capacitor in proper orientation—connect the positive terminal of the capacitor to the positive terminal of your power source, and the same for the negative terminals. Another thing to note while using capacitors is that any time there is a potential difference between the two conducting metal plates, there will be stray capacitance that will cause unwanted coupling between circuits. As a result, extreme care must be taken while laying out components on a printed circuit board.

8.7.1 Common Types of Capacitors

One of the most common types of capacitors used in electronic circuits is the *ceramic* capacitor that has dielectric material composed of either ceramic or porcelain soldered onto the conductive plates. Ceramic capacitors are constructed using two or more alternating layers of ceramic and metal layer that form the two electrodes. They come in fixed value of capacitance ranging from 1 pF to a few microfarads. Due to the small capacitance values, they are able to function satisfactorily at high frequencies, and hence are often used in circuits for electromagnetic-interference and radio-frequency-interference suppression. They are not polarized and so can be connected in any orientation. The multilayer ceramic capacitors, commonly known as MLCC, are most widely used in all electronic circuits and in various low-voltage applications.

The second most commonly used capacitor type is the *electrolytic* capacitor that consists of three subtypes: *tantalum*, *aluminum*, and *niobium*. The dielectric is made using tantalum, aluminum, or niobium plates to sandwich a semiliquid borax electrolyte paste. Among the three subtypes, the ones most commonly used are tantalum and aluminum. Tantalum capacitors generally reach up to 1000 µF, whereas aluminum capacitors can go as high as a few hundred millifarads. Tantalum capacitors, unlike aluminum capacitors, have more stable characteristics due to lower losses, and are often used in circuits with high-precision requirements. On the other hand, aluminum capacitors are typically used in power supplies and motor drive circuits for filtering, coupling, and bypassing signal noise.

Film capacitors are manufactured using thin insulating plastic film, for the dielectric, wrapped around in film-like sheets of metal foil that form the electrodes. Along with electrolytic capacitors, film capacitors are often used in AC and DC electronic circuits. They are not polarized and can be connected in a circuit in any orientation. They range from the thousand picofarads to hundred millifarads in capacitance.

8.7.2 Capacitors in Series

Capacitors in series add up with the following equation for equivalent capacitance, where n is an integer:

$$\frac{1}{C_{eq}} = \frac{1}{C_1} + \frac{1}{C_2} + \frac{1}{C_3} + \cdots + \frac{1}{C_N}$$

$$C_{eq} = \left(\frac{1}{C_1} + \frac{1}{C_2} + \frac{1}{C_3} + \cdots + \frac{1}{C_N} \right)^{-1}$$

(8.11)

Notice that the above expression for C_{eq} is similar to the equivalent resistance equation for resistors in a parallel connection. Refer to Figure 8.4a for the circuit configuration. Let us put some numbers to this circuit to make sense of the above expression. Assume $C_1 = 2.2$ µF, $C_2 = 4.7$ µF, and $C_3 = 10$ µF. The equivalent capacitance of the circuit can be calculated as follows:

$$C_{eq} = \left(\frac{1}{2.2\,\mu F} + \frac{1}{4.7\,\mu F} + \frac{1}{10\,\mu F} \right)^{-1} = 1.303\,\mu F$$

Remember to convert your capacitance into farads before doing any calculation. This is something that is very easy to miss because capacitance is generally denoted in microfarads.

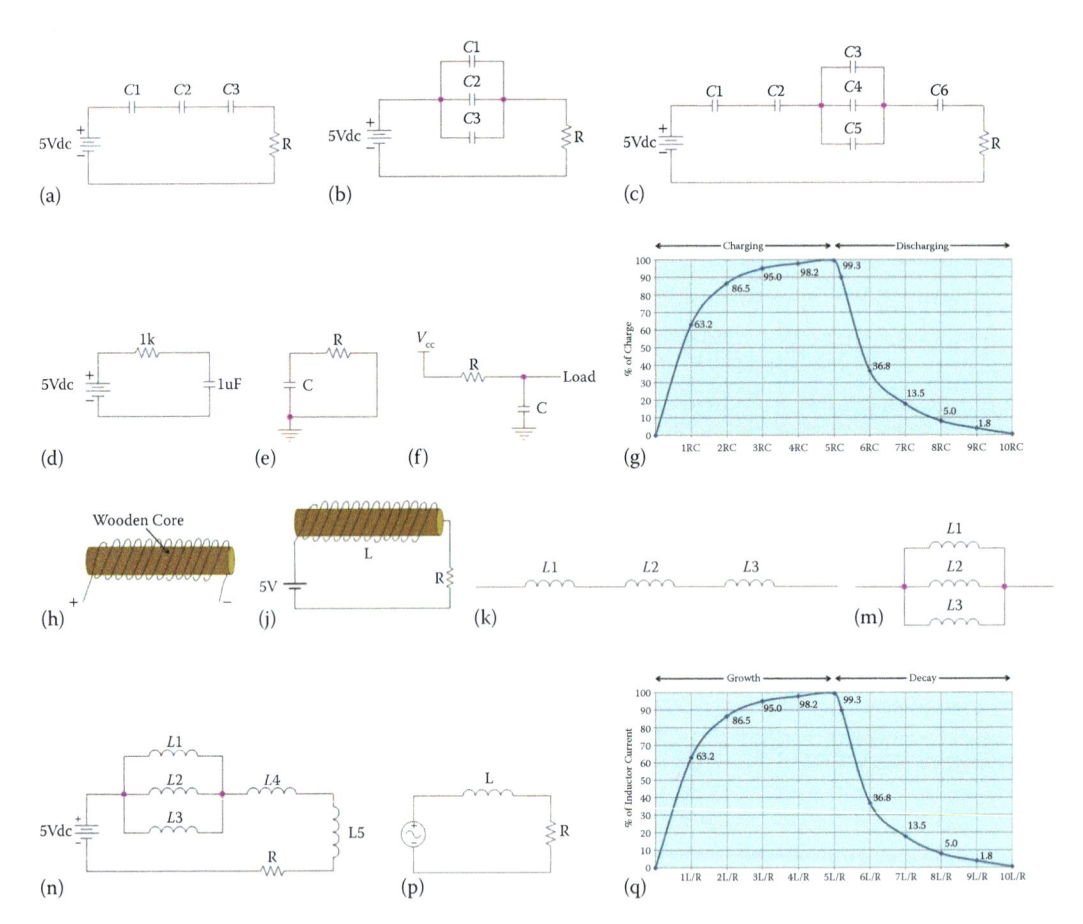

FIGURE 8.4
(a) Capacitors in series connection. (b) Capacitors in parallel connection. (c) Series and parallel combination for capacitors. (d) RC circuit. (e) Capacitor discharge circuit. (f) First-order RC filter circuit. (g) Graph for capacitor charging and discharging. (h) Basic inductor configuration with wooden core. (j) Implementing an inductor in a circuit. (k) Inductors in series connection. (m) Inductors in parallel connection. (n) Series and parallel combination for inductors. (p) First-order RL filter circuit. (q) Energy stored in an inductor.

8.7.3 Capacitors in Parallel

Similar to resistors in a series connection, capacitors in parallel add up arithmetically. For the circuit shown in Figure 8.4b, the equivalent capacitance in the circuit will be

$$C_{eq} = C_1 + C_2 + C_3 + \cdots + C_n \tag{8.12}$$

where n is an integer. For the capacitance mentioned in the previous section, total circuit capacitance will become: $C_1 + C_2 + C_3 = 2.2\ \mu F + 4.7\ \mu F + 10\ \mu F = 16.9\ \mu F$

8.7.4 Series and Parallel Combination

When you have to calculate the total capacitance for a combination of series and parallel capacitors, it is best to start with the equivalent capacitance in series connection first, and then add the remaining capacitance in parallel. For the circuit shown in Figure 8.4c total capacitance can be calculated as follows:

$$C_{parallel} = C_3 + C_4 + C_5$$

$$C_{total} = \left(\frac{1}{C_1} + \frac{1}{C_2} + \frac{1}{C_{parallel}} + \frac{1}{C_6} \right)^{-1}$$

Example 8.9

Assume $C_1 = 1$ µF, $C_2 = 2.2$ µF, $C_3 = 0.22$ µF, $C_4 = 0.1$ µF, $C_5 = 0.47$ µF, $C_6 = 1.47$ µF. What will be the total capacitance for the circuit shown in Figure 8.4c?

Solution:

$$C_{total} = \left(\frac{1}{1\,\mu F} + \frac{1}{2.2\,\mu F} + \frac{1}{0.22\,\mu F + 0.1\,\mu F + 0.47\,\mu F} + \frac{1}{1.47\,\mu F} \right)^{-1} = 0.294\,\mu F$$

8.7.5 RC Time Constant

As mentioned before, a capacitor stores electric charge when connected to a power source. In order for a capacitor to do that, it needs to get charged through a load, usually a resistor. Computing the *RC time constant* helps determine the rate at which the capacitor will be charged. By definition, the time constant (τ) of a circuit is referred to as the time required to be charged a capacitor to 63.2% of the power source voltage. Computing the time constant is easy; you simply multiply the resistor and capacitor values. Knowing the charging time is crucial to properly design your circuit with the appropriate capacitance. For example, if you are charging a 1 µF capacitor, by a 5 V power supply, through a 1 kΩ resistor, as shown in Figure 8.4d, the time constant for the circuit will be =(1000 Ω) × (0.000001 F) = 1 ms. Always remember to convert capacitance to farads before proceeding with time constant calculations. After the first 1 ms interval, the capacitor will be charged 63.2% of the source voltage = 3.16 V. In the next 1 ms duration, the capacitor will be charged an additional 63.2% of the difference between starting charge of 3.16 V, and supply voltage of 5 V = 3.16 V + ((5 − 3.16) × 63.2%) = 4.32 V. This cycle repeats itself until the capacitor is fully charged to approximately the source voltage. Table 8.4 shows capacitor charge percentage after five time constants. If you want to speed up the charging process of a capacitor, you need to reduce the RC time constant with either a less resistance or capacitance, or reduce both the resistor and capacitor values.

TABLE 8.4

Charge Stored % in Capacitor

τ Charging	%	τ Discharging	%
1	63.2	1	36.8
2	86.5	2	13.5
3	95.0	3	5.0
4	98.2	4	1.8
5	99.3	5	0.7

Similarly, for discharging a capacitor, you need a load connected in series, as shown in Figure 8.4e. The RC time constant refers to the amount of time needed for the capacitor to discharge 36.8% of supply voltage. Table 8.4 shows the capacitor discharge percentages for five time constant intervals. Figure 8.4g shows the timing diagram for a capacitor charging and discharging, and the equations are given as follows:

For charging

$$v_c = V\,(1 - e^{-t/\tau_c}) \tag{8.13}$$

For discharging

$$v_c = V e^{-t/\tau_d} \tag{8.14}$$

where:
τ_c is the charging time constant
τ_d is the discharging time constant

In order to determine how much energy is present in the capacitor while charging or discharging, additional math as explained in the next section.

8.7.6 Energy Stored in a Capacitor

The charging and discharging nature of a capacitor does not allow the current in a DC circuit. This is because a capacitor allows current when its conducting metal plates are either charging or discharging. Current is stopped when the capacitor is fully charged. Another way of saying this is that current is allowed through a capacitor only when it is allowed to continually charge and discharge, which can happen when the source itself is constantly changing, like an AC source. When the source is constant, in the case of a battery, the capacitor gets fully charged and then has no way of discharging, thereby blocking the current. A mathematical way to prove this has been shown later in this section. It is intuitive that charging and discharging of a capacitor happen through a resistive load, and is time dependent. The following equation gives the current through a conductor over time:

$$i = \frac{d}{dt}(q) \tag{8.15}$$

Then current through a capacitor becomes

$$i = \frac{d}{dt}(Cv) = C\frac{d}{dt}(v)$$

$$dv = \frac{1}{C}i\,dt \tag{8.16}$$

where C is the capacitance. Integrating the above expression for a time interval of t generates

$$v(t) = \frac{1}{C}\int_{-\infty}^{t} i(\tau)d\tau \tag{8.17}$$

$$v(t) = \frac{1}{C}\int_{t_0}^{t} i(\tau)d\tau + v(t_0) \tag{8.18}$$

where $v(t_0)$ can be assumed to be the initial charge present inside the capacitor prior to charging up for the time interval t_0 to t. From Equation 8.2, power delivered to a capacitor can be expressed as

$$p(t) = v(t)i(t) = Cv(t)\left(\frac{d}{dt}v(t)\right) \tag{8.19}$$

Hence, the energy stored in a capacitor due to the current-carrying conductor connected to a power source inducing the electric field can be expressed as

$$w_c(t) = \int_{-\infty}^{t} Cv(m)\left(\frac{d}{dm}v(m)\right)dm = C\int_{-\infty}^{t} v(m)\frac{dv(m)}{dm}dm \tag{8.20}$$

$$w_c(t) = \frac{1}{2}C\ [v^2(m)]_{m=-\infty}^{m=t} = \frac{1}{2}Cv^2(t) \tag{8.21}$$

and, substituting time varying expression of Equation 8.7 into Equation 8.19, we get

$$w_c(t) = \frac{1}{2}Cv^2(t) = \frac{q^2(t)}{2C} \tag{8.22}$$

The energy equation derived above shows that the energy stored in a capacitor will always be positive. When connected to a power source, a capacitor will absorb power from the circuit when storing energy, and release this stored energy back into the circuit when delivering power to the circuit. Because the energy stored in a capacitor is time dependent, the power source must be alternating in order for the current to flow through it. DC power sources, such as a battery, do not alternate with time and hence a capacitor acts as a blocking device for current in a DC circuit.

8.7.7 Designing a Low-Pass RC Filter

The intent of this section is to give you a brief overview on how a capacitor, along with a resistor, can be used to design a simple, first-order, low-pass filter in electronic circuits. By no means will this section provide you with all the details needed to design filters. If you are interested in learning more about digital filters, you should read textbooks about digital signal processing where you will get hardware and software perspective of filter design.

In a closed circuit with a power source and load, noise is always present in signal paths. Noise generally corrupts the performance of your design, and it is in your best interest to suppress noise as much as possible. One way of achieving this is to design a low-pass filter, which allows signal frequency lower than a certain threshold, known as the *cutoff frequency* (f_c), to pass through the load and attenuates all unwanted high-frequency signals above the cutoff limit. The attenuation of higher frequency signals is dependent on the filter design, which is beyond the scope of this chapter. For the circuit shown in Figure 8.4f, the capacitor essentially blocks low-frequency signals, and allows them to pass through to the load. For the high-frequency signals, however, the capacitor acts as a short circuit and the load does not see the effect of these frequencies. The cutoff frequency is determined using the RC time constant:

$$f_c = \frac{1}{2\pi\tau} = \frac{1}{2\pi RC} \tag{8.23}$$

The RC filter can be understood in terms of charging and discharging time of the capacitor. At frequencies lower than the cutoff point, the capacitor has enough time to charge up to a significant amount of the source voltage. In contrast, at high frequencies, the capacitor does not get enough time to charge before the direction of current is reversed. As a result, the output of the capacitor oscillates up and down by a small amplitude, which can be viewed as blocking signals with frequencies higher than the cutoff point.

Example 8.10

What will be the cutoff frequency of an RC low-pass filter that is designed with a resistor of 920 Ω and capacitor of 0.22 µF?

Solution:

$$f_c = \frac{1}{2\pi RC} = \frac{1}{2\pi(920\,\Omega)(0.22\times 10^{-6}\,F)} = 786\,Hz$$

Example 8.11

You want to design a low-pass filter with a cutoff frequency of 5 kHz. What resistor and capacitor value would you choose?

Solution:
RC time constant will be $(1/2\pi f_c) = [1/(2\times\pi\times 5000)] = 31.83$ µs
Choosing a 1 kΩ resistor will make the capacitor choice for $(31.83\,\mu s / 1\,k\Omega) = 0.032$ µF

8.8 Inductors

An *inductor* is a circuit component that is made of a conducting wire wrapped around a core in the form of a coil. In fact the circuit symbol of an inductor, as shown in Table 8.1, looks like a coil. There are different types of inductors that can be categorized based on the material composition of their cores, as shown in Figure 8.1b. The core can be made of iron or ferrite, air, or any nonmagnetic material. Iron core inductors are generally used in power supplies. Ferrite core inductors are heavily used in applications that operate with high frequencies. Air or nonmagnetic core inductors are widely used in filtering circuits, in combination with resistors, or capacitors, or both, depending on the filter design. Unlike capacitors that block abrupt change in voltage, inductors prevent instantaneous change of current in a circuit.

Inductors rely on *self-inductance*, often referred to as *inductance*, to achieve its functionality. Inductance is generally measured in millihenries (mH). To understand inductance, we need to take a step back and refresh the notion of *electromagnetic induction*. When you move a magnet next to a conductive wire, you induce current through that wire. Current always flows from higher potential to lower potential, and therefore creates a potential difference across the wire, also known as voltage. As the magnet induces this current through

Electric Circuits and Components

the wire, the current will generate its own magnetic field. If you wrap this wire around a nonmagnetic core for example, such as a cylindrical piece of wood, as shown in Figure 8.4h, the magnetic field would be intensified because now you are exposing more length of the wire to your magnet. Therefore, factors that impact the strength of the current-carrying wire's magnetic field are the number of turns around the core, amount of current through the wire, and material composition of the core. Note that a change in the magnetic flux of the magnet keeps the current flowing through the coil.

If you move the magnet faster over the coil, more current is generated. If you lower the speed of magnet movement, current will decrease in the coil. When you stop moving the magnet, current is halted in the coil. Self-inductance uses a similar concept, but now there is no external magnet present to induce current. Instead, the terminals of an inductor are connected in a circuit, as shown in Figure 8.4j, which provides for the current through the inductor coil. The coil carrying current creates a magnetic field around itself. As long as the current in the circuit is changing, the magnetic flux the coil induces will constantly change thereby inducing a voltage. The amount of voltage induced is dependent on the rate of current change in the circuit, and its polarity relies on whether the current is increasing or decreasing. This induced voltage across the coil will create a current that flows in the coil either with or against the flow of the circuit current. The direction of induced current depends on whether the circuit current is increasing or decreasing. If the circuit current is increasing, the induced current in the inductor coil will flow in the opposite direction to create a canceling effect, thereby slowing down the rate of current increase. If the circuit current is decreasing, the induced inductor coil current will flow with the circuit current to counteract the circuit current drop. At some point, if the circuit current becomes steady, self-inductance of the inductor will be halted. It is using this principle that an inductor is able to negotiate a sudden impulse of current in a circuit.

8.8.1 Series and Parallel Connection

Similar to resistors, inductors in series add up arithmetically, as shown in Figure 8.4k, through the following equation:

$$L_{eq} = L_1 + L_2 + L_3 + \cdots + L_n \tag{8.24}$$

where n is an integer.

When inductors are connected in parallel, as shown in Figure 8.4m, the equivalent circuit inductance is calculated using the following equation, where n is an integer. This is very similar to how resistors in parallel combine together for total resistance:

$$L_{eq} = \left(\frac{1}{L_1} + \frac{1}{L_2} + \frac{1}{L_3} + \cdots + \frac{1}{L_n} \right)^{-1} \tag{8.25}$$

For inductors connected both in series and parallel combination, as shown in Figure 8.4n, it is good practice to calculate the parallel inductance first and then add the series inductors shown below. It is essential that you convert inductance to henries before proceeding with calculations:

$$L_{\text{parallel}} = \left(\frac{1}{L_1} + \frac{1}{L_2} + \frac{1}{L_3} \right)^{-1}$$

$$L_{\text{total}} = L_{\text{parallel}} + L_1 + L_2$$

Note that the above expressions are only valid if the inductors are shielded; otherwise stray magnetic fields from nearby high current components, including other inductors, will affect them.

8.8.2 Energy Stored in an Inductor

The potential difference across an inductor is directly proportional to the rate of change of current in the circuit, and is given by the following expression:

$$v(t) = L\frac{d}{dt}i(t) \tag{8.26}$$

Integrating both side of the above equation, we obtain the expression for current in an inductor:

$$i(t) = \frac{1}{L}\int_{-\infty}^{t}v(m)d(m) = i(t_0) + \frac{1}{L}\int_{t_0}^{t}v(m)d(m) \tag{8.27}$$

The energy stored in an inductor can be derived from the expression of power delivered to the inductor:

$$p(t) = v(t)i(t) = i(t)L\frac{di(t)}{dt} \tag{8.28}$$

Integrating both sides over time yields the energy stored in the inductor in joules, shown as follows:

$$w_L(t) = \int_{-\infty}^{t}\left[L\frac{di(m)}{dm}\right]i(m)d(m) = \frac{1}{2}Li^2(t) \tag{8.29}$$

For a DC circuit, where the voltage does not vary with time, the voltage across the inductor can be considered to be zero. In other words, you can assume an inductor to be a short circuit when your power source is fixed, as in the case of a battery. Then you can ignore it in your DC circuit analysis.

Example 8.12

Current through a 10 mH inductor is given as 4 sin(244t) A. Determine the voltage across the inductor.

Solution:

$$v(t) = L\frac{di(t)}{dt} = (0.01)\frac{d}{dt}[4\sin 244t] = (0.01)(4)(244)\cos 244t = 9.76\cos 244t \text{ V}$$

Example 8.13

From the previous example, determine the energy stored in the inductor.

Solution:

$$w_c(t) = \frac{1}{2}Li^2(t) = \frac{1}{2}(0.01)(4\sin 244t)^2 = 0.08\sin^2 244t \text{ J}$$

8.8.3 RL Time Constants

An inductor requires a varying current through it, with a resistive load connected in series, as shown in Figure 8.4p, in order to store energy. The rate at which this can be done is dependent on the *RL time constant*, which indicates the amount of time the inductor needs to conduct 63.2% of the current when a voltage is applied across its terminals. RL time constant is calculated by

$$T = \frac{L}{R} \tag{8.30}$$

For example, if you have a 10 mH inductor in series with 100 Ω resistor, the RL time constant can be calculated as (0.01 H/100 Ω) = 0.1 ms. Table 8.5 shows the percentage of current through an inductor when connected to a resistive load, after each RL time constant interval. If you look at Table 8.4, you will notice the similarity with capacitor charging time intervals. A graph of the energy stored in an inductor is shown in Figure 8.4q.

8.8.4 Designing an RL Filter

While designing a filter with an inductor and resistor, as shown in Figure 8.4p, we need to take into account the *reactance* of the inductor. The reactance is a measure of the degree to which an inductor resists the current through it. Inductive reactance is measured in ohms using the following equation:

$$X_L = 2\pi f L \tag{8.31}$$

It can be seen from the above expression that the reactance of an inductor is directly proportional to the frequency of the AC signal. Thus, at higher frequencies the inductor will be more reluctant to allow current in this circuit. Let us verify this theory using a 10 mH inductor in a circuit with AC signal frequency 60 Hz. Inductive reactance will be as follows:

$$X_L = 2\pi \times 60\,\text{Hz} \times 0.01\,\text{H} = 3.77\,\Omega$$

If we increase the frequency of the signal by a hundred times, inductive reactance becomes

$$X_L = 2\pi \times (60 \times 100\,\text{Hz}) \times 0.01\,\text{H} = 377\,\Omega$$

The reactance increases a hundred times as well, indicating that for high-frequency signals the inductor will resist current even more. As a result, when designing a filter with an inductor, you must pay attention to the frequency of your alternating power supply. Reactance of a circuit is directly proportional to the input signal frequency.

TABLE 8.5

Energy Stored % in Inductor

RL Time Constant	%
1	63.2
2	86.5
3	95.0
4	98.2
5	99.3

8.9 Diodes

A *diode* is a two-terminal semiconductor device that allows the flow of current in one direction, as shown in Figure 8.1c. To understand how this is achieved, we need to look at how a diode functions. Because it is a semiconductor, it is made using a silicon crystal doped with elements such as boron and phosphorous. Figure 8.5a shows the silicon

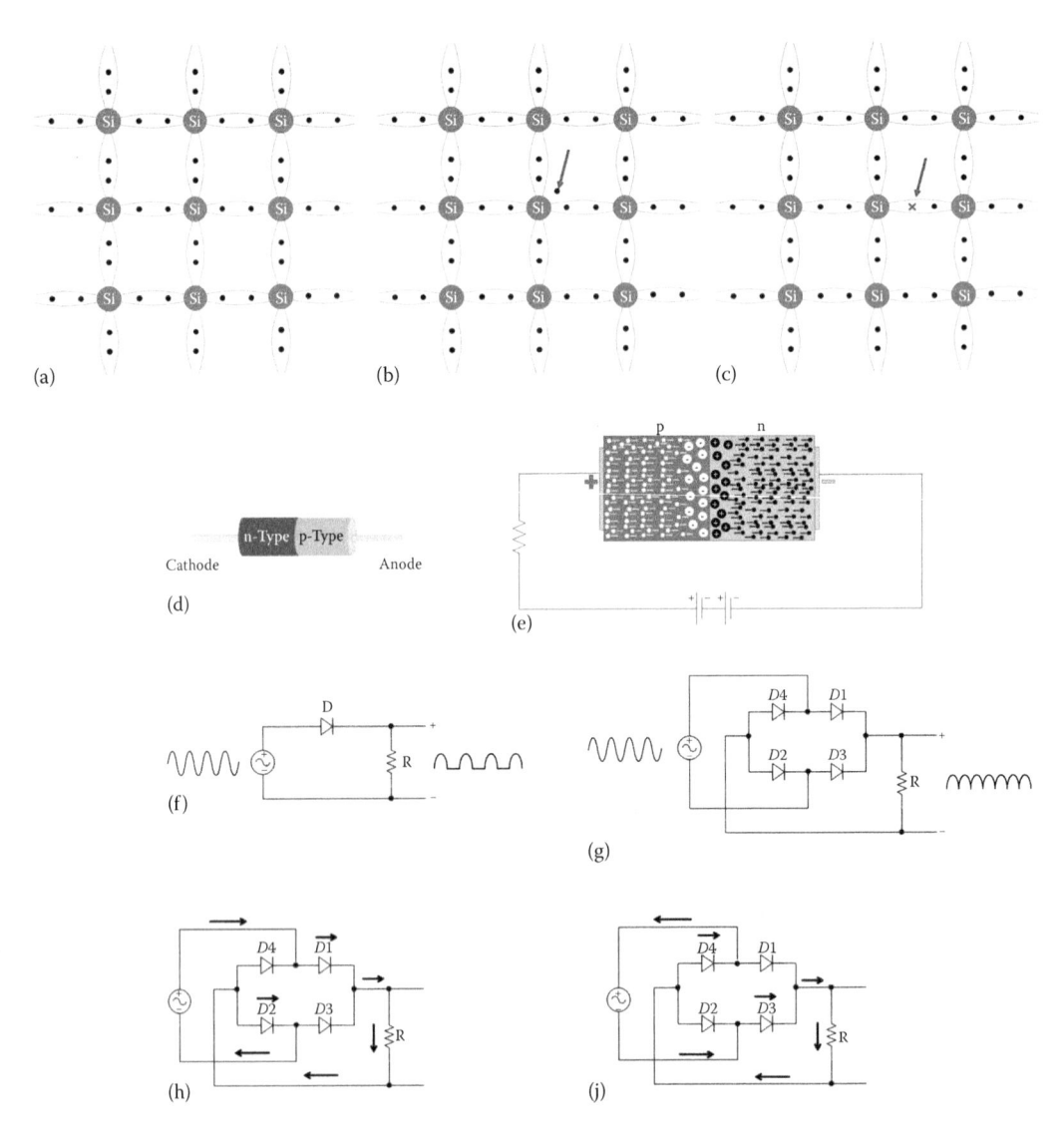

FIGURE 8.5
(a) Pure silicon crystal. (b) Silicon crystal doped with phosphorous having an extra electron. (c) Silicon crystal doped with boron having a deficit of electron. (d) Block diagram of a diode. (e) Forward biasing a diode. (f) Diode used for half-wave rectification in an AC circuit. (g) Diodes used for full-wave rectification in an AC circuit. (h) Direction of current during positive cycle of AC input for full wave rectification. (j) Direction of current during negative cycle of AC input for full-wave rectification.

crystal structure. Figure 8.5b and c shows how the crystal structure changes after the doping process. This would be a good stage to discuss in more detail the n-type and p-type semiconductors.

8.9.1 Doping of Silicon Crystal

From your chemistry classes you should remember that silicon has four electrons in its valence shell. These four electrons combine with four electrons of another silicon atom to form a solid Si–Si bond. The pattern is repeated and you get a silicon crystal (Figure 8.5a). A silicon crystal alone is strong and does not allow the flow of current. However, interesting things happen when you replace one silicon atom in the crystal with another element such as phosphorous. From the periodic table, you can find that phosphorous has five electrons in its valence shell. When it replaces a silicon atom in a silicon crystal, four of the five electrons in the phosphorous atoms' valence shell combine with four electrons of a nearby silicon atom forming an alternative crystal structure. The fifth electron does not have anything to bond with and hence remains free to move around the crystal structure, as shown in Figure 8.5b. When you replace additional silicon atoms with phosphorous atoms, more such free electrons emerge, which are available to flow freely. As you know, free-flowing electrons means current, and hence this modified Si–P crystal becomes an *n-type* semiconductor.

If you replace a silicon atom with boron, for instance, there will be one less electron per Si–B bond because boron has only three electrons in its valence shell, as shown in Figure 8.5c. This deficit of electron is called a *hole* and it behaves like a positive charge. Holes attract electrons because they are opposing charges, and an electron from a nearby Si–B bond moves over to fill this position of a hole. When this happens, the electron that just moved leaves behind a hole at its previous location, which gets filled up by another electron from a neighboring Si–B bond. This process continues with the addition of more boron atoms, and soon you have a number of electrons flowing through the crystal to fill up the hole spots, which is practically current. This may not be the most conventional way you would imagine current but it works. The modified Si–B crystal is called a *p-type* semiconductor.

8.9.2 Depletion Zone

We have established that n-type and p-type semiconductors can allow current. If we connect them together, we get a diode that allows current in one direction—from p-type to n-type. This is because when a voltage is applied across the diode terminals, right at the junction of p-type and n-type semiconductors, electrons move from the n-type side to the p-type side because they are attracted to the holes in the p-type side. This movement leaves behind a deficit of electrons in the n-type side, which themselves are essentially holes. It seems as if electrons and holes have swapped places at the junction. This electron and hole crossover region is known as the *depletion zone*.

The p-type side is known as the *anode*, and the n-type side is the *cathode*, as shown in Figure 8.5d. When a positive potential is applied at the anode, the depletion zone shrinks allowing more electrons to move from the n-type to the p-type side. This movement of electrons can be characterized as the flow of current from the anode to the cathode, and the phenomenon is known as *forward biasing*, as shown in Figure 8.5e. In contrast, if you apply positive potential at the cathode, the depletion zone will be expanded reducing the flow of electrons through the p–n junction. Reduction of electron flow can be viewed as blocking current.

8.9.3 Half-Wave Rectification

Now that we know how a diode is able to block current in one direction, we can find useful application for it in an AC circuit. When you plug an electrical device into a wall outlet, the device receives alternating current that switches at a frequency of 60 Hz (in Europe the frequency is 50 Hz). This means that current reverses direction every 1/60 s. The wave pattern looks similar to Figure 8.5f where voltage goes positive and negative. Not all electronic devices are protected for negative voltages. In fact, some devices may get permanently damaged due to negative voltage. In order to prevent this only positive voltage should be allowed to pass through such devices, and that can be achieved using a diode in-line with the power source. Such diodes are called *rectifying diodes*.

During the positive cycle of voltage, the anode gets positive polarity and cathode gets negative polarity allowing current through the circuit. When the input signal enters the negative cycle of voltage, the cathode gets positive polarity and anode gets negative polarity. As described in the previous section, this configuration resists the flow of electrons due to the expansion of the depletion zone inside the diode. Current is not allowed to flow in the circuit and the output voltage will be zero so long as the input voltage is in the negative cycle. This type of circuit is called *half-wave rectifier* because the circuit rectifies only the positive half of the input voltage and deletes the remaining half. In Figure 8.5f, you will notice that the amplitude of the output waveform is smaller than the input waveform. This is because there is a voltage drop associated with current in the diode. This drop is generally 600 mV across the diode, although it is dependent on the type of diode you are using in the circuit. The diode drop is also temperature dependent.

8.9.4 Full-Wave Rectification

In the half-wave rectification setup, you lose 50% of energy from the source. A better solution will be to design a circuit that will rectify the input voltage completely so that the electric appliances have access to maximum energy from the source without the worry of getting damaged. Such a circuit is known as a *bridge rectifier* shown in Figure 8.5g. During the positive voltage cycle of the input, $D1$ and $D2$ are forward biased, allowing current through and rectifying one-half of the input, as shown in Figure 8.5h. At this time, $D3$ and $D4$ are reverse biased blocking the current. In the negative voltage cycle, $D3$ and $D4$ are forward biased allowing current, and thereby rectifying the other half of the input, as shown in Figure 8.5j. At this time $D1$ and $D2$ are reversed biased, resisting current. The output waveform is now fully rectified and available to be implemented in electronic appliances. Notice that the amplitude of the output waveform is slightly lower than the input because of the diode drops. Because each half of the input signal goes through two diodes, the output waveform amplitude will be two diode drops below the input signal amplitude.

8.9.5 Zener Diodes

This type of diode is often used in circuits for voltage clamping. Normally, a rectifying diode can withstand a large reverse voltage to resist current. But at some point they will breakdown and allow current in the reverse direction, known as *avalanche breakdown*, resulting in their demise. However, Zener diodes are designed in such a way that they can

Electric Circuits and Components 215

withstand the avalanche region and operate properly. As a result they are able to let more and more current through them, which enables them to maintain a steady potential difference across them. This principle is used in circuits to regulate a node at a fixed voltage. For example, a resistive load in a circuit can handle up to 6 V maximum; however, its input signal could be as high as 10 V. In that case, you can clamp the input of the load with a 5 V Zener diode to regulate it. This will prevent the input signal from going above 5 V. Zener diodes are widely used in the industry as transient voltage suppressors (TVS) to protect downstream circuit components from a big inrush of current.

8.10 Transistors

A *transistor* can be viewed as a diode with an additional layer of either p-type or n-type semiconductor. As mentioned in the previous section, transistors are classified into two main categories—BJT and FET. For the purposes of this chapter, we will primarily focus on BJT. Transistors are sometimes called the miracle device of electronic components because they can perform various circuit functionalities within different design guidelines, as we will see through some examples. A transistor can come in two flavors: n–p–n or p–n–p. When p-type material is pressed in between two n-type layers, we get an n–p–n transistor, and when n-type material is sandwiched between two p-type layers, it is called a p–n–p transistor. These different sections are shown in Figure 8.6a and b. For the remainder of this section, all parameters defined will be based on n–p–n transistors for simplicity. The parameters will be of reverse polarity for p–n–p transistors as we will see in Section 8.9.5 with an example of a NOT gate construction.

The *collector* region is the biggest portion semiconductor material that is connected to the *base* that acts as the control switch for current to flow through the collector and *emitter*. Base is the middle, and the thinnest, region of the device that allows current through the device when a certain voltage is applied to it with reference to the emitter (V_{BE})—also known as *bias voltage*. Each of these regions has a terminal attached to it, which is why a transistor always has three leads attached to the device package. There are internal diodes present between base–emitter and base–collector junctions, as shown in Figure 8.6c for an n–p–n transistor and Figure 8.6d for a p–n–p transistor, which prevent current in the reverse direction. In a circuit, the reference designation for BJT starts with Q and for an FET it is M, as mentioned in Table 8.2.

8.10.1 Important Datasheet Parameters

When looking through the datasheet of a transistor, there are some parameters that you should pay special attention to because these parameters are crucial to proper functioning of the device:

- You can start by looking at *collector–emitter voltage* (V_{CEO}), which states the maximum voltage the transistor can handle across its collector and emitter terminals. V_{CEO} is generally about 45 V.
- The next parameter would be the *collector–base voltage* (V_{CBO}), which refers to the maximum voltage the transistor can handle across its collector and base. V_{CBO} is usually 50 V maximum.

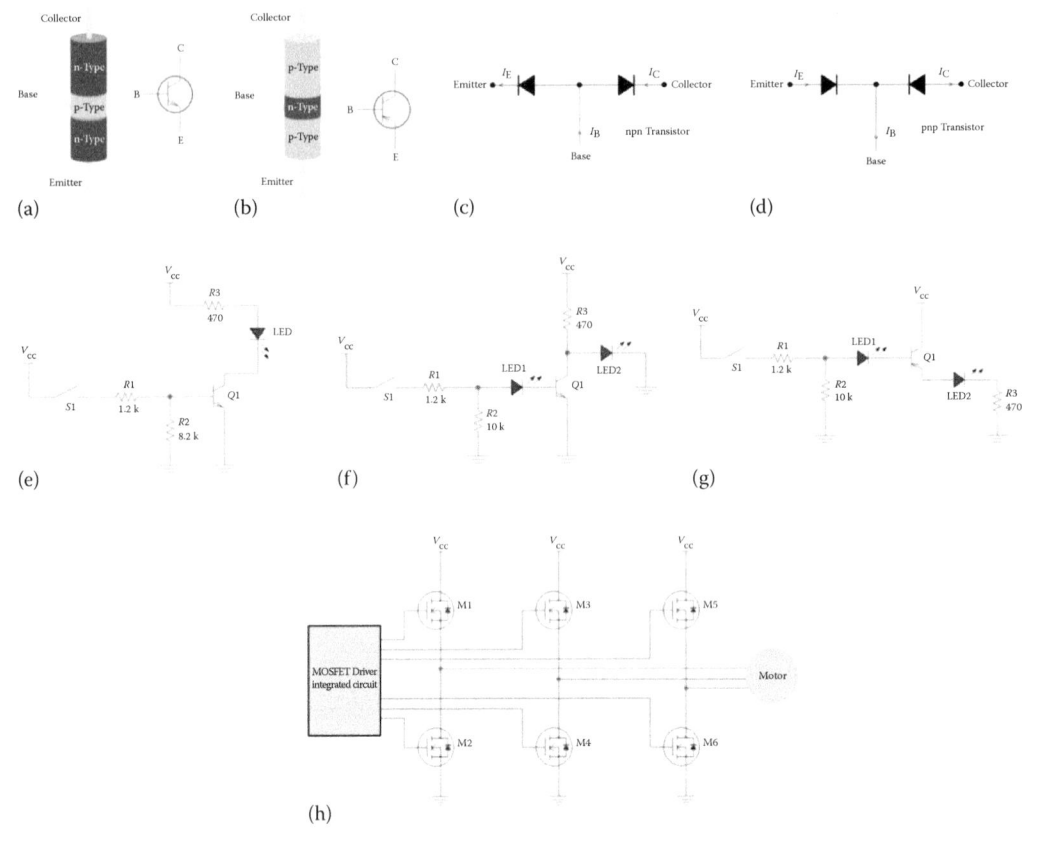

FIGURE 8.6
(a) n–p–n transistor. (b) p–n–p transistor. (c) n–p–n transistor simplified internal circuit connection. (d) p–n–p transistor simplified internal circuit connection. (e) n–p–n transistor used as a switch. (f) n–p–n transistors used to configure a NOT gate. (g) p–n–p transistors used to configure a NOT gate. (h) MOSFETs used to convert a DC circuit to an AC circuit and drive a motor.

- Next, you should look at the *emitter–base voltage* (V_{EBO}), which states the maximum voltage the transistor can tolerate across its emitter and base terminals. Most transistors will have this parameter around 5 V, indicating that only a small voltage needs to be applied to the base of the transistor for it to be functional. Circuits with transistors are designed such that a small voltage will be applied to the base. One piece of information to keep in mind is that an internal diode is connected between the base and emitter; the anode of the diode is connected to the base, and the cathode is connected to the emitter. Another way of looking at it could be that V_{EBO} refers to the maximum reverse bias voltage this internal diode can handle. This automatically makes the forward bias voltage (V_{BE}) 600 mV, often referred to as the diode drop. Base voltage will always be positive with respect to the emitter voltage.
- One of the most important parameters is the *collector current* (I_C), which refers to the maximum current that can flow through the collector–emitter path of the transistor. Depending on the type of transistor, I_C can be as high as a few hundred milliamperes, but often circuits will use a current-limiting resistor in-line with the collector to prevent a thermal event from happening to the transistor due to excessive current.

Electric Circuits and Components 217

- Another parameter you should notice is the *DC current gain* (h_{FE}), which refers to the ratio between base current (I_B) and collector current (I_C). h_{FE} is typically in the 100–250 range but can go as high as 600, depending on the type of transistor you choose. h_{FE} can be understood as the ability of the transistor to amplify current—the higher the number, the more amplification of current the circuit will receive from the transistor. This also implies that a transistor is a current-operated device, meaning a transistor can drive a high current load, but only requires a small amount of current at the base to be operational.
- And the last parameter is the *power dissipation* (P_D) of the transistor, which implies the total power a transistor can dissipate during continuous operation, usually a few hundred milliwatts. Sometimes P_D will not be explicitly mentioned in the datasheet; you will need to compute that from another parameter known as *thermal resistance* ($R_{\theta JA}$) using the following equation:

$$P_D = \frac{\theta_{amb}}{R_{\theta JA}} \tag{8.32}$$

where θ_{amb} is the ambient temperature of the transistor at worst-case circuit conditions.

8.10.2 Operating Regions

A BJT has the following regions of operation that depends on certain junction biasing factors. Depending on the circuit needs, a transistor may be forced to operate in a particular region for a prolonged time. Figure 8.1g shows a graphical representation of these operation regions.

8.10.3 Forward Active

In this region, the base–emitter junction is forward biased (approximately 600 mV, depending on the transistor) and base–collector junction is reverse biased implying that the collector–emitter voltage (V_{CEO}) will be greater than the base–emitter voltage (V_{BE}). When this happens, collector current (I_C) becomes almost proportional to base current (I_B) but many times larger for small variations in I_B due to the gain factor (h_{FE}).

8.10.4 Reverse Active

This region can be viewed as the opposite of the forward active region and is considered only in the case of a failsafe condition. Here, the base–collector junction is forward biased and the base emitter junction is reverse biased implying emitter–collector voltage (V_{EC}) which is greater than base–collector voltage (V_{BC}). We will not dwell in this region for too long because real-life application for a transistor in this region is limited.

8.10.5 Saturation

You would want to operate a transistor in this region for the most part because the *saturation* region corresponds to the transistor being fully ON (like a closed switch). Here, both the base–collector and the base–emitter junctions are forward biased, implying the base–emitter voltage (V_{BE}) is greater than or equal to collector–emitter voltage (V_{CE}).

8.10.6 Cutoff

This region is the exact opposite of the saturation region because very little current flows through the device implying that the transistor is OFF (like an open switch). Here, both the base–collector and base–emitter junctions are reverse biased. You should have your transistor in this region when you no longer want it to be functional in the circuit.

8.10.7 Avalanche Breakdown

A transistor reaches the breakdown region when a large potential is applied to the collector breaking down the base–collector junction diode. When that happens, the collector conducts current even when the base does not receive any current, and therefore, the transistor is no longer controllable. At this point it is said that the transistor has been permanently damaged. Breakdown region should be avoided by all means and to do that you need to make sure that the key parameter limits are taken into consideration while designing the circuit. A means to avoid the breakdown region is to implement an appropriate current-limiting resistor at the collector terminal.

8.10.8 Using a Transistor as a Switch

Figure 8.6e shows a circuit where the transistor is used as a switch to turn ON and OFF the LED. Assuming V_{cc} to be 5 V when S1 is closed, the bias voltage of the transistor will be about 4.36 V because R1 and R2 act as a voltage divider circuit. Calculated using Ohm's law, the base current (I_B) will be 4.17 mA, which is enough to turn ON Q1. At this time, the collector conducts current and the LED glows. Using Ohm's law, we can calculate the current through the LED to be 10.64 mA, which is clearly bigger than the base current. Hence, the transistor is effectively amplifying current. Now you might wonder why you need a transistor to light up the LED when you can directly connect the switch and the LED with an in-line resistor to achieve that. The purpose here is to show that a transistor can be used as a switch while amplifying the output current (something a regular ON–OFF switch cannot do).

8.10.9 Using an n–p–n Transistor as a NOT Gate

Figure 8.6f shows how a transistor can be used for inverting an input signal; assume $V_{cc} = 5$ V. When the circuit is functional, the LEDs will light up out of sequence, meaning when LED1 is glowing, LED2 will be OFF and vice versa. Initially, when S1 is open, no current is flowing through LED1 or the transistor's base. Therefore, LED1 and Q1 will be OFF. On the output side of Q1 current is flowing through LED2 lighting it up, and so we have established our first state—LED1 is OFF and LED2 is glowing.

When you close S1, current flows through LED1 and the base of Q1 lighting up LED1 and turning ON Q1. With Q1 ON, current flows through the collector–emitter, which essentially creates a short circuit across LED2. As a result, current bypasses LED2 and flows through R3 and Q1 to ground. LED2 goes dark, and we have established our second state—LED1 is glowing and LED2 is OFF. We have designed a NOT gate where the output of the circuit is the opposite of the input.

8.10.10 Using a p–n–p Transistor as a NOT Gate

All this discussion about n–p–n transistors could make p–n–p transistors feel left out. We should give the p–n–p transistors a chance to participate in the discussion because they are also widely used in electronic circuits. Figure 8.6g shows how a p–n–p transistor can be used for a NOT gate circuit; assume V_{cc} = 5 V. Before explaining this circuit, it is imperative to know that a p–n–p transistor works the opposite way from an n–p–n transistor. In order to turn ON Q1, a low-level ground voltage will be needed. When S1 is open, no current is flowing through LED1 and so it is dark. R2 pulls down the base voltage of Q1 to ground turning it ON in the process. On the output side of Q1 the collector is conducting current, essentially creating a short circuit between the emitter and collector terminals. Current now flows through LED2 making it glow. We have established our first state—LED1 is OFF and LED2 is glowing. The purpose of R3 is to limit the current through LED2.

Now as you close S1, current flows through LED1 making it glow, and switching OFF Q1. On the output side, no current flows through the emitter or LED2, and hence LED2 is dark. We have established the second state—LED1 is glowing and LED2 is OFF. We have successfully designed a NOT gate circuit using a p–n–p transistor.

8.10.11 Inverter Circuit with a MOSFET

A big portion of the transistor family is MOSFETs, which are used in fast switching circuits that often require high amounts of current, like for example an inverter circuit. A detailed discussion regarding a MOSFET will be a chapter in itself and outside the scope of this chapter. However, an example of a circuit with MOSFETs will do justice to this section. Without going into much detail, the circuit shown in Figure 8.6h switches the n-channel FETs, M1–M6, to drive a three-phase motor. The drain terminals of the upper FETs are connected to a 12 V supply; source terminals of the upper FETs are connected to the drain terminal of the lower FETs, and source terminals of the lower FETs are tied to ground. The gate terminals of all six FETs are connected to a driver IC that controls the switching time of each FET independently. The motor phase wires are connected to the source–drain junction of the FETs that can be called as the output of the circuit. The switching of FETs M1–M6 in sequence makes the motor draw current through the FETs, which results in an angular motion in a particular direction. Special attention is required while designing the timing sequence of the FETs because inaccurate timing could switch ON a pair of upper and lower FETs simultaneously, causing a direct short circuit between power source and ground, and damaging other circuit components. Such a scenario is often termed as shoot-through in a circuit. Depending on the size of the motor, the FETs need to source current in hundreds of amperes to the motor. However, the current the driver IC provides to the gate terminals of the FETs to turn them ON is only a tiny fraction of what the motor is drawing through the FETs. This is another design that shows how a transistor is able to amplify current by a large magnitude to drive a significant load. This example also serves as a basic design used in converting a DC circuit upstream to an AC circuit downstream.

8.11 Integrated Circuits

ICs are the fundamental building blocks to all electronic devices today. They consist of all the components mentioned in previous sections embedded on silicon wafer that gets packaged into a chip. ICs come in different package shapes, as shown in Figure 8.2n, and have a wide range in product portfolio, such as op-amps, timers, power supplies, sensors, and microcontrollers. In this section, we will only focus on a 555 timer and an op-amp design. Before we take a deeper dive, let us look at how an IC is manufactured that involves multiple high-precision fabrication steps.

8.11.1 IC Manufacturing Process

Typically the IC manufacturing process consists of more than a hundred steps, during which multiple copies of the IC are generated on a single piece of silicon wafer. Each IC manufacturer has their proprietary way of setting up their production lines for IC manufacturing, but generic steps include about 20 patterned layers into the silicon substrate for the finished product. This layering process is needed to create proper electrical connection inside and on the surface of the silicon wafer.

8.11.2 Wafer Production

First, purified silicon crystal is heated to the liquid state. A small piece of solid silicon is immersed into the molten silicon and then pulled out slowly, letting the liquid cool off to form an ingot of a single silicon crystal. The silicon crystal is then carefully sawed into very thin wafers that would be etched in a pre-programmed pattern, to create hundreds or thousands of IC.

8.11.3 Fabrication of Silicon Wafer

The fabrication process consists of a number of steps performed in sequence, which is described below. This is where the design of the IC meets reality; that is, the components and connections in the electrical design of the IC get imprinted on the silicon wafer. The fabrication process happens in a clean room within the manufacturer's facility. The clean rooms have extremely tight tolerances on environmental conditions such as temperature and humidity in order to prevent anomalies in the fabrication process. TSMC (Taiwan Semiconductor Manufacturing Company) is the largest independent semiconductor manufacturing company in the world.

8.11.4 Deposition

Before going into the etching process, where components get embedded onto the silicon, the wafers go through a two-step purification process. At first, the wafers get pre-cleaned via high-purity chemicals in order to increase yield of the IC. Next, under carefully controlled conditions, the wafers are cleansed with ultrapure oxygen, in diffusion furnaces, to form a uniform thickness of SiO_2 film on the wafer surface.

8.11.5 Masking

A mask can be visualized as having the blueprint of the circuit. Certain portions of the mask are transparent allowing light to pass, whereas other portions are opaque blocking light. In other words, masking is the process of covering up certain portions of a wafer while working on other areas, a process also known as *photolithography*. The mask is placed under a lamp that projects high-intensity ultraviolet light through the transparent sections of the mask to the silicon wafer. Prior to that, a light-sensitive film known as the *photoresist is applied to* the silicon wafer. Precise alignment of the wafer

Electric Circuits and Components 221

and the mask is crucial to this process, and hence automated photo aligners are used to ensure very tight tolerance on the alignment.

8.11.6 Etching

The high-intensity ultraviolet light etches the wafer portion present under the transparent section of the mask, leaving an impression of the photoresist. The wafer is then baked to harden the photoresist pattern. It is then exposed to a chemical solution that etches away the portions where no hardened photoresist is present. In this way, only required portions of the wafer that are representative of the IC design are now covered with the hardened photoresist layer. Next, the wafer is exposed to further chemicals or plasma in order to remove the hardened photoresist. The wafer is then checked to ensure the mask has imprinted the proper blueprint on the top surface of the wafer.

8.11.7 Doping

The wafer is doped with phosphorous and boron in the areas that the etch process had exposed, forming the n-type and p-type regions. Doping alters the electrical properties of silicon, as previously discussed, by providing additional electrons or holes in the crystal structure, shown in Figure 8.5b and c. If the IC design has multiple layers of circuit connections, the etching and doping processes are repeated for each layer in the design. As you can see, doping is not just an athletic activity. Depending on how it is done, it can prove to be very useful.

8.11.8 Passivation

Following the doping process, a series of metal depositions connects the individual parts of the circuit. Once interconnects have been established, a final dielectric layer is deposited on the wafer in order to prevent contamination and protect the integrity of the circuit. This process is called *passivation*. Wire bonds and probes then connect the top metal layer.

8.11.9 Electrical Testing

Now that the IC design has been impressed on the wafer, an automatic and computer-operated test runs through the system to ensure proper functionality of each chip on the wafer. The tests have upper and lower specification limits and if a chip on the wafer does not fall within the limits, it gets rejected.

8.11.10 Assembly

Robotic arms then carefully slice and visually inspect each chip on the wafer for anomalies. If a fault is detected, that piece of silicon wafer is rejected. Once everything looks fine, the chip gets assembled into a package, and the wire bonds get attached to the lead frames of the package for electrical connection. Each chip gets tested again before getting packaged in moisture proof containers and sent to customers.

8.11.11 IC Datasheets

When you are planning to implement an active or passive component in your circuit, you would always want to thoroughly review the accompanying *datasheet* from the manufacturer, especially for ICs. The datasheet contains valuable information about the component that will help you design your circuit accordingly. Many times you will find that even though an IC looks perfect for your application, parameters listed on the datasheet could electrically overstress the part in your specific design. You do not want to overstress your part in your circuit because it can cause a thermal event and result in permanent IC damage. This is very crucial when designing a circuit for safety critical applications such as airplanes, automotive vehicles, and hospital patient care equipment. This section highlights some key parameters in the datasheet that cannot be overlooked when implementing an IC in an electronic circuit. In addition to these parameters, you will need to look up other parameters that are specific to the functionality of the IC.

8.11.12 Supply Voltage

Every circuit needs a power source, whether AC or DC, and every IC needs power to operate. First you need to find the voltage rating of your IC for continuous operation. Generally, ICs require either 3.3 or 5 V supplies, but can operate at higher continuous voltages. If the IC you selected has a tight tolerance on supply voltage, you need to ensure that the power source to the IC is well regulated in order to prevent voltage transients. If your power source is noisy, it is good practice to use MLCCs or an RC filter to suppress voltage spikes.

8.11.13 Supply Current

Every IC draws some amount of current while performing its functionality. For example, if a 555 timer chip requires a maximum current of 15 mA to operate, then you need to make sure that the power supply can source more than 15 mA of current. In general, power supply ICs can source current in the hundred milliamp range, but you need to confirm this from your power supply IC datasheet.

8.11.14 Functional Description

IC manufacturers provide a lot of detailed information in the functional description section of the datasheet. You may already know about most of these features because you picked the IC for your design, but it is always good practice to review the functional description. You will be surprised how often you will find useful information in that section of the datasheet that you could have taken for granted. Most datasheets will provide a functional block diagram to accompany the description that details the inner circuit blocks of the IC. If you have specific questions regarding those functional blocks, it is highly recommended to contact the manufacturer for additional details (better safe with your design than sorry).

8.11.15 Characteristic Graphs

The majority of information you require for your specific design is hidden in graphs usually provided in the middle sections of the datasheet. Although the tables in the top half will provide information on numerous parameters, the information is generally based on specific test conditions that may or may not be relevant to your design. For example, you have a battery powered circuit that needs to draw little current and operate over a wide temperature range. You want to implement a magnetic sensor in your design and the sensor's datasheet specifies supply current to be 5 mA at 25°C. However, when you scroll down to the graphs portion of the datasheet, you find a graph of supply current over ambient temperature that

Electric Circuits and Components 223

implies that the sensor could draw current in the range of 1–10 mA over your desired temperature range. As a result you need to account for the 10 mA current draw in your design.

8.11.16 Absolute Maximum Ratings

This section, usually present within the first few pages of a datasheet, specifies parameter values for the survival of the IC. Under no condition can you exceed these parameter limits, or your IC will be damaged permanently. Read the instructions that accompany the absolute maximum ratings. They provide useful information, such as package thermal resistance, that will help you calculate maximum power dissipation rating (P_D) of the IC. The power dissipation of the device in your circuit must not exceed the rated P_D.

8.11.17 555 Timer

Out of all the ICs the 555 timer has been chosen because it is a widely used IC that provides a mix of both analog and digital circuits from a system perspective. The analog portion of the circuit, composed of a resistor and capacitor, is the input to the IC that determines how long the output of the device will stay active. The output is a digital signal that indicates whether the device is active or in reset mode. The 555 timer can be configured in two modes—*monostable* and *astable*. Figure 8.7a shows the pin assignment of this IC, and Figure 8.7d shows the functional block diagram inside the chip. As you can see, pin-2 is the trigger input and pin-6 is the threshold input to the chip. The three 5 kΩ resistor network provides a threshold limit of ⅔ V_{cc} and trigger limit of ⅓ V_{cc}. The comparator output will be high when the voltage at the (+) pin is greater than the voltage in the (–) pin; otherwise it will stay low. When the trigger level goes just below ⅓ V_{cc}, comparator B activates the latch that triggers a high output on pin-3. When the threshold level exceeds ⅔ V_{cc}, the latch is deactivated and the chip outputs a logic low signal. The external passive components control the timing between high and low of the chip.

8.11.18 Monostable (One-Shot) Multivibrator

The 555 timer IC can be configured to activate a small load for a certain duration and is determined using a pair of external resistor and capacitor, as shown in Figure 8.7b. A 5 V supply powers the chip, even though most 555 timers can handle up to 16 V depending on the manufacturer. When the IC receives a trigger signal, usually a voltage close to zero or ground, the output goes high activating a small load attached at pin-3 (not shown in Figure 8.7b). Prior to activation, known as the *stable state*, the input voltage to the IC at pin-2 is generally close to the supply voltage because, in most cases, the input pin would be connected to the power supply through a current-limiting resistor, as shown in Figure 8.7c. The duration for which the chip is activated, also known as the *pulse width* of the output, is determined using the following equation, measured in seconds:

$$t_{\text{pulse-width}} = 1.1 R_1 C_1 \tag{8.33}$$

Initially the output of the 555 timer is low meaning that the chip is not activated and the signal coming out of the chip is close to zero voltage. When the chip receives a low voltage signal at trigger pin, pin-2, the output goes high, close to the supply voltage, and C_1 starts to charge through R_1. The output of the chip stays high as long as the charge stored in the capacitor is approximately ⅔ V_{cc}. When this threshold is reached, the output of the

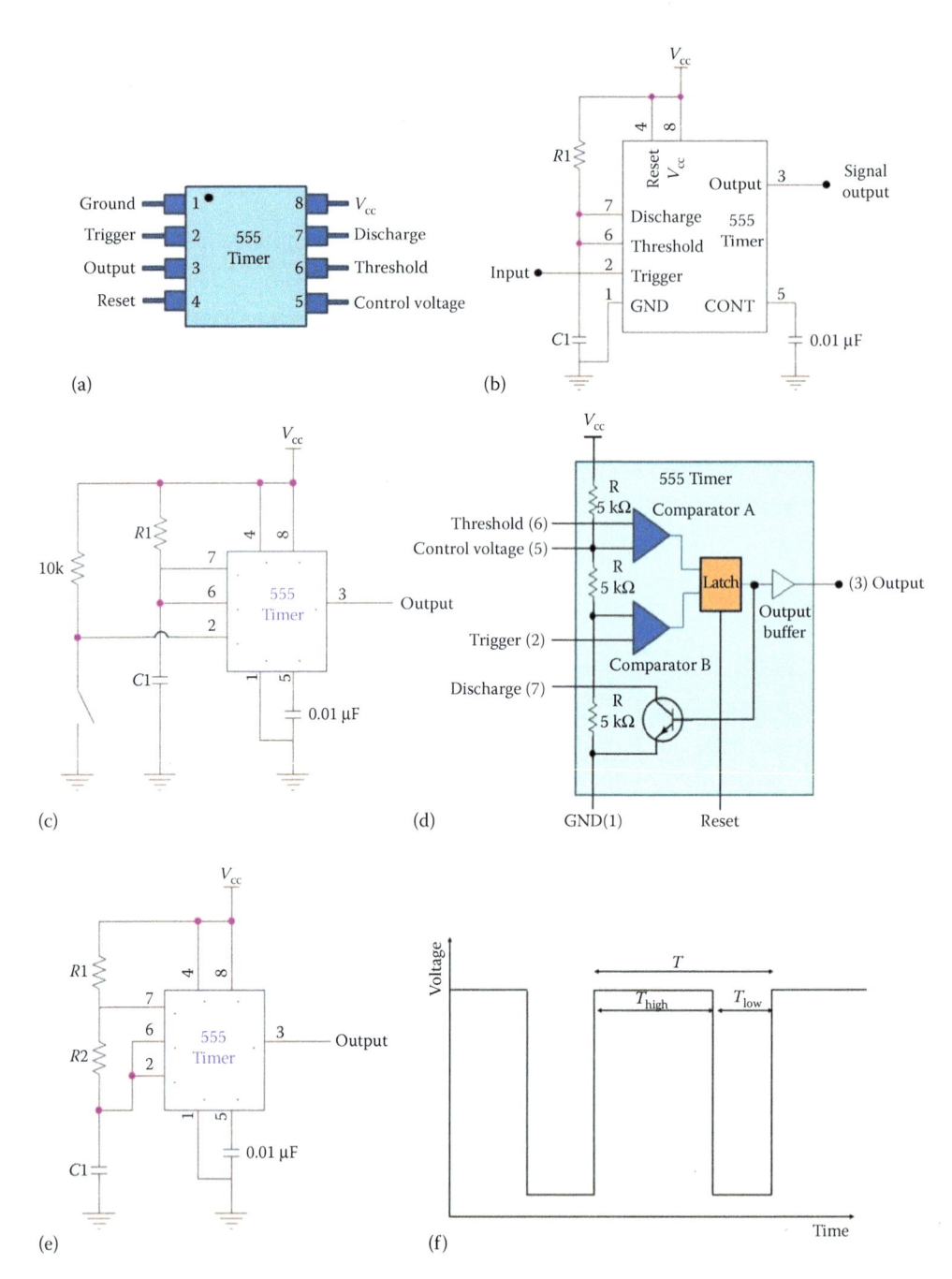

FIGURE 8.7
(a) Pin assignments of a 555 timer IC. (b) External resistor and capacitor used to configure a 555 timer IC as one shot. (c) 555 timer configured as a monostable (one-shot) multivibrator. (d) Internal functional block diagram of a 555 timer. (e) 555 timer IC configured as an astable multivibrator. (f) Astable multivibrator pulse durations of the unstable states.

chip goes low and the capacitor starts to discharge through an internal resistor of the chip. Changing the external resistor and capacitor values, according to the above equation, can alter the amount of time for which the output stays high. The constant parameter of 1.1 is specific to the 555 timer IC. This value will vary if you choose to use another device for one-shot configuration.

8.11.19 Astable Multivibrator

In this configuration, the 555 timer does not have the luxury of being in a stable state. It oscillates between two unstable states without the initiation of an external trigger. The circuit design is shown in Figure 8.7e. The external resistors and capacitor determine the duration the IC dwells in its unstable states, shown in Figure 8.7f. The two unstable states are denoted as T_{high} and T_{low}. They are computed using the following equations:

$$T_{high} = 0.7(R_1 + R_2)C_1 \tag{8.34}$$

$$T_{low} = 0.7 R_1 C_1 \tag{8.35}$$

$$T = 0.7(R_1 + 2R_2)C_1 \tag{8.36}$$

The percentage of time the output of the IC stays in the T_{high} region, compared to the total duration of the pulse T, as shown in Figure 8.7f, is called the *duty cycle* of the output and is expressed as

$$\text{Duty cycle} = \left(\frac{R_1 + R_2}{R_1 + 2R_2}\right) \times 100\% \tag{8.37}$$

In the astable configuration, C_1 charges through the two resistors, and oscillates between the charge levels ⅓ V_{cc} and ⅔ V_{cc}. When the charge level in C_1 is ⅓ V_{cc}, the 555 timer enters the unstable state T_{high} and when the C_1 charge level reaches ⅔ V_{cc}, the IC enters the unstable state T_{low}. The process gets repeated as the device oscillates within these unstable states.

8.11.20 Operational Amplifiers

In this section, the goal is to provide you with a high-level overview of an op-amp and the two basic configurations. The fundamental function of an op-amp is to amplify signals. A deep dive analysis of op-amps will require many more sections and even chapters. At that point you may begin to lose interest and move onto another textbook, which as an author, is not my intent. Back in the day, before the heavy use of electronics, amplifiers were used in telephone distribution systems to boost audio signals. This signal boosting feature was of particular interest to electrical engineers who found interesting ways to make an amplifier perform additional tasks, such as mathematical operations; hence the name *operational amplifiers*. In a nutshell, an op-amp IC consists of numerous transistors, resistors, capacitors, diodes, and so on packaged into a single chip to perform various tasks in different circuit configurations. Some op-amps require both positive (+V) and negative (–V) supply voltages from the power source, that typically range from ±5 to ±18 V. Other op-amps require only a single voltage supply and use the ground terminal as the reference voltage. We will only focus on op-amps that require one power source.

8.11.21 Noninverting Amplifier

In this configuration, as shown in Figure 8.8a, the op-amp amplifies the input signal via a gain factor shown in the following equation. The input signal (V_{in}) is applied to the noninverting (positive) terminal and the resistor network (R_{in} and R_f) is connected to the inverting (negative) terminal of the op-amp. R_s is used as a current-limiting resistor on the input signal path.

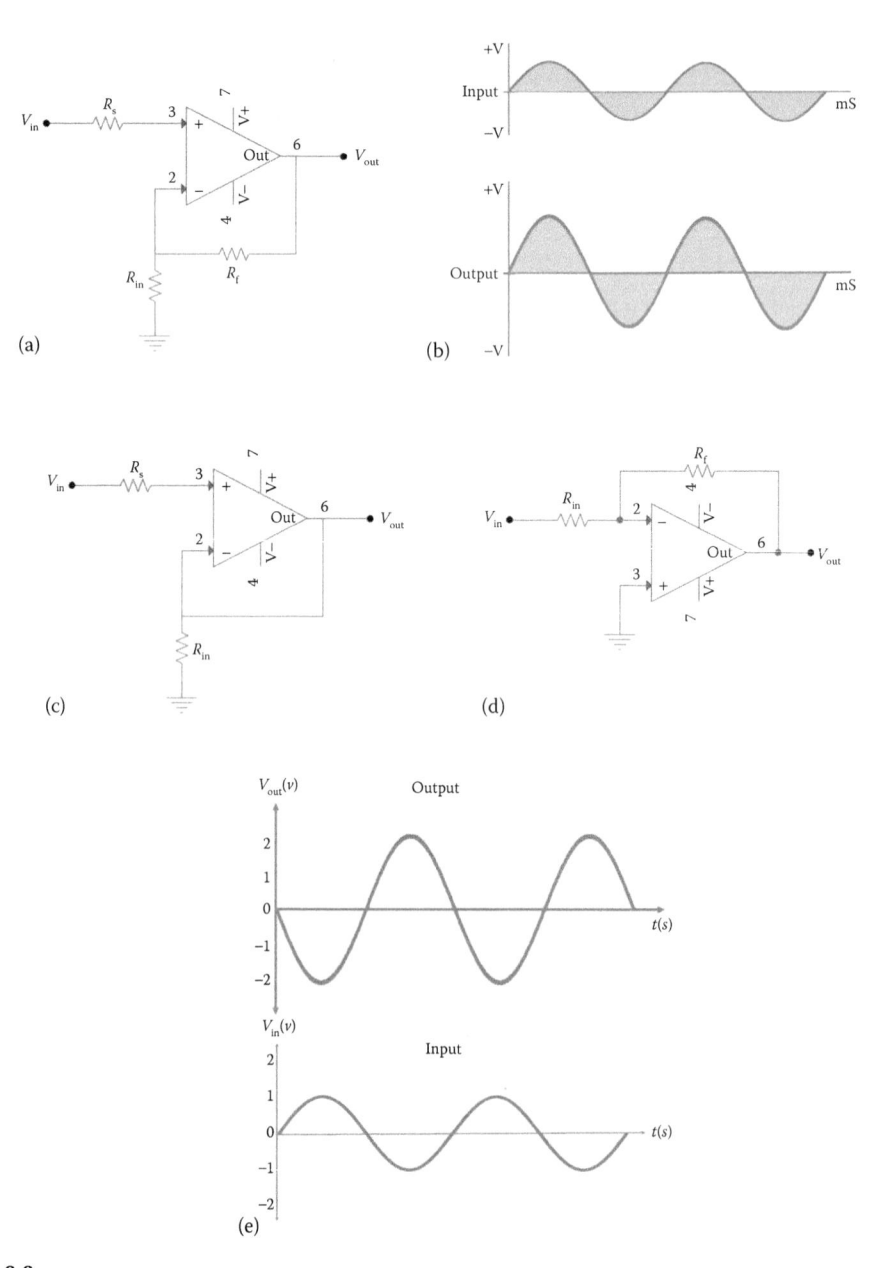

FIGURE 8.8
(a) Op-amp in noninverting configuration. (b) Output of an op-amp in noninverting configuration. (c) Unity gain buffer. (d) Op-amp in inverting configuration. (e) Op-amp output in inverting configuration with gain of 2.

Electric Circuits and Components

$$A_v = 1 + \frac{R_f}{R_{in}} \tag{8.38}$$

The output of the op-amp will be in phase with the input signal. The output voltage can be calculated using the following expression. Figure 8.8b shows the output voltage of an op-amp with respect to the input voltage, assuming gain of 2.

$$V_{out} = (V_{in})(A_v) \equiv V_{in}\left(1 + \frac{R_f}{R_{in}}\right) \tag{8.39}$$

The equation above implies that the closed-loop gain of a noninverting op-amp will always be greater than or equal to 1. If $R_f = 0$, the gain becomes 1, and the output will follow the input of the op-amp. Such a configuration is known as *voltage follower* or *unity gain buffer*, as shown in Figure 8.8c.

8.11.22 Unity Gain Buffer

You may be wondering what the purpose of the op-amp is if the input signal is not getting amplified. Why spend money to buy the op-amp IC and implement in the circuit if it is not going to do any signal boosting? These are valid questions, but there are special requirements for op-amp circuits with unity gain when a circuit requires signal isolation. The input impedance, measured in ohms, of the noninverting terminal of the op-amp is very large, in the range of 1 MΩ or 1×10^6 Ω, which means that the op-amp does not draw current from the input source. However, if a load is connected to the output of the op-amp, it can draw as much current as the op-amp can source, keeping the voltage level the same as the input to the op-amp. An application for such an op-amp configuration could be in a filter/rectifier circuit where the output is taken across a capacitor, and drawing current would drain the capacitor enough to bring voltage down, which would be undesired. You still want to maintain the same voltage level while drawing a small amount of current. In that case, you connect the output of the capacitor to the noninverting terminal of a unity-gain buffer that provides the same voltage level without drawing any current.

8.11.23 Inverting Amplifier

In this configuration, the input is connected to the inverting (negative) terminal of the op-amp, and a basic circuit design is shown in Figure 8.8d. The closed-loop gain is calculated as follows:

$$A_v = -\frac{R_f}{R_{in}} \tag{8.40}$$

The equation above implies that the output signal from the op-amp will be out-of-phase with the input signal. The output signal can be determined using the following equation:

$$V_{out} = (V_{in})(A_v) = -V_{in}\frac{R_f}{R_{in}} \tag{8.41}$$

Figure 8.8e shows an inverting op-amp output signal with respect to input and a gain of 2. Similar to the noninverting terminal, the inverting terminal of an op-amp has very high impedance, and therefore the op-amp does not draw any current from the source signal. If you set resistors R_{in} and R_f to the same value the gain of the op-amp will be –1. This is an interesting case because now the op-amp will keep the same amplitude as the input signal and invert the output. Such a configuration is also known as the *unity gain inverter circuit*.

8.12 Summary

- Electric charge is the presence of positive or negative charges such as electrons or protons, respectively. When present in space, they create an electric field. When a particle is in the vicinity of this field, it will experience a force that can be calculated using Coulomb's law.
- When there is a potential difference in a current-carrying conductor, electrons flow from the positive side of the wire to the negative side. This flow of electrons is called current. It should be noted that current flows from higher potential to lower potential in a wire or circuit, whereas electrons flow in the opposite direction.
- Electronic components that make up an electric circuit include resistors, capacitors, diodes, inductors, transistors, and various forms of ICs. They are labeled using specific letters and numbers, called reference designators, in a circuit that has been shown in Table 8.1. Different types of diodes, transistors, and ICs have been discussed.
- Resistor and inductors in series can be arithmetically added for total resistance and inductance in the circuit. Capacitors in parallel can be added in the same way to determine circuit capacitance. Resistors and inductors in parallel can be combined using Equations 8.6 and 8.25. Capacitors in series can be combined using Equation 8.11.
- Current through a resistor is proportional to the potential difference applied across it, according to Ohm's law. The proportionality constant is the resistance of the resistor.
- A capacitor blocks DC while allowing AC to flow through. This is shown using Equation 8.22, which implies that the energy stored in a capacitor is time dependent, and because DC does not vary with time, a capacitor will not allow DC to pass.
- A resistor and capacitor (RC) can be combined to form first-order low-pass filter circuit for a specific cutoff frequency that can be calculated using Equation 8.23. Any frequency above the cutoff frequency will be attenuated at the output of the filter. The capacitor acts as a short circuit for high-frequency signals letting them pass through to ground.
- A resistor and inductor can be combined to form a first-order LC circuit that will reject any high transient current thus protecting components sensitive to current spikes in a circuit.
- A diode allows current in only one direction. Diodes are often used in AC circuits for full-wave rectification because the negative amplitude of an AC source may damage electronic devices.

Electric Circuits and Components 229

- A transistor has five operating regions. It requires a small amount of current to turn ON, and enter saturation mode, and can drive a high current load thereby amplifying the input signal.
- MOSFETs are often used in high-frequency switching circuits that require a high current load to be driven by them. An example of an inverter circuit has been discussed in Section 8.10.11 where a DC circuit gets converted to an AC circuit to operate a three-phase motor.
- IC manufacturing process is a complex step-by-step sequence that requires high precision and accuracy in each step for greater yield. When designing a circuit with an IC, it is essential to read certain datasheet parameters, discussed in Section 8.10.2, to identify potential ways to protect the device and obtain maximum performance. ICs come in many different forms; two commonly used devices, a 555 timer and an op-amp, have been discussed.
- Op-amp comes in two different configurations: noninverting and inverting. They are generally used for amplifying the input signal but other circuit uses have been explained. Noninverting op-amp gain can be calculated using Equation 8.38 and inverting op-amp gain by Equation 8.40.
- In a noninverting op-amp circuit, if the feedback resistor is shorted, the circuit acts as a unity gain buffer that is often used in applications for device isolation. In an inverting op-amp circuit, if the feedback resistor is the same as the input resistor, the circuit acts as a unity gain inverter.

Problems

1. A charged particle placed in the vicinity of an electron will experience force according to Coulomb's law. Assuming the particle carries a charge of 6.53×10^{-19} C and they are placed 35 μm apart, determine the magnitude of force exerted between the particles. The charge of an electron is 1.6×10^{-19} C.
2. What will be the resistance and tolerance rating of a five-band resistor whose color stripes, in sequence, read: yellow, violet, black, orange, and gold?
3. Find the total resistance in the circuit shown in Figure 8.9a.
4. For the previous problem, assuming that V_{cc} is 5 V, what is the current through the circuit? *Hint*: use Ohm's law to compute current.
5. From the circuit shown in Figure 8.9b, determine the total circuit capacitance, and the RC time constant.

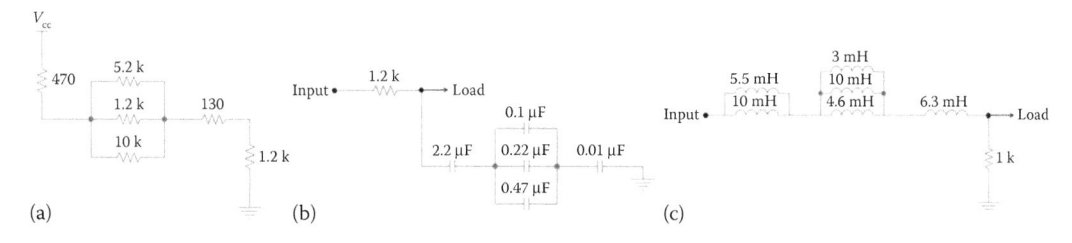

FIGURE 8.9
Figures pertaining to the problems in this section.

6. From the previous problem, compute the cutoff frequency of the first-order RC filter. How does the circuit behave when signal frequencies become higher than the cutoff frequency?
7. For the circuit shown in Figure 8.9c, determine the following parameters:
   a. Total inductance in the circuit
   b. The RL time constant
   c. The circuit reluctance for an input signal of frequency 50 Hz
8. You have a 15 mH inductor in a circuit with 5 cos(119t) A current. Determine the voltage across the inductor and the energy stored by the inductor.
9. A 555 timer IC has been configured as a monostable multivibrator. Determine the duration for which the IC stays in its unstable state when the external resistor and capacitor are set to 4.7 kΩ and 0.22 µF, respectively.
10. From Figure 8.7e, where a 555 timer is configured as an astable multivibrator, determine the following parameters when $R1 = 3.9$ kΩ, $R2 = 5.1$ kΩ, and $C1 = 1.2$ µF:
    a. T_{high} and T_{low}
    b. Duty cycle of the output
11. For noninverting op-amp configuration, what will be the output voltage if the feedback resistor (R_f) is 3.6 kΩ and input resistor (R_{in}) is 5.2 kΩ? Assume that the input voltage is 5 V.
12. An AC source has a signal amplitude of 6 V that gets connected to an op-amp in inverting configuration. The peak amplitude of the output signal is 2 V. What is one possible combination of resistance values you can choose for the feedback resistor (R_f) and input resistor (R_{in})?
13. You are given a 520 Ω resistor that needs to be connected to a 12 V supply. The power rating of the resistor is ¼ W. Is this resistor a good choice for the circuit? Explain your findings.

9
Engineering Economics

CHAPTER OBJECTIVE AND STUDENT LEARNING OUTCOMES

After completing this chapter, students will be able to

1. Define engineering economics terms, concepts, and terminologies.
2. Develop cash flow diagrams and use them to solve engineering economics problems. [ABET outcome e, see Appendix C]
3. Analyze multiple alternatives and make sound engineering economic decisions. [ABET outcome c, see Appendix C]
4. Use computational methods to solve problems to select the optimal solution. [ABET outcome k, see Appendix C]

9.1 Introduction

Engineers are involved in the management of projects ranging from single product development to complex projects that require financial decision making to maximize the economic value of the product or project. Knowledge and understanding of tools and techniques necessary for making economic decisions are essential for any engineering discipline. This chapter provides an overview of the economic decision-making process that requires formulating, estimating, and evaluating the economic outcomes of alternatives designed to accomplish a specific task or purpose. Engineering economics study involves problem identification, definition of the objective, cash flow estimation, financial analysis, and decision making. The final outcome or purpose of economic analysis is decision making that requires elements such as cash flow, interest, and time. The following step-by-step procedure can be used to solve engineering economics problems:

1. Define and understand the problem.
2. List information provided in the problem and information required in terms of variables.
3. The most commonly used variables are as follows:
   a. A = periodic payment or annual worth (AW) (starts at one period after P)
   b. P = present value (at time $t = 0$)

c. F = future value (at time $t = n$)
d. n = number of payment periods
e. i = interest rate for the payment period (PP) (monthly, yearly, etc.)

4. In most cases, four of the five variables will be provided and will require solution for the fifth variable.
5. Develop a cash flow diagram for the problem.
6. Select a solution method (formula, factor table, or Excel) to solve the problem.
7. For multiple alternatives, evaluate each alternative using an economic measure of worth (PW, AW, FW, or AW).
8. Evaluate each alternative and select the most economic solution.
9. Select the best economic alternative.

9.2 Nominal and Effective Interest Rate

Engineering economic study uses interest as annual rate unless specified otherwise. This section will familiarize students with interest rates and how interest rates change the monetary value of projects or investments. There are two types of interest rates that change the value of money: nominal and effective interest rates. The concepts of nominal and effective interest rates must be used when interest is compounded more than once each year.

Nominal interest rate: A nominal interest rate r is an interest rate that does not account for compounding. By definition,

$$r = \text{interest rate per time period} \times \text{number of periods} \tag{9.1}$$

A nominal rate may be calculated for any time period longer than the time period stated. After the nominal rate has been calculated, the compounding period (CP) must be included in the interest rate statement. For example, if $100 is invested at a nominal rate of 8% for five years, the final amount after five years will be $100 × 1.08 × 5 = $140.

Effective interest rate: An effective interest rate i is a rate wherein the compounding of interest is taken into account. Effective rates are commonly expressed on an annual basis as an effective annual rate; however, any time basis may be used:

$$\text{Effective rate per compounding period} = \frac{r\% \text{ per time period } i}{m \text{ compounding periods per } i} = \frac{r}{m} \tag{9.2}$$

where i is the interest period and m is the compounding frequency.

An effective rate may not always include the CP in the statement. If the CP is not mentioned, it is understood to be the same as the time period mentioned with the interest rate. For example, if $100 is invested at an effective rate of 8% for five years, the final amount after five years will be $100 (1 + 0.08)^5 = $100 × 1.4693 = $146.93. As can be observed, the effective interest rate increased the value of $100 from $140 to $146.93. In this example, the CP is not stated, and therefore CP is yearly as the interest rate is also per year.

9.2.1 Solving Engineering Economics Problems

Engineering economic analysis can be performed using three different methods: analytical, computational, and use of interest factor tables. All three approaches will provide similar solutions. It may be convenient or preferred by individuals to use one of these methods for some problems. However, it is recommended to use a second method to verify the accuracy of the solution:

1. *Analytical method:* Engineering economics problems can be solved using equations or formulas listed in the right column of Table 9.1. The symbols column lists the factors to solve problems using interest factor tables.
2. *Computational method:* Spreadsheets such as Excel provide financial functions that can be used to solve engineering economics problems. Financial functions are also available in several other computer programs. This section provides information about how spreadsheet functions can be used to solve engineering economics problems. A list of variables used in Excel is provided below to familiarize students with the spreadsheet functions (Table 9.2):

rate: interest rate per CP

nper: number of compounding periods

pmt: constant payment amount

PV: the present value. If PV is not specified, the function will assume it to be 0

FV: the future value. If FV is not specified, the function will assume it to be 0

TABLE 9.1

Engineering Economics Factors and Formulas

To Convert	Symbol	Formula
Present value (P) to future value (F)	($F/P, i\%, n$)	$(1+i)^n$
Future value (F) to present value (P)	($P/F, i\%, n$)	$(1+i)^{-n}$
Future value (F) to annual value (A)	($A/F, i\%, n$)	$\dfrac{i}{(1+i)^n - 1}$
Present value (P) to annual value (A)	($A/P, i\%, n$)	$\dfrac{i(1+i)^n}{(1+i)^n - 1}$
Annual value (A) to future value (F)	($F/A, i\%, n$)	$\dfrac{(1+i)^n - 1}{i}$
Annual value (A) to present value (P)	($P/A, i\%, n$)	$\dfrac{(1+i)^n - 1}{i(1+i)^n}$
Gradient value (G) to present value (P)	($P/G, i\%, n$)	$\dfrac{(1+i)^n - 1}{i^2(1+i)^n} - \dfrac{n}{i(1+i)^n}$
Gradient value (G) to future value (F)	($F/G, i\%, n$)	$\dfrac{(1+i)^n - 1}{i^2} - \dfrac{n}{i}$
Gradient value (G) to annual value (A)	($A/G, i\%, n$)	$\dfrac{1}{i} - \dfrac{n}{(1+i)^n - 1}$

TABLE 9.2

Spreadsheet Functions for Engineering Economics Problem

Name	Excel Format	Symbol
Present value (PV)	PV (rate, nper, pmt, FV)	$PV(i\%, n, A, F)$
Future value (FV)	FV (rate, nper, pmt, PV)	$FV(i\%, n, A, P)$
Annual value	PMT (rate, nper, PV, FV)	$PMT(i\%, n, P, F)$
Number of periods	NPER (rate, pmt, PV, NV)	$NPER(i\%, A, P, F)$
Interest rate	RATE (nper, pmt, PV, NV)	$RATE(n, A, P, F)$

3. *Interest factor tables:* Interest factor tables are available for different interest rates and can be used to solve engineering economics problems. This section provides three different interest factor tables (5%, 7%, and 10%) as an example to show how they can be used to solve problems. A step-by-step procedure to use interest factor tables is described as follows:

   a. From the problem statement, determine the interest rate, i.
   b. Determine the number of payment periods, n. The PP is sometimes mistaken as the annual period. For example, a car payment made monthly for four years has 48 payment periods and therefore, $n = 48$ months. If $n = 4$ is used as the number of payment periods, that answer will be incorrect. This is a common mistake observed in engineering economics problems.
   c. From Table 9.1 and the problem statement, determine what factor to use in the interest table. For example, if P is given in the problem and requires calculation of F, the factor (F/P) should be used as shown in row 1 of Table 9.1.
   d. For example, if $i = 5\%$ and $n = 6$, then $F/P = 1.3401$ as highlighted in Table 9.3 (see also Tables 9.4 and 9.5).

Example 9.1

Siwen Zhao, a mechanical engineering student at the University of Michigan–Flint, deposited $200 at the end of every year for seven years in an account that earned 5% annual effective interest. At the end of seven years, how much money will Siwen have in her account? Use (a) analytical, (b) computational, and (c) interest factor table methods.

Solution:

(a) Analytical methods: Use row 5 in Table 9.1:

$$F = (\$200)\left(\frac{F}{A}, 5\%, 7\right)$$

$$= (\$200)\frac{(1+0.05)^7 - 1}{0.05}$$

$$= (\$200)(8.1420) = \$1628.40$$

Engineering Economics

TABLE 9.3

Interest Factor Table for $i = 5\%$

	\multicolumn{8}{c	}{Interest Rate 5%}
	\multicolumn{2}{c	}{Single Payment}	\multicolumn{4}{c	}{Uniform Series Payments}	\multicolumn{2}{c	}{Arithmetic Gradients}
N	F/P	P/F	A/F	F/A	A/P	P/A	P/G	A/G
1	1.0500	0.9524	1.0000	1.0000	1.05000	0.9524		
2	1.1025	0.9070	0.48780	2.0500	0.53780	1.8594	0.9070	0.4878
3	1.1576	0.8638	0.31721	3.1525	0.36721	2.7232	2.6347	0.9675
4	1.2155	0.8227	0.23201	4.3101	0.28201	3.5460	5.1028	1.4391
5	1.2763	0.7835	0.18097	5.5256	0.23097	4.3295	8.2369	1.9025
6	*1.3401*	0.7462	0.14702	6.8019	0.19702	5.0757	11.9680	2.3579
7	1.4071	0.7107	0.12282	8.1420	0.17282	5.7864	16.2321	2.8052
8	1.4775	0.6768	0.10472	9.5491	0.15472	6.4632	20.9700	3.2445
9	1.5513	0.6446	0.09069	11.0266	0.14069	7.1078	26.1268	3.6758
10	1.6289	0.6139	0.07950	12.5779	0.12950	7.7217	31.6520	4.0991
11	1.7103	0.5847	0.07039	14.2068	0.12039	8.3064	37.4988	4.5144
12	1.7959	0.5568	0.06283	15.9171	0.11283	8.8633	43.6241	4.9219
13	1.8856	0.5303	0.05646	17.7130	0.10646	9.3936	49.9879	5.3215
14	1.9799	0.5051	0.05102	19.5986	0.10102	9.8986	56.5538	5.7133
15	2.0789	0.4810	0.04634	21.5786	0.09634	10.3797	63.2880	6.0973
16	2.1829	0.4581	0.04227	23.6575	0.09227	10.8378	70.1597	6.4736
17	2.2920	0.4363	0.03870	25.8404	0.08870	11.2741	77.1405	6.8423
18	2.4066	0.4155	0.03555	28.1324	0.08555	11.6896	84.2043	7.2034
19	2.5270	0.3957	0.03275	30.5390	0.08275	12.0853	91.3275	7.5569
20	2.6533	0.3769	0.03024	33.0660	0.08024	12.4622	98.4884	7.9030

TABLE 9.4

Interest Factor Table for $i = 7\%$

	\multicolumn{8}{c	}{Interest Rate 7%}
	\multicolumn{2}{c	}{Single Payment}	\multicolumn{4}{c	}{Uniform Series Payments}	\multicolumn{2}{c	}{Arithmetic Gradients}
N	F/P	P/F	A/F	F/A	A/P	P/A	P/G	A/G
1	1.0700	0.9346	1.00000	1.0000	1.07000	0.9346		
2	1.1449	0.8734	0.48309	2.0700	0.55309	1.8080	0.8734	0.4831
3	1.2250	0.8163	0.31105	3.2149	0.38105	2.6243	2.5060	0.9549
4	1.3108	0.7629	0.22523	4.4399	0.29523	3.3872	4.7947	1.4155
5	1.4026	0.7130	0.17389	5.7507	0.24389	4.1002	7.6467	1.8650
6	1.5007	0.6663	0.13980	7.1533	0.20980	4.7665	10.9784	2.3032
7	1.6058	0.6227	0.11555	8.6540	0.18555	5.3893	14.7149	2.7304
8	1.7182	0.5820	0.09747	10.2598	0.16747	5.9713	18.7889	3.1465
9	1.8385	0.5439	0.08349	11.9780	0.15349	6.5152	23.1404	3.5517
10	1.9672	0.5083	0.07238	13.8164	0.14238	7.0236	27.7156	3.9461
11	2.1049	0.4751	0.06336	15.7836	0.13336	7.4987	32.4665	4.3296
12	2.2522	0.4440	0.05590	17.8885	0.12590	7.9427	37.3506	4.7025
13	2.4098	0.4150	0.04965	20.1406	0.11965	8.3577	42.3302	5.0648
14	2.5785	0.3878	0.04434	22.5505	0.11434	8.7455	47.3718	5.4167

(*Continued*)

TABLE 9.4 (*Continued*)

Interest Factor Table for $i = 7\%$

				Interest Rate 7%				
	Single Payment		Uniform Series Payments				Arithmetic Gradients	
N	F/P	P/F	A/F	F/A	A/P	P/A	P/G	A/G
15	2.7590	0.3624	0.03979	25.1290	0.10979	9.1079	52.4461	5.7583
16	2.9522	0.3387	0.03586	27.8881	0.10586	9.4466	57.5271	6.0897
17	3.1588	0.3166	0.03243	30.8402	0.10243	9.7632	62.5923	6.4110
18	3.3799	0.2959	0.02941	33.9990	0.09941	10.0591	67.6219	6.7225
19	3.6165	0.2765	0.02675	37.3790	0.09675	10.3356	72.5991	7.0242
20	3.8697	0.2584	0.02439	40.9955	0.09439	10.5940	77.5091	7.3163

TABLE 9.5

Interest Factor Table for $i = 10\%$

				Interest Rate 10%				
	Single Payment		Uniform Series Payments				Arithmetic Gradients	
N	F/P	P/F	A/F	F/A	A/P	P/A	P/G	A/G
1	1.1000	0.9091	1.00000	1.0000	1.10000	0.9091		
2	1.2100	0.8264	0.47619	2.1000	0.57619	1.7355	0.8264	0.4762
3	1.3310	0.7513	0.30211	3.3100	0.40211	2.4869	2.3291	0.9366
4	1.4641	0.6830	0.21547	4.6410	0.31547	3.1699	4.3781	1.3812
5	1.6105	0.6209	0.16380	6.1051	0.26380	3.7908	6.8618	1.8101
6	1.7716	0.5645	0.12961	7.7156	0.22961	4.3553	9.6842	2.2236
7	1.9487	0.5132	0.10541	9.4872	0.20541	4.8684	12.7631	2.6216
8	2.1436	0.4665	0.08744	11.4359	0.18744	5.3349	16.0287	3.0045
9	2.3579	0.4241	0.07364	13.5795	0.17364	5.7590	19.4215	3.3724
10	2.5937	0.3855	0.06275	15.9374	0.16275	6.1446	22.8913	3.7255
11	2.8531	0.3505	0.05396	18.5312	0.15396	6.4951	26.3963	4.0641
12	3.1384	0.3186	0.04676	21.3843	0.14676	6.8137	29.9012	4.3884
13	3.4523	0.2897	0.04078	24.5227	0.14078	7.1034	33.3772	4.6988
14	3.7975	0.2633	0.03575	27.9750	0.13575	7.3667	36.8005	4.9955
15	4.1772	0.2394	0.03147	31.7725	0.13147	7.6061	40.1520	5.2789
16	4.5950	0.2176	0.02782	35.9497	0.12782	7.8237	43.4164	5.5493
17	5.0545	0.1978	0.02466	40.5447	0.12466	8.0216	46.5819	5.8071
18	5.5599	0.1799	0.02193	45.5992	0.12193	8.2014	49.6395	6.0526
19	6.1159	0.1635	0.01955	51.1591	0.11955	8.3649	52.5827	6.2861
20	6.7275	0.1486	0.01746	57.2750	0.11746	8.5136	55.4069	6.5081

(b) Computational method: Use Microsoft Excel
1. Select financial functions from the main menu.
2. Select the function FV from the financial functions.
3. An input box will appear on the screen requiring interest rate, number of payment periods (nper), and present value. Enter the values in the table.
4. Move to cell A1 and type = FV(5%, 7, 200, 0).
5. <Enter>.
6. Excel will show results of the function as $1628.40.

(c) Interest factor table method:
   1. Select Table 9.3 for the interest rate of 5%.
   2. Select $n = 7$ in row 1.
   3. Move to the right of the row with $n = 7$ to the fifth column (F/A column).
   4. Read the value in the intersecting cell of $n = 7$ and F/A. The value in the cell is 8.1420.
   5. Multiply the annual payment of 200 by the above factor of 8.1420 ($F = A \times F/A$):

$$\$200 \times 8.1420 = \$1628.4$$

The above examples demonstrate how all three methods provide similar solutions.

9.2.2 Effective Interest Rates for Annual and Any Time Period

The future worth (FW) F at the end of one year is the principal P plus the interest $P(i)$ through the year. The effective annual rate symbol i_a is used to write the relation for F with $P = \$1$:

$$F = P + Pi_a = 1(1 + i_a) \tag{9.3}$$

The effective rate i per CP must be compounded through all m periods to obtain the total effect of compounding by the end of the year. To determine monthly interest rate if annual interest rate is given, the annual rate must be divided by the number of months. For example, if the annual interest rate is 8%, the monthly interest rate can be calculated as $i = r/m = 8/12 = 0.75\%$. This means that F can also be written as $F = 1(1+i)^m$. Setting the two equations equal and solving for i_a yields the effective annual interest rate:

$$i_a = (1+i)^m - 1 \tag{9.4}$$

This calculates the effective interest rate per year from any effective rate over a shorter period of time. The equation can be generalized to determine the effective interest rate for any time period:

$$\text{Effective } i \text{ per time period} = \left(1 + \frac{r}{m}\right)^m - 1 \tag{9.5}$$

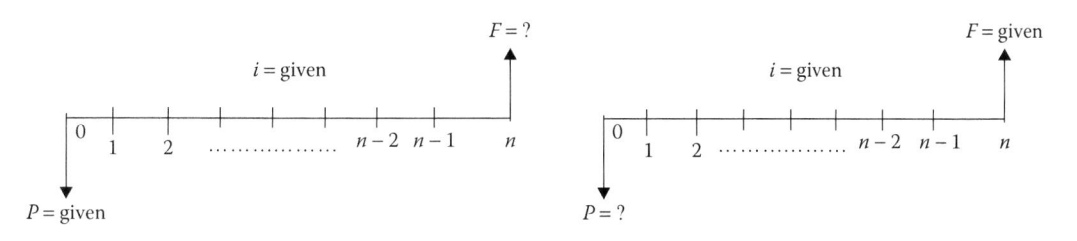

The effective interest rate over a CP is denoted by the term r/m, where m is the number of times that interest is compounded per time period. This generalization uses the symbol i instead of i_a to represent the effective interest rate.

Example 9.2

If a credit union pays 5.125% interest compounded quarterly (every three months), what is the effective annual interest rate?

Solution:

$r = 5.125\%$ is the nominal annual rate
   CP (how many times interest is compounded in a year), $m = 4$
   Use Equation 9.5:

$$i_a = (1+i)^m - 1 = \left(1 + \frac{0.05125}{4}\right)^4 - 1 = 0.05224$$

Example 9.3

A new engineering graduate contributes $300 per month to a 401(k) retirement account. The account earns interest at a nominal annual interest rate of 7%, with interest being credited monthly. What is the value of account after 35 years?

Solution:

$A = \$300$ per month

This problem requires knowledge of PP and compounding periods in financial transactions. PP is monthly, and therefore, both interest rate and number of payment periods must be converted to months. To convert interest rate from yearly to monthly, interest rate, r, needs to be divided by 12. The number of payment periods is found as $n = 35$ years $\times 12 = 420$ months.

Use Equation 9.2 to calculate the effective rate per month:

$$i = \frac{r}{m} = \frac{0.07}{12} = 0.00583$$

Because compounding is monthly, m is the total number of months:

$$m = (35\,\text{years})\left(12\,\frac{\text{month}}{\text{year}}\right) = 420\,\text{months}$$

Use Equation 9.13:

$$F = A\left[\frac{(1+i)^m - 1}{i}\right] = 300\left[\frac{(1+0.00583)^{420} - 1}{0.00583}\right] = \$539{,}802$$

9.3 Effect of Time and Interest on Money

Cash flow is central to economic analyses and may be viewed similar to a free body diagram in statics when analyzing forces and moments. Cash flows occur in numerous structures and quantities, including isolated single values, uniform series, and increasing or decreasing series by constant amounts or constant percentages. This section covers the derivations for the most commonly used engineering economy factors that take into account the time value of money.

9.3.1 Cash Flow Diagram

Because of the time value of money, the timing of cash flow over the life of a project is an important factor. Cash flow diagrams are an important part of the solution as it helps

Engineering Economics

visualization of financial transactions. Cash flow diagrams can be compared to *free body diagrams* used in solving statics or static equilibrium of forces problems. Solution steps in engineering economics problems require drawing an accurate cash flow diagram as the first step. The following guidelines can be used to draw a cash flow diagram. In a cash flow diagram, the horizontal axis (time) is marked as increments of period, up to the duration of the project. Cash received during any period is shown by an upward arrow and cash spent is shown as a downward arrow with arrow lengths proportional to the amount of cash flow. Two or more transactions in the same period are shown by upward and downward arrows:

Step 1: From the problem statement, determine the information provided and information required.

Step 2: List the given and required information as variables. The most commonly used variables are present value (*P*), future value (*F*), periodic payment (*A*), interest rate (*i*), number of payment periods (*n*), gradient (*g*), and so on.

Step 3: Draw the cash flow diagram showing the information about the variables that are provided and variable that is required.

Step 4: In a cash flow diagram, the following general rules must be followed unless otherwise stated in the problem:

    a. Present value (*P*) is at time zero (*t* = 0).
    b. Periodic payment (*A*) is at time (*t* = 1).
    c. Future value (*F*) is at the end of the last period of the transaction.
    d. Gradient (*G*) is one period after *A*.
    e. *P* is one period before *A* or *A* is one period after *P*.
    f. The last payment (*A*) and future value (*F*) both occur at the last period.

9.3.2 Single Amount Factors (*F/P* and *P/F*)

The fundamental factor converts present value (single amount of money invested now) to a future amount, *F*, the amount of money accumulated after *n* years from a single present worth (PW), *P*, with interest compounded one time per period. With an amount *P* invested at time *t* = 0, the amount F_1 accumulated after one year at the interest rate of *i* percent per year is

$$F_1 = P + P_i = P(1+i) \tag{9.6}$$

At the end of the second period, the amount accumulated is the amount after the first year plus the interest from the end of the first year to the second year over the entire F_1:

$$F_2 = F_1 + F_1 i = P(1+i) + P(1+i)i$$
$$= P(1+i+i+i^2) = P(1+i)^2 \tag{9.7}$$

The generalized formula to find *F* when given *P* is

$$F = P(1+i)^n \tag{9.8}$$

Conversely, the formula for finding P when given F is

$$P = F(1+i)^{-n} \tag{9.9}$$

Example 9.4

If the interest rate on an account is 10.5% compounded yearly, approximately how many years will it take to triple the amount (three times)?

Solution:

Draw a cash flow diagram.
   The future amount will be three times the present amount when the (F/P) factor is equal to 3.
   Use Equation 9.8:

$$F = P(1+i)^n$$

$$\frac{F}{P} = (1+0.105)^n$$

$$n \log 1.105 = \log 3$$

$$n = 11 \text{ years}$$

Example 9.5

Fifteen years ago $1000 was deposited in a bank account, and today it is worth $2500. The bank pays interest semiannually (every six months). What was the effective rate per period?

Solution:

Draw a cash flow diagram with $P = 1000$, $F = 2500$, and $t = 15$ years:

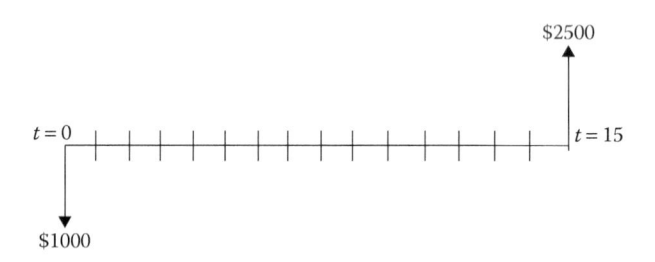

$$P = \$1000$$

$$n = (15 \text{ years})(2 \text{ compounding periods per year}) = 30 \text{ compounding periods}$$

Use Equation 9.8:

$$F = P(1+i)^n$$

$$\frac{F}{P} = (1+i)^n$$

$$\frac{2500}{1000} = (1+i)^{30}$$

$$i = 0.031$$

Engineering Economics 241

9.3.3 Uniform Series PW Factor and Capital Recovery Factor (*P/A* and *A/P*)

An expression for the PW can be derived by considering each *A* value as a FW *F* and calculating each PW and then summing the results. The equation to calculate the equivalent *P* value in year zero for uniform end-of-period series of *A* values beginning at period 1 and ending at period *n* is

$$P = A\left[\frac{1}{(1+i)^1}\right] + A\left[\frac{1}{(1+i)^2}\right] + A\left[\frac{1}{(1+i)^3}\right] + \cdots + A\left[\frac{1}{(1+i)^{n-1}}\right] + A\left[\frac{1}{(1+i)^n}\right]$$

$$P = A\left[\frac{1}{(1+i)^1} + \frac{1}{(1+i)^2} + \frac{1}{(1+i)^3} + \cdots + \frac{1}{(1+i)^{n-1}} + \frac{1}{(1+i)^n}\right]$$

$$\frac{P}{1+i} = A\left[\frac{1}{(1+i)^2} + \frac{1}{(1+i)^3} + \frac{1}{(1+i)^4} + \cdots + \frac{1}{(1+i)^n} + \frac{1}{(1+i)^{n+1}}\right]$$

$$\frac{1}{1+i}P = A\left[\frac{1}{(1+i)^2} + \frac{1}{(1+i)^3} + \frac{1}{(1+i)^4} + \cdots + \frac{1}{(1+i)^n} + \frac{1}{(1+i)^{n+1}}\right] \quad (9.10)$$

$$P = A\left[\frac{1}{(1+i)^1} + \frac{1}{(1+i)^2} + \frac{1}{(1+i)^3} + \cdots + \frac{1}{(1+i)^{n-1}} + \frac{1}{(1+i)^n}\right]$$

$$\frac{-i}{1+i}P = A\left[\frac{1}{(1+i)^{n+1}} - \frac{1}{(1+i)^1}\right]$$

$$P = \frac{A}{-i}\left[\frac{1}{(1+i)^n} - 1\right]$$

$$P = A\left[\frac{(1+i)^n - 1}{i(1+i)^n}\right] \quad i \neq 0$$

When P is given and A needs to be determined, the reverse equation is

$$A = P\left[\frac{i(1+i)^n}{(1+i)^n - 1}\right] \tag{9.11}$$

Example 9.6

A new engineering graduate decided to buy a car with a purchase price of $25,000. She saved money to make a down payment of $5000 and borrowed the remaining balance from a bank at 5% annual interest, compounded monthly for five years. Calculate the nearest value of the required monthly payments to pay off the loan.

Solution:

Draw a cash flow diagram of the problem:

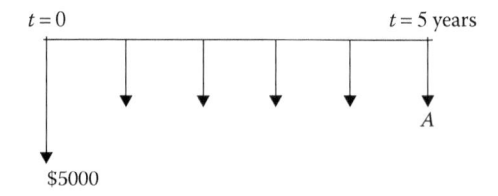

$$P = \$25,000 - \$5,000 = \$20,000$$

$$i = \frac{5\%}{12 \text{ compounding periods per year}} = 0.417\%$$

$$n = (5 \text{ years}) \times (12 \text{ months per year}) = 60$$

Use Equation 9.11 to compute the annual value:

$$A = P\left[\frac{i(1+i)^n}{(1+i)^n - 1}\right] = 20,000\left[\frac{0.00417(1+0.00417)^{60}}{(1+0.00417)^{60} - 1}\right] = 20,000(0.01887) = \$377.46$$

9.3.4 Sinking Fund Factor and Uniform Series Compound Amount Factor (A/F and F/A)

If P from Equation 9.3 is substituted into Equation 9.9, the following formula results:

$$A = F\left[\frac{1}{(1+i)^n}\right]\left[\frac{i(1+i)^n}{(1+i)^n - 1}\right]$$

$$A = F\left[\frac{i}{(1+i)^n - 1}\right] \tag{9.12}$$

The above equation is the A/F or sinking fund factor. It determines the uniform annual series A that is equivalent to a given future amount F.

Engineering Economics 243

The above equation can be rearranged to find F for a stated A series in periods 1 through n:

$$F = A\left[\frac{(1+i)^n - 1}{i}\right] \tag{9.13}$$

The term in brackets is called the uniform series compound amount factor, or the F/A factor. When multiplied by the given uniform annual amount, A, it yields the FW of the uniform series:

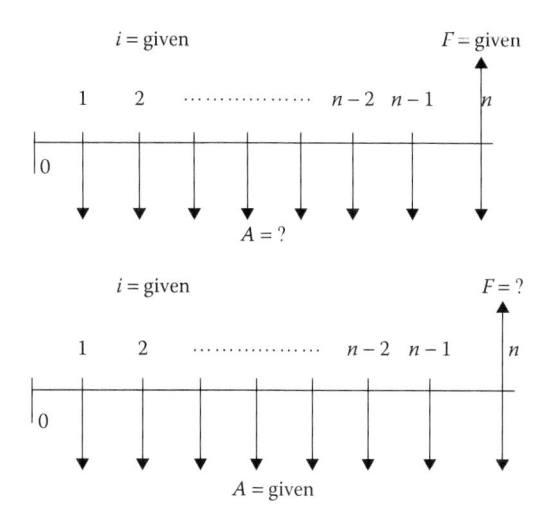

Example 9.7

A civil engineer plans to save $500 each month in a savings account for 10 years at 10% annual interest, compounded monthly. The engineer had a family emergency after two years when she needed to withdraw money from the account. How much money will be available after two years?

Solution:

Draw a cash flow diagram. In the diagram, periodic payment, A, starts in period one, while last payment (A) and future value (F) are in the same period $t = 24$:

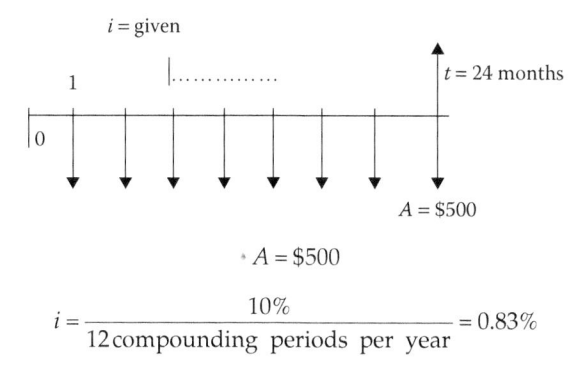

$$i = \frac{10\%}{12 \text{ compounding periods per year}} = 0.83\%$$

$$n = (2 \text{ years}) \times (12 \text{ compounding periods per year}) = 24 \text{ compounding periods}$$

Use Equation 9.13:

$$F = A\left[\frac{(1+i)^n - 1}{i}\right] = \$500\left[\frac{(1+0.0083)^{24} - 1}{0.0083}\right] = \$13218.26$$

9.3.5 Arithmetic Gradient Factors (P/G and A/G)

In some instances, cash flow is nonuniform as the payment increases or decreases from one period to the next. When the change of amount is constant, it is defined as arithmetic gradient. The amount of change is called the gradient. The cash flow at the end of period one is the base amount of the cash flow series. Define the symbols G for gradient and CF_n for cash flow in year n as follows.

G is the constant arithmetic change in cash flow 3 from one time period to the next:

$$CF_n = \text{base amount} + (n-1)G \tag{9.14}$$

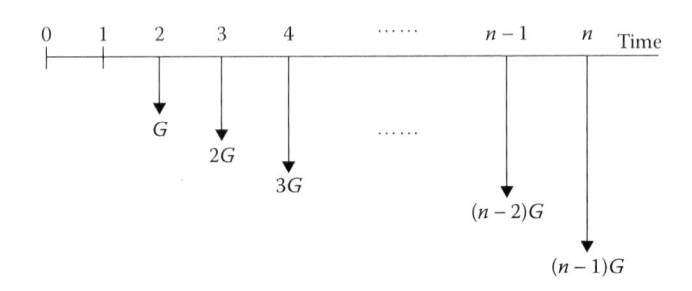

Conventional gradient is the gradient beginning between years 1 and 2. The total PW P_T is the addition of the PW of the uniform series and the present of the gradient series only:

$$P_T = P_A \mp P_G \tag{9.15}$$

The corresponding equivalent AW A_T is the total amount of the base amount series and gradient series AW:

$$A_T = A_A \mp A_G \tag{9.16}$$

The P/G factor is used for PW, the A/G factor for annual series, and the F/G factor for FW. The PW at year 0 of only the gradient is equal to the sum of the PW of the individual cash flows, where each value is considered a future amount. The present value of the gradient series equation is shown as follows:

$$P_G = \frac{G}{i}\left[\frac{(1+i)^n - 1}{i(1+i)^n} - \frac{n}{(1+i)^n}\right] \tag{9.17}$$

The gradient series AW equation is as follows:

$$A_G = G\left[\frac{1}{i} - \frac{n}{(1+i)^n - 1}\right] \tag{9.18}$$

Engineering Economics

9.3.6 Geometric Gradient Series Factors

A geometric gradient series is a cash flow series that either increases or decreases by a constant percentage each period. The uniform change is called the rate of change.

The relation to determine the total PW P_g for the entire cash flow series may be derived by multiplying each cash flow in the below figure (a) by the P/F factor, and the geometric gradient series PW factor for values of g not equal to the interest rate i:

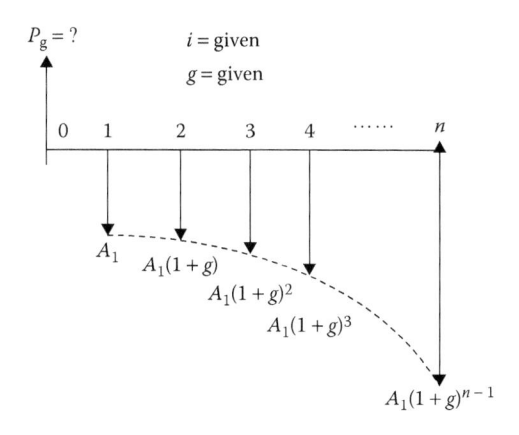

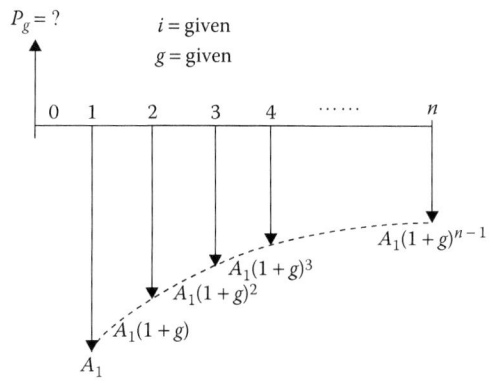

The equation for P_g and the geometric gradient series PW factor $(P/A, g, i, n)$ formula are as follows:

$$P_g = A_1(P/A, g, i, n)$$

$$(P/A, g, i, n) = \begin{cases} \dfrac{1-\left[(1+g)/(1+i)\right]^n}{i-g} & g \neq i \\ \dfrac{n}{1+i} & g = i \end{cases} \quad (9.19)$$

Example 9.8

The maintenance cost for a car this year is expected to be $500. The cost will increase $50 each year for the subsequent nine years. The interest is 8% compounded annually. What is the approximate PW of maintenance for the car for this time?

Solution:

To calculate the PW of the maintenance costs of nine years, both the uniform gradient and the uniform series factors must be used:

$$P = A(P/A, 8\%, 10) + G(P/G, 8\%, 10)$$

$$= 500(6.7101) + 50(25.9768) = \$4654$$

9.4 PW, AW, and FW Analysis

The evaluation and selection of economic proposals require cash flow estimates over a stated period of time, mathematical techniques to calculate the measure of worth, and a guideline for selecting the best proposal. For terminology purposes, each viable proposal is called an alternative.

9.4.1 PW Analysis

PW comparison of alternatives with equal lives is straightforward. The present worth P is renamed PW of the alternative. If the alternatives have the same capacities for the same time period (life), the equal-service requirement is met. Calculate the PW value at the stated minimum attractive rate of return (MARR) for each alternative.

The PW of the alternatives must be compared over the same number of years and must end at the same time to satisfy the equal-service requirement. For cost alternatives, failure to compare equal service will always favor the shorter-lived mutually exclusive alternative, even if it is not the more economical choice, because fewer periods of costs are involved:

$$\text{PW} = P = \text{PV}$$

9.4.2 FW Analysis

FW of an alternative may be determined directly from the cash flows, or by multiplying the PW value by the FP factor, at the established MARR. Analysis of alternatives using FW values is especially applicable to large capital investment decisions:

$$\text{FW} = F = \text{FV}$$

9.4.3 AW Analysis

The AW method offers a prime computational and interpretation advantage because the AW value needs to be calculated for only one life cycle. The AW value determined over one life cycle is the AW for all future life cycles:

$$\text{AW} = A = \text{PMT}$$

9.5 Rate of Return Analysis

The rate of return (ROR) is the rate paid on the unpaid balance of borrowed money, or the rate earned on the unrecovered balance of an investment so that the final payment or receipt brings the balance to exactly zero with interest considered.

The ROR value is determined in a generically different way compared to the PW or AW values for a series of cash flows, so we can define the ROR as the interest rate that makes the PW or AW of a cash flow series exactly equal to 0. This can make cash outflows for the present value equal to cash inflows:

$$0 = PW$$

$$PW_0 = PW_1$$

Because it is the same for AW, then

$$0 = AW$$

$$AW_0 = AW_1$$

9.6 Decision Making

Decision making is the key learning outcome of this chapter as engineers are often required to make economic decisions to select alternative projects, products, and services. The following is a list of engineering economics examples:

1. To acquire engineering analysis (computational fluid dynamics or finite element analysis) software, economic analysis is required to compare price, terms, and purchase or lease decisions.

2. When a new product is designed and developed, manufacturing cost, material cost, and labor cost must be analyzed to determine whether to make the components in house or outsource them.

3. To improve manufacturing efficiency through automation, initial cost, maintenance cost, and life of different equipment, systems need to be analyzed using engineering economic analysis.

Example 9.9

Consider the purchase of two vehicles. One is a 2015 Toyota Camry and the other is a 2015 Honda Civic Sedan. Assume that the annual interest rate is 10%.

	Toyota Camry	Honda Civic Sedan
Initial cost	$23,795	$20,110
Annual fuel cost	$1000	$700
Salvage value	$18,995 after 3 years	$15,000 after 3 years

Using the equivalent uniform annual cost method, decide which vehicle should be purchased.

Solution:

This problem requires converting all transactions to uniform annual cost (A).

a. Convert initial cost (P) of $23,795 to A using analytical methods:

$$A(1) = P\left[\frac{i(1+i)^n}{(1+i)^n - 1}\right] = \$23,795\left[\frac{0.1(1+0.1)^3}{(1+0.1)^3 - 1}\right] = \$9568.32$$

b. The annual fuel cost of $1000 per year is provided in the problem as the uniform annual cost (A), and therefore, no conversion is needed; A(2) = 1000.
c. Convert the salvage value (F) of $18,995 to uniform annual cost (A):

$$A(3) = F\left[\frac{i}{(1+i)^n - 1}\right] = \$18,995\left[\frac{0.1}{(1+0.1)^3 - 1}\right] = (-)\$5738.67$$

The salvage value is an earning or positive cash flow. Therefore, A(3) is deducted from A(2) and A(1).

Total uniform annual cost:

$$A = A(1) + A(2) + A(3)$$
$$= \$9568.32 + \$1000 - \$5738.67 = \$4829.65$$

For Honda Civic Sedan:

a. Convert P to A:

$$A(1) = P\left[\frac{i(1+i)^n}{(1+i)^n - 1}\right] = \$20,110\left[\frac{0.1(1+0.1)^3}{(1+0.1)^3 - 1}\right] = \$8086.53$$

b. Annual fuel cost, A(2) = $700/year
c. Convert salvage value (F) to A:

$$A(3) = F\left[\frac{i}{(1+i)^n - 1}\right] = \$15,000\left[\frac{0.1}{(1+0.1)^3 - 1}\right] = (-)4531.72$$

Total uniform annual cost:

$$A = A(1) + A(2) + A(3)$$
$$= \$8086.53 + \$700 - \$4531.72 = \$4254.81$$

The uniform annual cost of Toyota Camry is 4829.65 compared to $4254.81 of Honda Civic.

Therefore, Honda Civic has lower cost and is an economically better choice (Table 9.6).

Engineering Economics

TABLE 9.6
Commonly Used Symbols and Terms

Term	Symbol	Definition
Annual amount or worth	A or AW	Periodic payment or payment at regular intervals
Annual operating cost	AOC	Annual expenses that are related to the operation of a device or equipment
Cash flow	CF	The movement of money into or out of a business, project, or activity
Compounding frequency	m	Number of compounding periods in a year
Initial cost	P	Total initial cost
Future amount or worth	F or FW	The value of an asset at a specific date in the future
Gradient, arithmetic	G	Constant amount of change between payment periods
Gradient, geometric	G	Constant rate of change (in percent) in each time period
Interest	i	Amount earned or paid based on amount invested or borrowed
Interest rate	i	Interest expressed in percent of the original amount for each time period
Life (estimate)	n	Number of periods or years for which an alternative or asset will be evaluated
Net present value	NPV	Another name for present worth (PW)
Present amount or worth	P or PW	Amount of money at current time. Current time is at $t = 0$
Salvage/market value	S or MV	Expected value of an asset after a time period when the asset is sold or disposed

Problems

Draw a cash flow diagram for each problem and solve using at least two of the three methods described in this chapter:

1. An interest rate of 12% per year, compounded monthly, is equivalent to what effective rate per quarter?

2. You borrow $4500 for one year from a bank at an interest rate of 13% per year. How much money will you pay back?

3. Jane put $3000 into an empty saving account with a nominal interest rate of 6%. No other contributions are made to the account. With monthly compounding, how much interest will have been earned after 10 years?

4. A company puts $30,000 down and will pay $2000 every year for the life of a machine (8 years). If the salvage is zero and the interest rate is 10% compounded annually, what is the present value of the machine?

5. Tom plans to invest all money earned into a saving account earning 8% interest, compounded quarterly. He hopes to have $5000 at the end of 12 months. How much money will he have to save?

6. $10,000 will be needed in 5 years to buy a condo. How much money will you need to deposit into a bank at the end of each year with an interest rate of 10% compounded annually?

7. Calculate the present worth of a geometric gradient series with $10,000 in year 1, which increases by 4% each year through year 5. Use 10% per year as the interest rate.

8. A bonus package pays an employee $2000 at the end of the first year, $2500 at the end of the second year, and so on, for the first 10 years of employment. What is the present worth of the bonus package at 5% interest?

9. An engineer is considering two materials for use in a bend. According to the following table, determine what material should be used based on the present worth comparison.

	Material A	Material B
Initial cost	$20,000	$35,000
Maintenance and operating	$7000/year	$5000/year
Salvage	$4000 after 5 years	$15,000 after 5 years

10. The manager of a manufacturing factory needs to decide between two different machines. Machine A will have an initial cost of $40,000, an annual operating cost of $30,000, and a service life of 4 years. Machine B will cost $50,000 to purchase, an annual operating cost of $18,000 with 4-year life. At an interest rate of 10% per year, which should be selected on the basis of a present worth analysis?

11. An industrial engineer is considering two robots for purchase, Robot 1 and Robot 2. Which should be selected on the basis of a future worth comparison at an interest rate of 10% per year.

	Robot 1	Robot 2
Initial cost	$50,000	$80,000
Maintenance and operating	$10,000/year	$8000/year
Salvage	$25,000 after 3 years	$45,000 after 3 years

12. A manufacturing company wants to evaluate two CNC machines. Which should be selected on the basis of an annual worth comparison at an interest rate of 10% per year?

	Machine X	Machine Y
Initial cost	$300,000	$475,000
Maintenance and operating	$30,000/year	$40,000/year
Salvage	$35,000 after 3 years	$45,000 after 5 years

10
Probability and Statistics

CHAPTER OBJECTIVE AND STUDENT LEARNING OUTCOMES

After completing this chapter, students will be able to

1. Define probability and statistics with examples.
2. Apply probability theorems in real-world applications. [ABET outcome a, see Appendix C]
3. Solve problems using probability theorems. [ABET outcome e, see Appendix C]
4. Perform statistical analysis using standard deviation and normal distribution. [ABET outcome e, see Appendix C]

10.1 Introduction

Why are probability and statistics important for engineers to know? This is because engineers are required to determine the reliability of a product and predict the occurrence of failure.

Our quality of life depends on statistics. Every day we come across circumstances where we have to make choices. These choices depend on how well we can assess the situation based on our or others' past experience. In manufacturing, statistics is used to determine whether or not products meet customers' expectations. It is not the intent of the authors to present statistics in a deep fashion, but rather basic concepts of probability and statistical values are most applicable in quality. Statistics is used in quality engineering to determine how many parts are within tolerance, what is the mean dimension of a part and variation from the mean dimension, etc.

If things were done right just 99.9% of the time, then we would have to accept

- One hour of unsafe drinking water per month
- Two unsafe plane landings per day at O'Hare International Airport in Chicago
- 16,000 pieces of mail lost by the US postal service every hour
- 20,000 incorrect drug prescriptions per year
- 500 incorrect surgical operations each week
- 22,000 checks deducted from the wrong bank accounts per hour
- 32,000 missed heartbeats per person per year

10.2 Probability

We are often surprised when we meet people who share the same birthday with us. This example can be expanded to pretty much anything: brand or color of the car, shoe size or the height of the person, and so on. In fact, as the number of people we meet increases, the chance of two people in a group having the same birthday increases dramatically. Try verifying this by asking your classmates, and then extend the sample size to all freshmen level students.

Probability is the likelihood of something happening. In other words, it is the chance that an event will take place. For example, the probability of a coin toss resulting in either *heads* or *tails* is 1 (100%), because there are no other options, assuming that the coin lands flat.

10.2.1 Probability Theorems

Probability affects all aspects of our lives. Intuitively we understand things like the chance it will snow, the probability of there being slow traffic on the commute to school or work, or the chance that our favorite player will be injured during the match. Probability is the chance that something will happen. However, having a probability attached to an event does not mean that the event will definitely take place.

Back to the coin tossing example: a fair coin is tossed and the winner is the person who correctly predicts on which face the coin will land. If we toss the coin number of times, we will start capturing a pattern: approximately 50% of the time the coin lands on heads and another 50% of the time it lands on tails, resulting in 100% certainty that the coin will land either on heads or tails.

Probability plays an important role in industry, be it a manufacturing or service industry. There is a chance that a tool will break, the assembly line will stop, the equipment will go out of order, a person will be late for work, or the plane will take off with delay due to a long line at the counters.

The probability of an occurrence is written as $P(A)$ and is equal to

$$P(A) = \frac{\text{number of occurences}}{\text{total number of possibilities}} = \frac{s}{n}$$

Example 10.1: Determining Probabilities

$$P(\text{opening a 600-page book to page 221}) = \frac{1}{600}$$

There is only one successful case possible to open page 221 out of 600 total pages.

$$P(\text{heads}) = \frac{\text{coin lands head side up}}{\text{two sides of a coin}} = \frac{1}{2}$$

Or consider a shipping container with 500 computers being loaded to a ship, with seven televisions mixed in the same computer container. During the delivery, the probability of randomly selecting one of the TVs is

$$P(\text{selecting a TV}) = \frac{7}{500}$$

Or 166 students enrolled to an Industrial Engineering program in Fall 2016, of whom 57 are female students. What is the probability of meeting one female student at the departmental orientation for engineering students?

$$P(\text{of meeting a female engineering student}) = \frac{57}{166}$$

Theorem 10.1: Probability Is Expressed as a Number between 0 and 1: $0 \leq P(A) \leq 1$

If a probability of an event happening is 1, then there is a certainty that it will happen (100%). On the other hand, if the probability of an event occurring is 0, then there is a certainty that it will never occur. In between the certainties that an event will definitely certainly occur or not occur exist probabilities defined by their ratios of desired occurrences to the total number of occurrences, as seen in Example 10.1.

Theorem 10.2: The Sum of the Probabilities of Events in a Situation Is Equal to 1.00: $\sum Pi = P(A) + P(B) + \cdots + P(N) = 1.00$

Example 10.2: The Sum of the Probabilities

A manufacturer of oil filters receives raw materials from three different suppliers. The stockroom currently contains 20 filters from supplier A, 30 filters from supplier B, and another 50 filters from supplier C. From these 100 filters in the inventory, a worker will encounter the following probabilities in selecting a filter for the next job:

P(filter from supplier A) = 20/100 = 0.20 or 20%
P(filter from supplier B) = 30/100 = 0.30 or 30%
P(filter from supplier C) = 50/100 = 0.50 or 50%

According to Theorem 10.2:
P(filter from supplier A) + P(filter from supplier B) + P(filter from supplier C) = 0.20 + 0.30 + 0.50 = 1.00

Theorem 10.3: If $P(A)$ Is the Probability That Event A Will Occur, Then the Probability That Event A Will Not Occur Is $P(A') = 1 - P(A)$

Example 10.3: Determining the Probability That an Event Will Not Occur

Currently, the stockroom contains oil filters from only suppliers A and B. There are 20 filters from supplier A and 30 filters from supplier B. If the filter selected was from supplier A, what is the probability that a filter from supplier B was not selected?

$$P(A') = 1 - P(A)$$

where

$$P(A) = P(\text{selecting a filter from supplier A}) = 0.40$$

Then

$$P(A') = P(\text{selecting a filter from supplier B}) = 1 - 0.40 = 0.60$$

If the events cannot occur simultaneously, they are considered *mutually exclusive*. One event occurs and that prevents the other event from happening, therefore mutually exclusive events can happen only one at a time. An example of mutually exclusive events is rolling a die and getting a 6 because this prevents all other values from happening.

Theorem 10.4: For Mutually Exclusive Events, the Probability That Either Event A or Event B Will Occur Is the Sum of Their Respective Probabilities: $P(A \text{ or } B) = P(A) + P(B)$

This theorem is also called *the additive law of probability*. The *or* in the probability is represented as the "+" sign.

Example 10.4: Probability of Mutually Exclusive Events I

At an engineering orientation you and another fellow student are tied in a competition for the door prize. The organizers announced that each of you should select two values on a single die. If one of your values comes up, as an outcome from die rolling, then you will be the winner. You select 3 and 5 as your numbers. What is the probability that either a 3 or a 5 will appear on the die when it is rolled?

$P(\text{rolling a 3 or a 5 on a die}) = P(3) + P(5)$
$P(3) = 1/6, P(5) = 1/6$
$P(\text{rolling a 3 or a 5 on a die}) = 1/6 + 1/6 = 1/3$

Example 10.5: Probability in Mutually Exclusive Events II

At the oil filter manufacturing plant in Example 10.2, a worker visits the inventory. If 20% of the filters comes from supplier A, 30% from supplier B, and 50% from supplier C, what is the probability that the worker will randomly select a filter from either supplier A or supplier C?

$P(\text{filter from supplier A}) = 0.20$
$P(\text{filter from supplier B}) = 0.30$
$P(\text{filter from supplier C}) = 0.50$

Because the choice of a filter from supplier A precludes choosing supplier C, these events are mutually exclusive. According to Theorem 10.4, we have

$$P(\text{filter from A or filter from C}) = P(A) + P(C) = 0.20 + 0.50 = 0.70$$

Theorem 10.5: When Events A and B Are Not Mutually Exclusive Events, the Probability That Either Event A or Event B or Both Will Occur Is $P(A \text{ or } B \text{ or both}) = P(A) + P(B) - P(\text{both})$

Example 10.6: Probability in Nonmutually Exclusive Events I

A computer-aided design class at the local university consists of 32 students, 12 female and 20 male. The professor asked, "How many of you are out-of-state students?" Eight of the women and five of the men raised their hands. What is the probability that a student selected at random will be female, out-of-state, or both?

$$P(\text{female or out-of-state}) = P(F) + P(O) - P(\text{both})$$

Calculation shows

$$P(\text{female}) = \frac{12}{32}$$

$$P(\text{out-of-state}) = \frac{13}{32}$$

$$P(\text{both}) = \frac{8}{32}$$

$$P(\text{female or out-of-state}) = \frac{12}{32} + \frac{13}{32} - \frac{8}{32} = \frac{17}{32}$$

If students who are both out-of-state and female are not subtracted, then the total of 25/32 students would be inaccurate. This happens because those people would be counted twice, once for each group.

Example 10.7: Probability in Nonmutually Exclusive Events II

The secretary of the engineering department is trying to determine what routes students take the most. Knowing this will help the secretary post announcements accordingly. The secretary collected the following information:

P(post printed in color) = 0.12
P(post printed in grayscale) = 0.29

Students that look at a color printed post may also look at a grayscale printed announcement. Therefore, these two events are not mutually exclusive, and P(student reads both color and grayscale print) = 0.07.

The secretary would like to know the probability that the student will read either color or grayscale print. According to Theorem 10.5:

P(student looks at color printed) + P(student looks at grayscale printed) − (student looks at both prints) = 0.12 + 0.29 − 0.07 = 0.34

Theorem 10.6: If A and B Are Dependent Events, the Probability That Both A and B Will Occur Is $P(A \text{ and } B) = P(A) \times P(B|A)$

This theorem indicates that the occurrence of event B is dependent on the outcome of event A. This relationship between A and B is denoted by $P(B|A)$. The vertical bar is translated as *given that*. The probability that both A and B will occur is the probability that A will occur multiplied by the probability that B will occur, given that A has already occurred.

Example 10.8: Probability in Dependent Events I

Let us consider the out-of-state student example observed above. What is the probability that the selected student is female and she is also out-of-state? This is a conditional probability because the answer will depend on the condition whether or not the selected student is a female. The equation in Theorem 10.6 will be as follows:

$$P(B|A) = \frac{P(A \text{ and } B)}{P(A)}$$

$$P(\text{out-of-state} | \text{female}) = \frac{P(\text{female and out-of-state})}{P(\text{female})}$$

The probability of selecting one of the 12 females in the class is equal to P(female) = 12/32. We discussed that out-of-state is conditioned on a female student being selected. Therefore, the probability of being both female and out-of-state [P(female and out-of-state)] = 8/32:

$$P(\text{out-of-state} | \text{female}) = \frac{8/32}{12/32} = 2/3$$

Example 10.9: Probability of Dependent Events II

To attend an ABET symposium in Atlanta, a professor in Flint must first fly to Chicago and board a connecting flight. Because of the nature of the flights, the traveler has very little time to change planes in Chicago. On his way to the airport, he ponders the possibilities. On the basis of past traveling experience, he knows that there is a 70% chance that his plane will be on time in Chicago [P(Chicago on time) = 0.70]. This leaves a 30% chance that he will arrive late in Chicago. He realizes that there is a 20% chance that he will be late to his Chicago plane but on time to the Atlanta plane [P(Chicago late and Atlanta on time) = 0.20]. Suppose that the professor arrives late in Chicago. What is the probability that he will reach Atlanta on time given that he arrived late in Chicago?

$$P(\text{Atlanta on time} | \text{Chicago late}) = \frac{P(\text{Chicago late and Atlanta on time})}{P(\text{Chicago late})}$$

From the problem statement above, we know

$$P(\text{Chicago on time}) = 0.70$$

$$P(\text{Chicago late}) = 1 - 0.70 = 0.30$$

$$P(\text{Chicago late and Atlanta on time}) = 0.20$$

$$P(\text{Atlanta on time} | \text{Chicago late}) = \frac{0.20}{0.30} = 0.70$$

If the occurring events are *independent* of each other, then the occurrence of one event does not influence the second one. Therefore, the outcome of one event will be unaffected by the other one. Hence, we can say $P(A|B) = P(A)$ and $P(B|A) = P(B)$, which results in $P(A \text{ and } B) = P(A) \times P(B)$.

Example 10.10: Probability in Independent Events

A car manufacturing firm employs 200 people in the drive train and transmission departments. Of all 80 female employees, 32 work in the drive train and 48 are in the transmission. Additionally, of all 120 male employees, 48 are employed in the drive train and 72 in the transmission. If an employee is selected randomly from 200 total participants, what is the probability that the selected employee works in the drive train department?

$$P(\text{drive train}) = \frac{80}{200} = 0.4$$

What is the probability of selecting an employee who is both female and works in the drive train department?

$$P(\text{female and drive train}) = P(\text{female}) * P(\text{drive train}) = \frac{80}{200} * \frac{80}{200} = \frac{4}{25}$$

The problem already tells us that of the 200 people at the company, there are 32 female employees in the drive train department. 32 of 200 is the same as 4 of 25, so you can see how it works.

Independence can be established if the following are true:

$$P(\text{drive train} \mid \text{female}) = P(\text{drive train})$$

and

$$P(\text{female} \mid \text{packaging}) = P(\text{female})$$

Theorem 10.7: If A and B Are Independent Events, the Probability That Both A and B Will Occur Is $P(\text{A and B}) = P(\text{A}) \times P(\text{B})$

Sometimes this is also referred to as *joint probability*, meaning that both A and B can occur at the same time.

Example 10.11: Joint Probability I

A company purchases headlight bulbs from two different suppliers. U&M Corporation supplies 60% of the bulbs and 40% of the headlight bulbs come from C&H Manufacturing. Both suppliers are going through quality problems. A total of 95% of the bulbs coming from U&M Corporation perform as expected. However, only 80% of the bulbs from C&H Manufacturing perform.

What is the probability that a bulb selected at random will be from U&M Corporation and that it will perform as expected?

Because the bulbs are coming from independent manufacturers, we can apply the equation in Theorem 10.7:

$$P(\text{U\&M and perform}) = P(\text{U\&M}) \times P(\text{perform})$$

$$= 0.60 \times 0.95 = 0.57$$

We can also calculate the probability that a randomly selected bulb is supplied by C&H Manufacturing and it works as follows:

$$P(\text{C\&H and perform}) = P(\text{C\&H}) \times P(\text{perform})$$

$$= 0.40 \times 0.80 = 0.32$$

$$= 0.40 \times 0.80 = 0.32$$

Example 10.12: Joint Probability II

The company is interested in determining the probability that any one of the bulbs selected at random, regardless of the supplier, will not perform.
   We know that the sum of probabilities is 1.00. Additionally, we calculated the probabilities of the performing bulbs for each supplier. To calculate all nonperforming bulbs, we only need to subtract all performing bulbs from 1. Therefore,

All performing bulbs = $P(\text{C\&H and perform}) + P(\text{U\&M and perform}) = 0.32 + 0.57 = 0.89$

$$\text{All nonperforming} = 1 - 0.89 = 0.11$$

10.2.2 Permutations and Combinations

As the number of ways that a particular outcome may occur increases, so does the complexity of determining all possible outcomes. For instance, if a part must go through four different machining operations and for each operation there are several machines available, then the scheduler will have a number of choices about how the part can be scheduled through the process. Permutations and combinations are used to increase the efficiency of calculating the number of different outcomes possible.
   A *permutation* is the number of arrangements that n objects can have when r of them are used:

$$P_r^n = \frac{n!}{(n-r)!}$$

It is important to remember that permutation is used when the order of arrangement is important.

Example 10.13

An industrial engineer needs to calculate the number of permutations possible for the following machining operations to take place:

Centering (four machines): The part must go to two different centering machines, each set up with different tools.
Grinding (three machines): Grinding the part takes only one grinder.
Milling (five machines): The part must go through three different milling machines, each set up with different tools.

The engineer understands that order is important, and number of permutations for each work center must be calculated. These calculated values will then be multiplied together to determine the total number of possible outcomes.

Centering (four machines selected two at a time):

$$P_r^n = \frac{n!}{(n-r)!} = \frac{4!}{(4-2)!} = 12$$

Grinding (three machines selected one at a time):

$$P_r^n = \frac{n!}{(n-r)!} = \frac{3!}{(3-1)!} = 3$$

Milling (five machines selected three at a time):

$$P_r^n = \frac{n!}{(n-r)!} = \frac{5!}{(5-3)!} = 60$$

Because there are 12 different permutation outcomes for centering, 3 operations for grinding, and 60 different outcomes for milling machines, the total number of permutations that the engineer will have to list is $12 \times 3 \times 60 = 2160$.

When the order of objects in the calculation is not important, the number of possibilities can be calculated by using the equation for *combination*:

$$C_r^n = \frac{n!}{r!(n-r)!}$$

Example 10.14: Calculating a Combination

An interdisciplinary group project is assigned for the mechanical and industrial engineering students. There are seven members present from the industrial engineering and three of them must be in the group. The number of different combinations of members from industrial engineering is

$$C_r^n = \frac{n!}{r!(n-r)!} = \frac{7!}{3!(7-3)!} = 35$$

There are five members from the mechanical engineering department and two of them must join the group. The number of different combinations of members from mechanical engineering is

$$C_r^n = \frac{n!}{r!(n-r)!} = \frac{5!}{2!(5-2)!} = 10$$

Multiplying the two outcomes to obtain the total number of different combinations from industrial and mechanical engineering students, we receive

$$\text{Total} = C_3^7 \times C_2^5 = 35 \times 10 = 350$$

10.3 Statistics

Even though you may not have realized it, you probably have made some statistical statements in your everyday conversation or thinking. Statements like "I study about 6 hours per class every week" or "You are more likely to pass the exam if you start preparing earlier" are actually statistical in nature.

Statistics, the collection, tabulation, analysis, interpretation, and presentation of numerical data, provide a viable method of supporting or clarifying a topic under discussion. Misuses of statistics have led people to distrust them as the quotations shown below:

> Do not put your faith in what statistics say until you have carefully considered what they do not say.
>
> William W. Watt

There are two kinds of statistics, the kind you look up and the kind you make up.

Rex Stout

Correct use of statistics helps to extract meaning from the presented information to make key decisions in the business. This chapter shows how to use probability and statistics to illuminate the understanding of a situation, process, or product.

We often see disease statistical reports in the news. If the reported data are simply shown to the population, chances are that people will ignore the data just because of a simple fact that the reported data on the number of infected people from this disease may not represent threat. However, if the reported data show how the trend of infection increased or decreased over a period of time and how it invaded the geolocation, people may have a better idea of how that disease may affect them. For example, studies have shown that 85% to 95% of lung cancers are smoking related. The statistic should tell you that almost all lung cancers are related to smoking and that if you want to have a good chance of avoiding lung cancer, you shouldn't smoke. Furthermore, if the genes of a human are affected by the cigarette, then the diseases can potentially be passed on to their children. Statistics are critical in determining the chances of a new baby being affected by the disease.

10.3.1 Statistical Difficulties

Newspaper and magazine editors urge their readers to *click on our website poll* and then they publish the results in the next issue. The newspaper may learn about their readership for marketing purposes. However, these survey results may be biased as 1936 presidential election survey showed. *The Literary Digest* magazine, one of the most respected magazines at that time, predicted that Landon would win the election by a large margin while the real election results turned out to be the opposite. *The Literary Digest* magazine polled over 10 million people and received 2.4 million responses. Those who responded were the magazine's readers and were mostly upper class people who owned telephones. The survey did not take into consideration that the voters will not be composed only of *The Literary Digest* magazine readers with higher income. Therefore, the survey results favored Landon with 57% versus Roosevelt with 43% of the votes. The actual results were Landon 38% and Roosevelt 62%.

10.4 Statistics in Organizations

Statistics can help organizations to support their safety, quality, delivery, and efficiency goals. Most organizations implement tools such as Six Sigma, lean manufacturing, or continuous improvement process. Accurately collected data and analysis can be used to understand and predict whether the organization is on the right track. From the examples above we realized that collection of data, analysis, and accurate interpretation is crucial in presenting to the audience. Once gathered, data are presented to the organization leadership, and can then be utilized in decision making.

10.4.1 Descriptive and Inferential Statistics

The study of statistics can be branched into two main categories: descriptive and inferential. *Descriptive statistics* give information that describes the data in some manner. For instance, suppose two suppliers provide brake clutches for various vehicles. If 100 clutches were shipped and 40 of 100 were sold to car manufacturer A, then one description of the data on the sold clutches would be that 40% of clutches went to manufacturer A. If the same supplier keeps record of how many clutches sold every month, then it can calculate the mean and report the average sales. This average can be used to determine how many clutches the supplier should produce every month to decrease inventory expenditure. Now suppose you need to collect data on a very large number of people, such as average height of male students at your university with a population of 40,000. It would take a long time to measure every single person. Some people might not even want to participate in your data collection process. This is where inferential statistics comes into play. *Inferential statistics* make inferences about populations using a small portion of the data drawn from the population.

10.4.2 Sample versus Population

Statistical data can be analyzed through population and sample data sets. The population includes all possible measurements, outcomes, or objects of interest to us in a particular study. The sample refers to a portion of the population that represents the population from which it is selected. A simple difference between the population and the sample is that the population includes each element from the set of observations, whereas the sample consists only of observations drawn from the population. All manufactured tractors in an organization would represent the population of tractors. Only models that were manufactured in the year 2015 would represent a sample of the population (entire family of tractor population). If you can statistically analyze the entire population and present the data to the team, your report would be ideal. However, in the industry, cost is the main driving force. Representing the entire tractor population will become time consuming and costly (this exercise would involve visiting every customer that purchased the product, whether or not they still have it or the tractor is still functional). Therefore, taking a sample of tractors to represent the business might be cost effective and less time consuming as well as easier to interpret.

Example 10.15: Taking a Sample

A tractor manufacturing firm has just received a shipment of 1,000 tires in a semitruck. The firm had ordered 800 front (28.7 major diameter) and 200 rear (52.6 major diameter) tires. The shipping department wishes to check that there actually are 20% of rear tires (RT) and 80% front tires (FT). He does not want to count all of the tires, so he decided to sample the population. Table 10.1 shows the results of 10 random samples of 10 tires each.

A greater number of RT are found in some samples than in others. However, when the results are compiled, the RT comprise 19%, very close to the desired value of 20%. The shipping department is pleased to learn that the samples have shown that there are approximately 20% RT and 80% FT.

From the above example, we see that the sample represented the population. However, for the sample to be unbiased, it needs to be a *random sample* (each item in the population has the same opportunity to be selected). Example 8.2 illustrates how nonrandom sampling can lead to incorrect inferences about the outcomes. Therefore, when reading statistical reports and their validity, it is crucial to understand the source of the data collected and how they were collected to avoid bias. To avoid bias, you must find answers to the following questions:

- How was the situation defined?
- Who was surveyed?
- How many people were contacted?
- How was the sample taken?
- How were the questions worded?
- Is there any ambiguity?

TABLE 10.1

A Sampling of Tires

Sample Number	Sample Size	Number of FT	Number of RT	Percentage of RT
1	10	8	2	20
2	10	7	3	30
3	10	8	2	20
4	10	9	1	10
5	10	10	0	0
6	10	7	3	30
7	10	8	2	20
8	10	9	1	10
9	10	8	2	20
10	10	7	3	30
Total	100	81	19	19

10.5 Data

Two types of statistics exist: deductive and inductive. Also known as descriptive statistics, deductive statistics describe a population or complete group of data. When describing a population using deductive statistics, the investigator must study each entity within the population. This provides a great deal of information about the population, product, or process, but gathering the information is time consuming and costly. Inductive statistics deal with a limited amount of data or a representative sample of the population as discussed earlier. Once samples are analyzed and interpreted, predictions can be made concerning the larger population of data. In quality control, two types of numerical data can be collected. Variables data, those quality characteristics that can be measured, are treated differently from attribute data, those quality characteristics that are observed to be either present or absent, conforming or nonconforming.

10.5.1 Measures of Central Tendency

Averages, medians, and modes are the statistical values that define the center of a distribution. Because they show the place where the data tend to be gathered, these values are commonly called the measure of central tendency. Mean and median are the two most common such measures. Mean, denoted as $\bar{x}$, is the simple arithmetic average. The *mean* is determined by adding the values together and then dividing this sum by the total number of values. The mean of this set of numbers {1, 1, 5, −1} is $\bar{x} = ((1+1+5-1)/4) = 6/4 = 1.5$. When this value is calculated for the population, it is referred to as the population mean and is signified by μ.

Algebraic equation for the sample mean is

$$\bar{x} = ((x_1 + x_2 + \cdots + x_n)/n) = 1/n \sum_{i=1}^{n} x_i$$

where:
x_i denotes variable measurements on each observation
Σ denotes the summation operator
Bar notation denotes the average of a sample ($\bar{x}$)
n means number of readings

Example 10.16: Clutch Plate: Determining the Mean

Calculate the mean clutch thickness received by a gauge: {0.0625, 0.0626, 0.0624, 0.0625, 0.0627}.

$$\bar{x} = \frac{(0.0625 + 0.0626 + 0.0624 + 0.0625 + 0.0627)}{5} = \frac{0.3127}{5} = 0.0625$$

The *median* is the value that divides an ordered series of numbers so that there are an equal number of values on either side of the center, median, value. The series of values are ordered from highest to lowest; then the median is determined as the value that is located at the center of the ordered series. If the number of observed values is odd, the median will be the center of the values. If the number of observed values is even, then the median will be the average of the two centered values. From the above example, the median of the given five values will be 0.0626.

The *mode* is the most frequently occurring number in a group of values. In a set of numbers, a mode may not occur. A set of numbers may also have two or more modes. If a set of numbers or measurements has one mode, it is called unimodal. If it has two numbers appearing with the same frequency, it is said to be bimodal. Similarly, distributions with more than two modes are called multimodal. The mode in the clutch example is 0.0625 because it is the most frequently occurring value.

Measures of central tendency describe the center position of the data. They show how the data tend to cluster around a center value. If the distribution is symmetrical, the mean, mode, and median values are equal. A comparison of mean, median, and mode determines whether a distribution is skewed and, if it is, in which direction.

In the right-skewed distribution, the mean is greater than the median. Similarly, the left-skewed distributions correspond to the mean value that is less than the median value. In a symmetric distribution, the mean is approximately (sometimes exactly) equal to the median.

Example 10.17: Clutch Plate: Relationship

The average, median, and mode of the clutch plate data provide information about the symmetry of the data. Equal mean, median, and mode will indicate that the data are symmetrical. From the previous example, the values for the clutch plate are as follows:

Mean = 0.0625 inch
Median = 0.0626 inch
Mode = 0.0625 inch

The plate data show that the mean and mode are equal to each other and the median is different. From Figure 10.1, we can observe that the greater median than the mean corresponds to the negatively skewed curve (left skewed).

10.5.2 Spread: Measures of Dispersion

The observer can determine the spread of the data with two measurements, range, and standard deviation. Dispersion measurements reveal the relation of the data to the mean, or the center point of the accumulated values. *Range* represents the difference between the highest and smallest values in a series of values. Because the range describes how far the

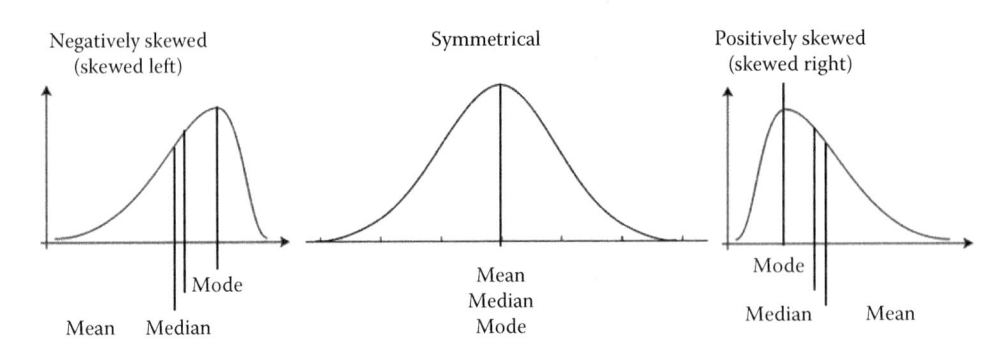

FIGURE 10.1
Comparison of mean, median, and mode.

data are spread, all values in the series (population or sample) will fall within these two maximum and minimum values. Range can be calculated as

$$R = X_h - X_l$$

where:
R is the range
X_h is the highest value in the series
X_l is the lowest value in the series

Example 10.18: Clutch Plate: Calculating Range Values

Range value can be calculated for the Clutch Plate thickness example. The series of values for various thicknesses were as follows: 0.0625, 0.0626, 0.0624, 0.0625, and 0.0627.
Range can be calculated by subtracting the lowest value (0.0624) from the highest value (0.0627) in the series:

$$R = X_h - X_l = 0.0627 - 0.0624 = 0.0003 \text{ inches}$$

The range shows where each end of the distribution is located. However, it does not show how data are grouped within the distribution. The *standard deviation* shows the dispersion of the data within the distribution. It describes how the individual values fall in relation to their means, the actual amount of variation present in a set of data. Standard deviation uses all of the measurements and provides more reliable information about the dispersion of the data. The standard deviation can be calculated with the following equation:

$$\sigma = \sqrt{\frac{\sum_{i=1}^{n}(X_i - \mu)^2}{n}}$$

where:
σ is the standard deviation of the population (*variance*)
μ is the mean value of the series of measurements (population)
X_i is the values of each observation
n is the number of observations

Usually the large number of participants (population size) makes it difficult to calculate the population standard deviation, referred to as *variance*. Therefore, with the use of a computer, a smaller portion (sample) from the population is drawn to make inferences about the participants. Calculation of sample standard deviation is almost similar to the equation above. However, in short, the standard deviation is simply the square root of the variance:

$$s = \sqrt{\frac{\sum_{i=1}^{n}(X_i - \bar{X})^2}{n-1}}$$

where:
s is the standard deviation of the sample
$\bar{X}$ is the average value of the series of measurements (sample)
X_i is the value of each observation
n is the number of observations

Example 10.19: Clutch Plate: Determining the Standard Deviation of a Sample

Calculate the sample standard deviation from the set of values drawn from the population:

$$s = \sqrt{\frac{\sum_{i=1}^{n}(X_i - \bar{X})^2}{n-1}}$$

$$= \sqrt{\frac{(0.0624 - 0.0625)^2 + 2(0.0625 - 0.0625)^2 + (0.0626 - 0.0625)^2 + (0.0627 - 0.0625)^2}{5-1}}$$

$$= 0.0001$$

Example 10.20: Gas Mileage: Determining the Standard Deviation of a Sample

A customer is interested in testing the gas mileage for seven different cars to decide which of the cars will be the most economical to commute to school/work. A sample of a much larger production run was driven under the normal circumstances to determine the number of miles per gallon the cars got. The following mpg readings were obtained: 36, 35, 39, 40, 35, 38, 41.
Calculate the sample standard deviation.
We need to compute the average value of the data set first:

$$\bar{X} = \frac{36 + 35 + 39 + 40 + 35 + 38 + 41}{7} = 37.7 \approx 38$$

Using the sample standard deviation equation, we obtain

$$s = \sqrt{\frac{(36-38)^2 + 2(35-38)^2 + (39-38)^2 + (40-38)^2 + (35-38)^2 + (38-38)^2 + (41-38)^2}{7-1}}$$

$$= 2.45$$

10.5.3 Central Limit Theorem

The central limit theorem states that the distribution of the sum (or average) of a large number of independent, identically distributed variables will be approximately normal, regardless of the underlying distribution. The importance of the central limit theorem is hard to overstate; indeed it is the reason that many statistical procedures work. So what is the *large* number? The answer depends on two factors:

- Requirements of accuracy: The more closely the sampling distribution needs to resemble a normal distribution, the more sample points will be required. In other words, the larger the sample size, the more accurate the information is.
- The shape of the underlying population. The more closely the original population resembles a normal distribution, the fewer sample points will be required.

In practice, a sample size of $n = 30$ is large enough when the population distribution is roughly bell-shaped.

10.5.4 Normal Frequency Distribution

Because this chapter briefly talks about probability and statistics, we will not be considering normal distributions in detail. The normal frequency distribution is the familiar bell-shaped curve, commonly called the *normal curve* (Figure 10.1).

The normal frequency distribution has six distinct features:

1. A normal curve is symmetric about μ, the central value.
2. The mean, mode, and median are all equal.
3. The curve is unimodal and bell shaped.
4. Data values concentrate around the mean value of the distribution and decrease in frequency as the values get further away from the mean.
5. The area under the normal curve equals 1. One hundred percent of the data are found under the normal curve, 50% on the left-hand side, 50% on the right-hand side.
6. The normal distribution can be described in terms of its mean and standard deviation by observing that 99.73% of the measured values fall within ±3 standard deviations of the mean ($\mu \pm 3\sigma$). Additionally, 95.5% of the data fall within ±2 standard deviation of the mean, and 68.3% of the data fall within ±1 standard deviation of the mean.

As the data fall away toward the horizontal axis, the curve flattens. The tails of the normal distribution approach the horizontal axis indefinitely and never cross it.

10.5.5 Standard Normal Probability Distribution: Z Tables

If the investigator knows the mean and standard deviation values, the area under the normal curve can be determined with Z tables using the following equation:

$$Z = \frac{X_i - \bar{X}}{s}$$

where:
  X_i is the individual value of interest
  X-bar is the sample average
  s is the sample standard deviation

Probability and Statistics 267

We can equally adopt this equation to determine the area under the normal distribution curve for the population:

$$Z = \frac{X_i - \mu}{\sigma}$$

where:
X_i is the individual value of interest
μ is the population mean
σ is the population standard deviation

The population standard deviation σ can also be calculated as $\sigma = \sigma/\sqrt{n}$.

The percentages of proportion of the observations are provided in the Z table to find the area under the curve along with the Z value. The sign of the Z value represents whether the interested data are on the left side of the center (less than mean) or on the right side (greater than mean). The positive value of Z shows that it is on the right side of the curve and the negative value indicates that it is on the left side of the curve.

Determination of the area under the bell curve using the Z table involves calculating the Z score from the above equation. Next look up the obtained Z value on the table. Notice that only one decimal place is shown under the "Z" column. You will need to follow the horizontal number set to match the calculated value in the Z table. Intersection of the column and the row will reveal the percentage area under the curve. Let us take a look at an example:

Example 10.21: Clutch Plate: Using Standard Normal Probability Distribution

The engineering team working with the clutch plate thickness data have determined that their data approximate a normal curve. They would like to determine what percentage of parts from the samples taken is below 0.0624 (Figure 10.2).

1. From the collected data they calculated an average of 0.0625 and a standard deviation of 0.0001. They used the Z tables to determine the percentage of parts under 0.0624 inch thick:

$$Z = \frac{0.0624 - 0.0625}{0.0001} = -1$$

From Appendix (Z table): −1 corresponds to 0.1587, or 15.87% of the parts are thinner than 0.0624 inch.

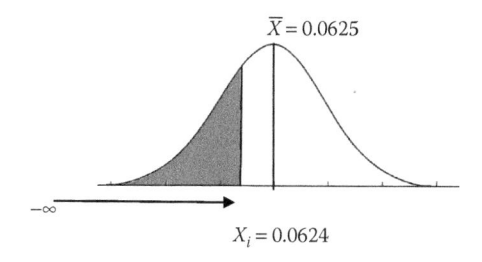

FIGURE 10.2
Area under curve, $X_i = 0.0624$.

10.6 Summary

$$P(A) = \frac{\text{number of occurences}}{\text{total number of possibilities}} = \frac{s}{n}$$

Theorem 10.1: Probability Is Expressed as a Number between 0 and 1: $0 \leq P(A) \leq 1$

Theorem 10.2: The Sum of the Probabilities of the Events in a Situation Is Equal to 1.00:
$\sum Pi = P(A) + P(B) + \cdots + P(N) = 1.00$

Theorem 10.3: If *P(A)* Is the Probability That Event A Will Occur, Then the Probability That A Will Not Occur Is $P(A') = 1 - P(A)$

Theorem 10.4: For Mutually Exclusive Events, the Probability That Either Event A or Event B Will Occur Is the Sum of Their Respective Probabilities: $P(A \text{ or } B) = P(A) + P(B)$

Theorem 10.5: When Events A and B Are Not Mutually Exclusive Events, the Probability That Either Event A or Event B or Both Will Occur Is $P(A \text{ or } B \text{ or Both}) = P(A) + P(B) - P(\text{both})$

Theorem 10.6: If A and B Are Dependent Events, the Probability That Both A and B Will Occur Is $P(A \text{ and } B) = P(A) \times P(B|A)$

Theorem 10.7: If A and B Are Independent Events, the Probability That Both A and B Will Occur Is $P(A \text{ and } B) = P(A) \times P(B)$

Permutations

$$P_r^n = \frac{n!}{(n-r)!}$$

Combinations

$$C_r^n = \frac{n!}{r!(n-r)!}$$

Problems

1. In a raffle at a school event, the probability of drawing the name of sophomore from the participating group is 0.35, the probability of drawing the name of a junior is 0.46, the probability of drawing the name of a senior is 0.15, and the probability of drawing a freshman is 0.04. What is the probability that a junior or a freshman will be a winner?

2. At a local car dealership, the officials would like to give a prize to 100 people selected at random from those attending the end of year event at the dealership. As of the closing day, 12,500 people have attended the event and completed the entry form for the prize. What is the probability that an individual who attended the event and completed the entry form will win a prize?

3. At a kindergarten, a class contains eight kids with yellow T-shirts numbered 1 to 8, six children with orange T-shirts numbered 1 to 6, and ten kids with gray T-shirts numbered 1 to 10. What is the probability of selecting a child with an orange T-shirt numbered with a 5? Of choosing a child with an orange T-shirt? Of choosing a child with a number 5?

4. If there are five different parts to be stocked but only three bins available, what is the number of permutations possible for five parts taken three at a time?
5. If a manufacturer is trying to put together a sample collection of the product and order is not important, how many combinations can be created with 15 items that will be placed in packages containing five items? How many permutations can be made if order is important?
6. A tractor manufacturing organization receives its GPS navigators from two different suppliers: 75% come from HMZ Co. and 25% come from MRM Navigators Co. The percentage of GPS navigators from HMZ that perform according to specification is 95%. The GPS navigators from MRM Navigators perform according to specification only 80% of the time. What is the probability that any one GPS navigator received by the manufacturer performs according to specifications?
7. If one of the GPS navigators from problem 6 performed according to specifications, what is the probability that it came from HMZ Co?
8. In the airport tower, the employees control air traffic within the area with as many as 500 planes. Communicating with the pilots and navigating their routes through different air zones accordingly requires highly skilled personnel. In zone A, there are 300 planes, of which 150 captains have little experience (2–5 years of flight), 75 have moderate experience (5–10 years of flight), and 75 are veterans (more than 15 years of flight). Five percent of the captains with little experience may need additional assistance, while only 1% of each of the other two categories of captains will need additional assistance. What is the probability that when communicating with the air traffic control, that particular captain will need additional assistance in the area?
9. Suppose a car seat supplier makes seats in four different styles from three different fabrics. Use the table to calculate the probability that a seat selected at random will be made from fabric 1. If the seat is style 1, what is the probability that it will be made from fabric 2? What is the probability that a seat of style 4 will be selected at random? If a seat is made from fabric 3, what is the probability that it is a style 3 seat?

Style	Fabric 1	Fabric 2	Fabric 3	Total
Style 1	150	55	100	305
Style 2	120	25	70	215
Style 3	80	60	85	225
Style 4	110	35	110	255
Total	460	175	365	1000

10. Describe the concepts of a sample and a population.
11. Describe the central tendency of data.
12. Describe measures of dispersions.
13. Calculate mean, median, mode, range, and standard deviation for the following data: 225, 226, 227, 226, 227, 228, 228, 229, 222, 223, 224, 226, 227, 228, 225, 221, 227, 229, 230.
14. At the Information Technology Service department of a university, students answer phone calls to troubleshoot problems that callers may have. Engineering students are interested in how long calls typically last. Use the following information for *call duration* to calculate mean, median, mode, range, and standard deviation.

Customer	Call Duration	Customer	Call Duration
1	4	7	40
2	15	8	45
3	34	9	30
4	22	10	10
5	25	11	8
6	36	12	3

15. For problem 14, calculate the percentage of calls that last for over 15 minutes. The average length of the call is 8 minutes, with a standard deviation of 3 minutes.

16. If the average wait time is 12 minutes with a standard deviation of 3 minutes, determine the percentage of callers who wait less than 15 minutes for their problem to be resolved.

17. Two brothers run a restaurant. All orders are served at the restaurant. The brothers need to do some planning to optimize timing. If the mean time for an order and service is 45 minutes and the standard deviation is 10 minutes, what percentage of the customers will receive their order in less than 65 and more than 35 minutes? The data are normally distributed.

18. The life of a motorcycle battery is normally distributed with a mean of 800 days and a standard deviation of 45 days. What fraction of these batteries would be expected to last beyond 1000 days?

CASE STUDY 10.1

Part 1

Angie's company has recently told her that her pension money can now be invested in several different funds. They have informed all their employees that it is the employee's decision which of the six risk categories to invest in. The six risk categories include conservative, secure, moderate, balanced, ambitious, and aggressive. Within each risk category, there are two fund choices. This gives each employee 12 funds in which to invest capital.

As the names suggest, each fund category attempts to invest the person's capital in a particular style. A conservative fund will find investments that will not expose the investor's money to risk. This category is the *safest*. Money invested here is not exposed to great potential losses, but it is not exposed to great gains either. The aggressive fund is for investors who believe that when nothing is ventured, nothing is gained. It takes the greatest risks. With this fund, investors have the potential to make great gains with their capital. Of course, this is also the fund that exposes the investor to the potential for great losses. In between, the other four categories mix the amount of risk an investor is subject to.

Angie is rather at a loss. In the packet from the investment company, she has received over 150 pages of information about each of the accounts. For the past three nights, she has been trying to read it all. Now all the information is running together in her head, and she has to make up her mind soon.

Just as she is about to give up and throw a dart at the dart board, she discovers an investment performance update (Table 10.1) among the many pages of data. Remembering her statistical training, she decides to study the performance of the different funds over the past 11 years. To fully understand the values she calculates, she decides to compare those values with the market performance of three market indicators: treasury bills, the Capital Markets Index, and the Standard & Poor's 500 Index. Investors use these indicators to compare the performance of their investments against the market.

Assignment

From Table 10.2 calculate the average of each of these funds and market performance indicators.

Part 2

As her next step, Angie is going to interpret the results of her calculations. To make sure that she has a comfortable retirement, she has decided that she would like to have her money grow at an average of 15% or greater each year.

TABLE 10.2

Funds and Market Indices, 1993–2003

	93	94	95	96	97	98	99	00	01	02	03
Conservative											
C1	10.4	12.8	18.4	15.0	3.0	8.2	11.3	8.0	13.2	7.3	10.7
C2	10.9	14.0	24.2	15.9	2.5	8.7	13.7	10.2	17.9	8.0	11.6
Secure											
S1	18.2	10.2	23.7	18.9	7.8	12.0	23.0	4.4	21.2	8.0	8.2
S2	16.3	9.3	27.8	19.1	0.5	12.8	15.3	1.4	27.9	2.7	13.1
Moderate											
M1	14.5	10.0	29.7	23.2	11.6	16.0	19.7	0.8	29.5	7.6	7.5
M2	21.9	14.4	27.2	17.6	6.4	15.7	28.2	4.5	27.6	6.6	4.5
Balanced											
B1	27.2	9.2	24.7	17.3	2.3	21.8	22.8	1.4	20.8	7.7	7.5
B2	21.5	8.9	23.2	13.2	7.8	12.7	24.8	2.3	25.6	9.2	11.9
Ambitious											
A1	36.2	6.3	32.4	11.8	2.6	14.7	22.9	3.5	38.6	10.8	15.8
A2	27.3	2.6	33.5	14.7	3.4	15.1	32.8	26.1	24.1	13.8	25.0
Aggressive											
G1	23.7	7.6	33.4	25.8	5.5	13.3	28.6	6.9	51.5	4.0	24.1
G2	22.3	0.3	38.3	19.7	6.0	14.1	37.0	22.1	57.6	6.4	21.2
Treasury Bills											
	9.2	10.3	8.0	6.3	6.1	7.1	8.7	8.0	5.7	3.6	6.7
Capital Market Index											
	17.7	8.3	28.6	16.5	3.4	13.2	21.6	1.0	25.0	8.4	13.4
Standard & Poor's 500											
	22.60	6.30	31.70	18.60	5.30	16.60	31.60	23.10	30.40	7.60	10.10

Assignment
Use the average values you calculated to determine which funds will average greater than 15% return on investment per year.

Part 3
Angie has narrowed her list to those funds averaging a 15% or greater return on investment per year. That still leaves her with several to choose from. Remembering that some of the funds are riskier than others, Angie is quick to realize that in some years she may not make 15% or more. She may add 25% or higher to her total amount of pension, or she may lose 5% of her pension. An average value does not guarantee that she will make exactly 15% year. An average reflects how well she will do when all of the good years (gains) and bad years (losses) are combined.

Angie has 20 years to work before she retires. While this gives her some time to increase her wealth, she is concerned that a few bad years could significantly lower the amount of money that she will retire with. To find the funds that have high gains or high losses, she turns her attention range and standard deviation.

Assignment
Calculate the range and standard deviation associated with each of the funds and market indicators with an average of 15% or greater return on investment.

Part 4
For all of the funds that she is interested in, Angie has noticed that the spread of data differs dramatically. One of the ranges is almost 60 points! She has also noticed that the riskier the fund, the greater the range of values.

Because the ranges may have been caused by an isolated year that was particularly good or particularly bad, Angie turns her attention to the standard deviations, which also increase as the riskiness of the fund increases.

Decision time! Comparing the average return on investment yielded by the stock market (Standard & Poor's 500) during this 11-year period, Angie notices that several of the investment funds have brought in return that were less than the Standard & Poor's 500 market indicator. She decides to take them off her potential funds list. This removes M2 and B1 from consideration. She also removes from consideration any fund that has a large range and standard deviation. Because she is willing to take some risk, but not a dramatic one, she decides not to invest in the aggressive fund. This removes G1 and G2 from consideration. Now she can concentrate on a smaller group of funds: M1, A1, and A2. So now, instead of having to devour hundreds of pages of information, Angie can look at the pages detailing the performance of these three funds and make her decision about which one to invest in.

To help make the choice between these funds, Angie has decided to use the Z tables to calculate the percentage of time that each fund's return on investment fell below 12% that she feels she absolutely has to make each year.

Assignment
Assume a normal distribution and use the Z tables to calculate the percentage of each of the remaining funds' return on investments that fell below 12% in the past 11 years. Do not perform this calculation for the fund indicators.

11
Computer Programming

CHAPTER OBJECTIVE AND STUDENT LEARNING OUTCOMES

1. Establish a detailed algorithm to perform a task or solve a problem. [addresses ABET Criteria 3(k)]
2. Recognize the basic structures of programming languages.
3. Compose algorithms in the form of pseudocode.
4. Use computer programming techniques to solve engineering problems. [addresses ABET Criteria 3(k)]
5. Select appropriate programming techniques to perform specific tasks. [addresses ABET Criteria 3(k)]

11.1 Importance of Programming

Computers, cell phones, modeling software, and other programmable devices are integral in the work of all engineers regardless of their specific discipline. It is very important that engineers not only have the ability to use these devices but also have the ability to understand and control them.

Computers can be programmed to do all sorts of things. The process of writing sequential instructions to enable the computer to complete a specific task is called *computer programming*. Computer programming has experienced tremendous developments since its inception. Computer languages continue to evolve from once archaic forms into a more comprehensible and capable code. This evolution has enabled more people to master the art of computer programming.

However, computer program means more than just feeding instructions to a device. There are a number of invaluable skills that you get from programming that infiltrate into other areas of an engineer's responsibilities. The following are just a few examples of the benefits to understanding programming:

- Programming teaches problem solving. As you develop *algorithms*, step-by-step instructions that solve a particular problem, you will learn strategies to problem solving such as divide and conquer, recursion, and heuristics, which can be used in attempts to model and solve *any* type of problem.

- When you program, you will undoubtedly hone your troubleshooting skills. Let us face it; things go wrong from time to time, and we do not always instantly know what the problem is. In programming, you will learn to test and systematically isolate the cause(s) of the problem, and the same techniques can be used in troubleshooting all sorts of systems.
- Even if you do not become a programmer, you will be able to communicate with programmers. It is essential that the computer programmer have the communication skills to understand and clarify the desires of the customer and/or other project stakeholders.

As you can see, even if you were to decide that programming is not for you, there are still benefits in obtaining a basic understanding of what programming is and how it is accomplished.

11.2 Languages and Applications

Each computer language has its own *syntax*, or a set of rules, that must be strictly followed when writing a program. The syntax rules dictate what elements must be included in each instruction and how those instructions are formatted. A computer language's syntax can be compared to the way we structure our sentences and paragraphs when writing a paper. There are certain elements that every sentence is required to have, namely a noun and a verb, but there are several ways to relay the same information ranging from basic to elegant phrasing. Just like every written language has different grammatical rules, each computer language has a different syntax that must be learned before successful communication can occur.

11.2.1 High-Level Languages

There are several types of languages that a programmer can learn. *High-level languages* are frequently used because they are the most similar to human language. High-level languages also tend to be portable across various CPUs and are easier to edit and maintain. The languages C, C++, Java, and Fortran are examples of this type of programming language.

In a high-level programming language, a program is composed of many individual instructions called *statements*. A programming statement can be composed of keywords, operators, punctuation, and other allowable programming elements and must be arranged in the proper sequence to perform an operation. Simple statements are executed in sequence, or one after another, until all the statements are executed, but programming languages also include control structures that will alter the consecutive sequence executed in a predefined, structured manner. All computer programming languages utilize the simple statements and control structures discussed in this chapter, and once you become familiar with their basic functionality, switching from one programming language to another becomes more of a matter of learning new syntax instead of grasping entirely new concepts.

11.2.2 Machine Code and Assembly Language

Machine code is the basic low-level programming language designed to be recognized by a computer. The language is composed entirely of binary code or long sequences of 0s and 1s that represent on and off electric impulses. Each binary instruction is composed of multiple binary digits. A program can have hundreds of thousands or even millions of binary instructions, and writing such a program would be tedious and time consuming. Misplacing a single 0 or 1 digit will cause an error in the program.

Although a computer's CPU only understands machine code, it is very difficult and impractical to write entire programs in machine code. For this reason, *assembly languages* were created in the early days of computing as an alternative. Assembly languages use short words, typically referred to as *mnemonics*, instead of using binary numbers for instructions. For example, in assembly language, the mnemonic *add* typically means to add numbers, *div* typically means to divide numbers, and *mov* typically means to move a value to a specified location in memory.

Example:

High-Level Programming with C++	Assembly Language	Machine Code
	# f-> $s0; g->$s1; h->$s2	1001010110010100111
if (k==0) f = i + j;	# i-> $s3; j->$s4; k->$s5	1101010011011100101
else if (k==1) f = g + h;	bne $s5, $0, L1	1001010110010100111
else if (k==2) f = g - h;	add $s0, $s3, $s4	1101010011011100101
else if (k==3) f = i - j;	j Exit	1001010110010100111
OR	L1: addi $t0, $s5, -1	1101010011011100101
select (k) {	bne $s0, $0, L2	1001010110010100111
case 0: f = i + j; break;	add $s0, $s1, $s2	1001010011011100101
case 1: f = g + h; break;	j Exit	1111010111010100111
case 2: f = g - h; break;	L2: addi $s0, $s5, -1	1101010011011100101
case 3: f = i - j; break;	bne $s0, $0, L3	1001010110010100111
}	sub $s0, $s1, $s3	1101010011010000101
	j Exit	1111010110010100111
	L3: addi $s0, $s5, -3	1101010011011100101
	bne $s0, $0, Exit	1001010111110100111
	sub $s0, $s3, $s4	1101010111111100101
	Exit:	1001010110010100111
		...

11.2.3 Compiling and Executing a Program

Ultimately, all code must be translated into machine code for the computer to understand. For high-level programs, a *compiler* is a program used to translate the high-level language program into machine code. Similarly, an *assembler* is a program designed to translate programs written in assembly code into machine code. These translation programs create a file typically referred to as an *executable*. Once the executable has been created, the programmer can then give the command to run the program.

In some cases, such as with the Python language, an *interpreter* is used instead of compiling and executing the program. An interpreter is a program that both translates and executes instructions in a high-level language program in the same operation. The interpreter reads each individual instruction in the program one by one, converts it to machine code, and then immediately executes it. This process repeats for every instruction in the program. While combining the translation and execution into one step works well when debugging the code, once the program is debugged and finalized, it still undergoes the translation each time the program runs, which delays execution.

11.3 Algorithm Development

When programmers are given a task to program, they must first think about what they are trying to do and develop an algorithm in which to accomplish that task. An *algorithm* is a set of step-by-step instructions that perform a specific task. In general, an algorithm is a detailed procedure to solve a problem.

One of the most challenging aspects of computer programming is understanding that computers, with all their power and possibilities, are fundamentally unintelligent. They can only do exactly what the programmer tells them to do—nothing more and nothing less. When we communicate with others, we typically make basic assumptions that we cannot make when we program. For example, if you have dinner with your family one evening and ask a sibling to hand you the salt, you have made the assumption that your sibling knows what salt is, where to find it, and how to pass it to you. In programming, this is not the case. If you ask the computer to retrieve a piece of data, at the very least, you must tell the computer what to call the data, what type and size it is, where to get it and how to get it to you. While this is a different way of thinking than we are used to, there are some methods, tools, and general guidelines that we can use to help make this process easier.

11.3.1 General Guidelines for Creating Simple Algorithms

The first and most important step to creating a successful algorithm is to understand the task at hand. The programmer should start by defining the following:

- The program *outputs*—determine what answers or results are desired from the program.
- The program *inputs*—consider what data is going to be provided to the program and from where it will be obtained.

The second step is to make sure that you have all background information necessary to complete the task. For example, if you intend to calculate the area of a tetrahedron, you may need to look up the formula to make that calculation.

Thirdly, think about the types of tasks that are involved in producing the final result and split the algorithm into smaller, more manageable segments. In most cases, it is easier to tackle the smaller sections than to think of the program as a whole when starting to program.

Computer Programming

The fourth step involves choosing a format to express your algorithm and then writing it. In this chapter, we discuss two formats, flowcharts and pseudocode, but there are others that you may choose to use individually or in conjunction with another format.

Be sure to review your algorithm, looking closely at each step. If a step can be broken down into two or more smaller steps, adjust your algorithm to include these steps. By following these steps, you should be able to test your algorithm by hand to make sure that it operates as desired.

11.3.2 Flowcharts

Flowcharts are graphical representations of the steps of an algorithm. One advantage of using flow charts to write algorithms is that the flow chart symbols and their meanings are fairly standardized and tend to have the same meaning for most users. Flowcharts are particularly helpful when trying to communicate to others how the program is intended to operate. It can also be easily followed to aid in analyzing the functionality of the algorithm.

You will find a list of the most frequently used flowchart symbols and their meanings in Table 11.1. Flowcharts are written and read from top to bottom or left to right. The flow chart symbols are linked together with arrows showing the process flow direction. The purpose of all flowcharts is to communicate how a process works or should work without any technical or group-specific jargon.

As an example, consider a program that will determine whether to build a pool based on the user's yard size and the radius of the pool that the user hopes to install. The following shows an algorithm represented as a flowchart:

TABLE 11.1

Frequently Used Flowchart Symbols

Shape	Symbol	Name and Function
Oval	⬭	Start/end—denotes the beginning or end of a program
Parallelogram	▱	Input/output—denotes an input or output operation
Rectangle	▭	Process—denotes a step or operation to be performed
Diamond	◇	Decision—denotes a decision and a corresponding route based on the result of a condition
Directional arrows	→	Flow line—arrowheads that connect and show the sequence of process steps

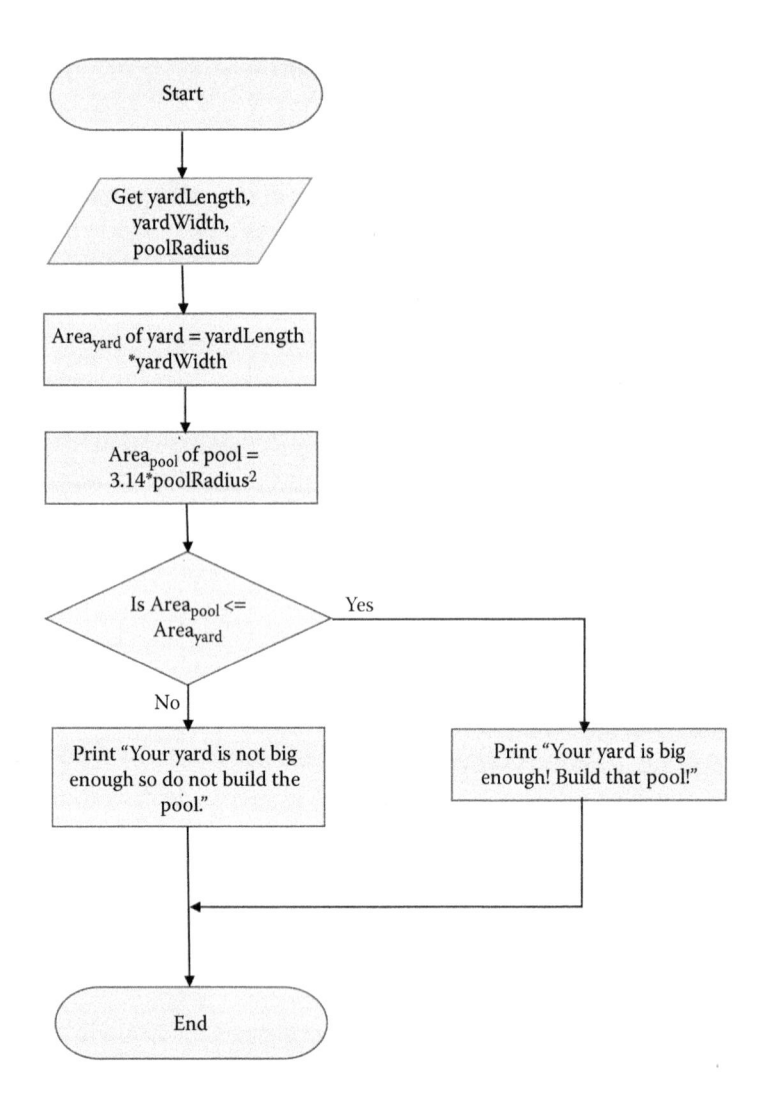

```
            ╭─────────╮
            │  Start  │
            ╰────┬────╯
                 ▼
         ╱─────────────╱
        ╱ Get yardLength,╱
       ╱  yardWidth,    ╱
      ╱   poolRadius   ╱
     ╱────────┬───────╱
              ▼
    ┌──────────────────┐
    │ Area_yard of yard = yardLength │
    │      *yardWidth  │
    └────────┬─────────┘
             ▼
    ┌──────────────────┐
    │ Area_pool of pool = │
    │  3.14*poolRadius² │
    └────────┬─────────┘
             ▼
         ◇ Is Area_pool <= ◇  Yes
         ◇    Area_yard    ◇─────►
             │ No
             ▼
```

Area$_{yard}$ of yard = yardLength *yardWidth

Area$_{pool}$ of pool = 3.14*poolRadius^2

Is Area$_{pool}$ <= Area$_{yard}$

Print "Your yard is not big enough so do not build the pool."

Print "Your yard is big enough! Build that pool!"

End

There are, however, a few disadvantages to using flowcharts. In situations where the logic of a task becomes complex, the flowchart quickly becomes cluttered and, at times, challenging to follow. Because flowcharts consume a lot of physical space, reading the flowchart may require that the user follows graphics across multiple screens or pages. Depending on the software used to create flowcharts, editing may also be difficult.

Still, many people believe that flowcharts play a vital role in programming and have proven to be very beneficial when trying to understand the logic of long, complicated programs. Flowcharts also facilitate communication between programmers and any potential stakeholders that lack a strong, technical background, and they typically make the debugging process easier.

11.3.3 Pseudocode

Instead of focusing on the syntax of a specific programming language, this introduction will employ pseudocode to demonstrate the basic concepts of computer programming. The word *pseudo* means *false*, and, thus, pseudocode means *false code*. *Pseudocode* is an informal way of expressing the details of a computer program or algorithm and can be easily read. Pseudocode is typically used in the planning phase of programming, and allows the programmer to develop a step-by-step algorithm that focuses on the logic involved with the task instead of language-specific syntax. When the time comes to write the code in the appropriate programming language, it should be a simple task of translating each line of pseudocode into the proper syntax.

In general, there are no strict rules to follow when writing pseudocode, but, for the purpose of this chapter, the following general guidelines will be adhered to:

- Ending Programming Statements
  - In the English language, a period is used to indicate that we are finished with a statement. Similarly, many computer languages require that a line of code ends with some sort of punctuation. In the case of our pseudocode, we will use a semicolon to indicate the end of each line.
- Blocks
  - A *block of code* is two or more statements that are grouped together. Some languages use curly brackets to indicate the beginning and the end of a particular block of code. In our pseudocode, we will use the words **BEGIN** and **END** to represent the beginning and the end of a code block, respectively.
- Indentation
  - Every time a new block begins, the instructions contained in that block should be indented one additional tab length. Likewise, every time the program encounters the **END** of a code block, the program indentation is decreased by one tab. Complying with an indentation schema is not actually required by the computer, but this should help ensure better readability of our code and aid in debugging efforts.

As we continue through this chapter and introduce you to additional programming concepts, you will find additional guidelines for writing pseudocode. Please apply these guidelines whenever you are asked to write pseudocode in this course.

Program Structure Example

The following program asks the user to input an item number, quantity, and cost of each item and then calculates the total cost of ordering those items. If the order cost is over $50.00, then $14.99 is added to cover shipping costs. Otherwise, the shipping cost is $9.99. The total cost, including shipping, is then relayed to the user:

```
BEGIN #begin program
   #declarations go here
   DECLARE itemNumber as REAL;
   DECLARE costEachItem as REAL;
   DECLARE qtyOrdered as REAL;
   DECLARE orderTotal = 0.0 as REAL;

   PRINT "Please enter the item number, the cost for each item and the
quantity that you would like to order.";
   GET itemNumber, costEachItem, qtyOrdered;
   orderTotal = qtyOrdered * costEachItem;
   IF (orderTotal > 50.00)
```

```
    BEGIN IF
           PRINT "Your shipping costs are $14.99.";
           orderTotal = orderTotal + 14.99;
    END IF
    ELSE
    BEGIN ELSE
           PRINT "Your shipping costs are $9.99.";
           orderTotal = orderTotal + 9.99;
    END ELSE
    PRINT "Your order total is ", orderTotal;
END         #end program
```

11.4 Comments and Documentation

Almost all high-level programming languages allow the programmer to include comments within their code. The purpose of comments is to add details and/or explanation of how the code is intended to function. Comments are purely for the programmer's benefit and are not executed or processed by the computer at all. They are discarded when the program is compiled and do not exist in the machine code. Depending on the language, a special character or character sequence indicates the beginning of a comment section. In some cases, an end of the comment section is also indicated, but many languages assume that once the character or character sequence has been encountered, the comment continues until the end of that line in the code.

For the purpose of our pseudocode, we will use the *pound symbol (#)* to indicate that a comment has begun. Comments may occupy their own line, or an instruction may appear at the beginning of a line with a comment to the right of the instruction. Once a comment symbol is encountered, the computer will consider everything on that line and to the right of the comment symbol to be a comment (Figure 11.1).

11.5 Reserved or Keywords

Each high-level language has its own set of predefined words that the programmer must use to write a program. The words that make up a high-level programming language are known as *keywords* or *reserved words*. Each keyword has a specific meaning and cannot be used for any other purpose. In our pseudocode, we will represent these reserved words entirely with capital letters. Table 11.2 lists reserved/keywords addressed in this chapter.

11.6 Variables, Constants, and Their Data Types

Every piece of data used in a computer program has a data type. The data type lets the computer know how much memory to allocate in storing that data, how to retrieve

Example 1
```
# This is a comment that appears on a line of its own
# Once the # sign is encountered, anything that follows on
# the same line is considered a comment
```
Because the above comment spans three computer lines, the programmer needs to use the comment symbol (#) at the beginning of each line to continue the comment onto the next line. For example, if the # sign were omitted in the third line of the comment, the computer would experience an error as it tries to compile the text as code.

Example 2
```
DECLARE orderQuanitity as INTEGER # I can say anything after the comment symbol
```
When a programmer combines executable statements and comments on the same line, the executable statement always precedes the comment. Even in the few languages where it is acceptable to put a comment at the beginning of a line followed by an executable statement, it is highly recommended that a programmer place the executables before the comment to maintain consistency and ease in code maintenance.

Example 3
```
# Calculate the total price # totalPrice = orderQuanity * pricePerItem;
```
If the programmer wanted to execute this statement, this would be *incorrect*. Once the comment symbol appears on a line, the remainder of that line is considered a comment. In this example, *totalPrice = orderQuanity * pricePerItem* is part of the comment and is never executed.

FIGURE 11.1
Examples Using Comments in Pseudocode

TABLE 11.2
A List of Reserved Pseudocode Words That Will Be Used in This Chapter

Reserved Pseudocode Words								
DECLARE	SET	REAL	INTEGER	CHAR	STRING	ARRAY	BOOLEAN	GET
READ	PRINT	WRITE	FOR	IF	THEN	ELSE	CASE	SELECT
DO	WHILE	UNTIL	AND	OR	NOT	BEGIN	END	FUNCTION
PROCEDURE	ELSEIF	BREAK						

Note: These words should always be capitalized and cannot be used for any other purpose in our program.

that data, and what operations can be performed with that data. The basic data types are as follows:

- An *integer* is any whole number and includes both positive and negative numbers.
- A *real* or *floating-point number* is a number with a decimal point.
- A *boolean* is a variable that can only be true (represented by a 1) or false (represented by a 0).
- A *character* is any single ASCII letter, number, or punctuation mark.
- An *array* is a list of related data all of the same type. It can be an array of any type listed above.
- A *string* is a special type of array composed of characters.

A *constant* is a value that is needed by the program but never changes during the execution of the code. A *variable* is a value that can be changed throughout the course of the program. Before they can be used, both constants and variables need to be *declared* with their data type and assigned a name, also called an *identifier*. The identifier tells the computer to set aside a location in memory to store that piece of data. If you were to use an analogy to understand how this works, think about information that you might need to obtain from a library. When your roommate first moved into town, you had to tell him the mailing address of the library so that he could locate it. However, now that your roommate knows where the library is, you simply call the library by name knowing that your roommate will be able to locate you if necessary. Additionally, because no two libraries have the exact same information available at a given time, the available content at that library would be the value of the data stored at that location.

In pseudocode, if we wanted to use a constant floating-point number to represent an hourly pay rate, a variable floating-point number to represent the number of hours worked, and another variable floating-point number to represent the gross pay earned for working for that duration of time, it would look like the following:

```
DECLARE HourlyRate = 10.50 as REAL CONSTANT;
DECLARE hoursWorked as REAL;
DECLARE grossPay as REAL;
```

Assigning identifiers to constants and grouping declaration statements in the program is a good programming practice. Alternatively, it is acceptable to hard code the actual value of the constant wherever it is needed in the program. However, if there comes a time when we want to change that value, it is harder to find. Furthermore, it becomes tedious to locate and change all instances of that constant, especially in programs where the constant is used multiple times. It is helpful to declare and set constants in the same location within the code to make finding and changing values even easier.

Similarly, depending on the scope of a variable, it is also a good programming practice to group variable declarations together at the beginning of each program, function, and/or procedure. Technically, variables can be declared and initialized anywhere within the program as long as it is before that variable is needed, but grouping them together at the beginning of a block of code ensures that the variable will be available for use and facilitates finding and maintaining that variable in the future.

11.6.1 Initialization

Once your variables have been declared, you will need to assign them values. Variables are considered *undefined* if a value has not been assigned. Undefined variables are unpredictable, and it is a good practice to *initialize* all variables to avoid unnecessary problems. In the gross pay calculation example above, we might decide to initialize both *hoursWorked* and *grossPay* to 0.0. We could do this using a **SET** command as follows:

```
SET hoursWorked = 0.0;
SET grossPay = 0.0;
```

Another way to do this is to combine the declaration and the initialization into one step. The pseudocode for this would look as follows:

```
DECLARE hoursWorked = 0.0 as REAL;
DECLARE grossPay = 0.0 as REAL;
```

It is important to note that only the second method works with constants. Once a variable is declared to be a constant, its value cannot be changed during the life of the program. Thus, the value of a constant must be set at the time it is declared.

By initializing these variables, we are ensuring that the computer does not have some unexpected value stored in the location of that variable and that the value is of the correct data type. Otherwise, there could be any type and size of data in that location, and if it happens to be of an incompatible data type or size, it may cause errors and possibly crash the program when it is executed.

11.6.2 Naming Conventions

As long as reserved words and illegal characters are not used, naming a variable or constant identifier is left up to the programmer. There are numerous naming conventions that programmers use, and unless mandated by an employer, the convention that a programmer chooses is merely a matter of preference. In general, variable and constant identifiers should be a combination of letters and/or numbers and should not contain any spaces or special characters, with the exception of the underscore character. Variable names should be as descriptive as possible while using the fewest characters possible. Some variables simply cannot be adequately identified with a single word, yet distinguishing words within a name that contains no spaces can be difficult. One naming convention that increases the readability of such identifiers uses the underscore character to replace spaces in descriptive variable names. Thus, declaring a variable to store the number of vacation days that an employee has earned might be named *vacation_days*.

Another common convention is known as *CamelCase*. In CamelCase variables, the first word of a variable name begins with a lowercase letter while the first letter of any additional words is capitalized. Thus, the variable that would be used to store the number of vacation days that an employee has earned would be named *vacationDays*.

For the purpose of consistency, we need to define a naming convention to use in our pseudocode. CamelCase naming will be adhered to where the first word of a variable name starts with a lowercase letter and the first word of a constant begins with a capital letter. Thus, *hoursWorked* would be an identifier for a variable and *HoursWorked* would represent a constant.

It is important to note that, in the majority of programming languages, variable and constant names are *case sensitive*. Thus, if the letters of a variable are not all in the same case, they are considered to be different variables assigned to different memory locations. Thus, the variable named *hoursWorked* is stored at a different memory location and has a different value than the variable identified as *HoursWorked*. While these are acceptable as independent variables, it is recommended that the programmer avoid using variable names within a program that differ only by case.

While it is acceptable to assign nondescriptive identifiers to variables and constants, it is strongly recommended to use more descriptive identifiers so that the programmer can easily infer their purposes in the program. For example, if *num1* and *num2* were used in place of *hoursWorked* and *grossPay* in our example above, it would be easy for the programmer to forget or lose track of the intended purpose of the variables. The numbers 1 and 2 are located right next to each other on the keyboard increasing the likelihood that a typo might occur, and locating errors of this type makes debugging more challenging. The more detail portrayed in a variable name, the better. When dealing with data that has units, some people even recommend representing units in a variable's name. Thus, if the height of a telephone pole is measured in feet, the variable might be named *poleHeightFeet*.

Frequently, the naming convention that a programmer uses is defined by management or by the project team. When a programming project has multiple contributors and/or consists of multiple components that work together, it is especially important for all contributors to agree on the naming convention that should be adhered to.

11.7 Input and Output

Every programming language has the ability to load data into the program and allow several different options for handling the output of a program. This is commonly referred to as the *input* and *output* of a program.

11.7.1 PRINT and WRITE

Output is required at various points during the program execution. Among other things, it can be used to provide the user additional instruction and/or clarification, display results from the program operations, or store data for later use. Two of the most basic and frequently used forms of output are displaying information on a computer or device screen and writing the data to a file so that it can be opened or accessed again at a later time. The reserved word, **PRINT,** will be used as our pseudocode keyword that represents displaying the data on a computer or device screen, and the reserved word, **WRITE,** to represent writing the data to an external file. Both **PRINT** and **WRITE** should be followed by the data to be displayed or written. Characters enclosed in double quotes ("") will be displayed exactly as typed. When a variable name follows either of these keywords, the value stored in that variable name will appear in the output. If you want multiple items to appear in the output, they should be separated by commas.

11.7.2 GET and READ

Similarly, the two most common ways of getting data values are by asking the user to enter the data via the keyboard and loading the data from a separate data file. For our purposes, we will use **GET** as our pseudocode keyword that represents prompting the user for data values, and **READ** to represent reading data from a file. Both **READ** and **GET** should be followed by variables that have been declared to store the correct type of data. Thus, when asking for the number of days until Christmas, the variable should be declared as an integer. When reading multiple pieces of data, the variable names in which to store that data should be separated by commas. While the **GET** command tells the computer to wait for an input from the user, it does not provide any instruction or guidance to the user as to what data is expected. Thus, the **GET** command is commonly used in conjunction with a **PRINT** command that asks the user to enter data.

Keep in mind that there are many *digital sensors* that exist and can be used in producing and supplying data to a program. Thermometers, light sensors, scales, and measuring devices are examples of *digital sensors*. In practice, the process of getting data into the program from these devices varies, but for the purposes of this book, we will assume that you can obtain this information in the same manner as you would obtain input from the keyboard.

Input–Output Example

```
PRINT "Please enter the day of the month";
GET dayOfMonth;
PRINT "This is day", dayOfMonth, "of the month."
```
This set of code will prompt the user to enter the day of the month. Assuming that the user enters the number 10, it will then display the following:
`This is day 10 of the month.`

11.8 Assignment and Arithmetic Operators

Most high-level programming languages come equipped with the ability to perform basic operations. An *operator* performs an action on one or more *operands*. The common arithmetic operators are shown in Table 11.3.

The *assignment* operator is the only operator from the list in Table 11.3 that has the built-in ability to work with all data types. It instructs the computer to store the resulting value in the memory location referred to by the identifier. The general form of the assignment statement is

`Identifier = expression;`

Be careful not to confuse the assignment operator with equality. The single = sign indicates assignment, while a double =, or ==, indicates a comparison and will be discussed later in this chapter.

The other operator that you might not be familiar with is %. The % sign is the symbol for the modulo operator. The modulus is simply the remainder after a division operation has been performed. For example, 13 divided by 4 is 3 with a remainder of 1. Thus, the modulus is 1. One practical application of the modulo operator is to determine if a particular number is an even or an odd number. In order to be even, the number has to be divisible by 2 with no remainder. A modulo operation with no remainder is 0, and thus, the program would declare a number, called *numToTest*, to be even if *numToTest* % 2 = 0.

TABLE 11.3

Basic Arithmetic Operators

Operation	Symbol
Assignment	=
Addition	+
Subtraction	−
Multiplication	*
Division	/
Modulus	%
Exponents	^

TABLE 11.4

Operation Precedence in Mathematical Calculations

Operation	Symbol
Parenthesis	()
Exponentiation	^
Multiplication, division, and modulus	* / %
Addition and subtraction	+ −

11.8.1 Order of Execution

Arithmetic operators work the same way as normal mathematical operations with which you are already familiar. They all produce the desired mathematical results and are all subject to rules of precedence when determining the order of execution. Understanding these few simple rules on how values are calculated is essential to accurately predicting the output of a program that requires arithmetic calculation (Table 11.4).

When operators of equal value are to be executed, the order of execution is left to right. If you find that you are unsure of the execution order, add parenthesis to mandate in what order the operations will be executed.

11.9 Conditional Expressions

A *conditional expression* is a type of expression that tests your data to see if it meets a condition, and then takes an action depending on the result. Both selection statements and iterative loops rely on the results of conditional expressions to determine their behavior.

11.9.1 Relational Operators

Relational operators check the relationship between two operands. If the relation is true, it returns value 1 and if the relation is false, it returns value 0 (Table 11.5):

TABLE 11.5

Basic Relational Operators

Operation	Meaning	Example of True	Example of False
==	Is equal	8 == 8	8 == 2
!=	Is not equal	5 != 4	5 != 5
>	Greater than	3 > 2	2 > 3
>=	Greater than or equal to	18 >= 1	18 >= 21
<	Less than	2 < 3	3 < 2
<=	Less than or equal to	1 <= 1	1 <= 0

Computer Programming

TABLE 11.6

Basic Logical Operators

Operation	Meaning	Alternative Symbol
AND	Evaluates as true if and only if both conditions are true	&&
OR	Evaluates as true if at least one condition is true	||
NOT	Evaluates as true only if the condition is false	!

11.9.2 Logical Operators

Logical operators are used to combine expressions containing relation operators. In most programming languages, the logical operators shown in Table 11.6 are predefined and available for use.

When we describe the results of a logical operation, we frequently use *truth tables*. Truth tables list all possible inputs and the result of the comparison, or the output. The truth tables for the logical operators discussed in this chapter can be found as follows:

AND Truth Table			OR Truth Table			NOT Truth Table	
A and B Are Inputs			A and B Are Inputs			A Is the Input	
C Is the Output			C Is the Output			B Is the Output	
A	B	C	A	B	C	A	B
True	True	True	True	True	True	True	False
True	False	False	True	False	True	False	True
False	True	False	False	True	True		
False	False	False	False	False	False		

There are numerous examples in our day-to-day lives where we use the above logic. While booleans can be thought of as true or false values, they can really be thought of in any terms where there can only be one of two results. Thus, yes/no can also be used to represent boolean values. To help us better understand the logic, consider the following analogies:

AND Analogy			OR Truth Table			NOT Truth Table	
Your mother asks you to stop at the store to pick up milk **AND** eggs			Your mother asks you to pick up chocolate ice cream **OR** vanilla ice cream from the store			You reach a fork in the road	
Did You Get Milk?	Did You Get Eggs?	Is Mom Happy?	Did You Get Vanilla?	Did You Get Chocolate?	Is Mom Happy?	Do I Go Left?	Do I Go Right?
Yes	Yes	Yes	Yes	Yes	Yes	Yes	No
Yes	No	No	Yes	No	Yes	No	Yes
No	Yes	No	No	Yes	Yes		
No	No	No	No	No	No		

Examples of Conditional Expressions

Assume that you are using numbers to represent days of the week. Number 1 will be used to represent Sunday, number 2 used to represent Monday, all the way to number 7, which will be used to represent Saturday. The user will be asked to enter the day of the week as a number and the value will be stored in the variable *dayOfTheWeek*.

a. Write a conditional expression that will evaluate to true if the *dayOfTheWeek* is a weekday.
   ANSWER: ((*dayOfTheWeek* >= 2) **AND** (*dayOfTheWeek* <= 6))
   EXPLANATION: Because the numbers 2–6 represent the days Monday–Friday, if the user enters any number from 2 to 6, then the expression will evaluate to true.
b. Write a conditional expression that will evaluate to true if the *dayOfTheWeek* is a weekend day.
   ANSWER: ((*dayOfTheWeek* == 1) **OR** (*dayOfTheWeek* ==7))
   EXPLANATION: Because numbers 1 and 7 represent the days Sunday and Saturday, respectively, if the user enters either number 1 or number 7, then the expression will evaluate to true
c. Evaluate the following conditional expression:
   ((*Dayoftheweek* == 1) **AND** (*dayOfTheWeek* ==7))
   ANSWER: Regardless of what the user enters, this will always be false
   EXPLANATION: Because it is impossible for it to be both Saturday AND Sunday on a particular day, this will always evaluate to false
d. Evaluate the following conditional expression:
   (*Dayoftheweek* = 1)
   ANSWER: Regardless of what the user enters, this will always be true
   EXPLANATION: Because a single equal sign is used in this expression, it is assigning the number 1 to *dayOfTheWeek* instead of comparing it to the number 1, and any number other than zero always evaluates to true.

11.10 Selection Statements

Selection statements use a condition to determine which set of instructions get executed. There are two types of selection statements, **IF-THEN** statements and **SELECT-CASE** statements. Both can accomplish the same task, and the choice of which to use depends on the situation and programmer's style.

11.10.1 IF-THEN

The simplest type of selection statement is the **IF…THEN** statement. This statement tests a condition, and if it is true, then the program executes the following instruction or set of instructions. Otherwise, the instructions immediately following the **IF-THEN** statement do not get executed and instruction execution resumes outside of the **IF-THEN** statement.

The **IF-THEN** statement is written in the following form:

```
IF (condition) THEN
BEGIN IF
     STATEMENTS…
END IF
```

Computer Programming

IF-THEN-ELSE and **IF-THEN-ELSE IF** are extensions of the **IF-THEN** clause that most high-level programming languages support. With **IF-THEN-ELSE**, the action to be performed when the condition is false can be specified after the **ELSE**. In an **IF-THEN-ELSE IF** clause, multiple conditions can be checked for within the same **IF** statement.

The following is the typical form of an **IF-THEN-ELSE** clause:

```
IF (condition) THEN
BEGIN IF
      STATEMENTS...
END IF
ELSEIF (condition 2)
      BEGIN ELSEIF
      STATEMENTS...
END ELSEIF
ELSE
      BEGIN ELSE
      STATEMENTS...
END ELSE
```

11.10.2 SELECT...CASE

The **SELECT...CASE** statement determines which code to execute based on the value of a variable. This allows the programmer to specify the program behavior for each possible value that a variable can have. Basic **IF...THEN** statements can only respond by executing the code if the condition is true or by not executing the **IF-THEN** statements if the condition is false. The **SELECT-CASE** tests a single variable to see if it is equal to specific values. It takes the following form:

```
SELECT (variable)
     CASE value1:
            STATEMENTS....
            BREAK;
     CASE value2:
            STATEMENTS...
            BREAK;
     # can add more CASE blocks as necessary
     DEFAULT:
            STATEMENTS;
     END CASE;
```

Note the use of the keyword **BREAK** after each.

Selection Structure Examples

The following examples produce the same results:

IF...THEN...ELSE	SELECT...CASE

`IF (number == 1) THEN`
`BEGIN IF`
   `PRINT "The number is 1.";`
   `number = number +1;`
`END IF`
`ELSEIF (number == 2) THEN`
`BEGIN ELSEIF`
   `PRINT "The number is 2.";`
   `number = number +1;`
`END ELSEIF`
`ELSEIF (number == 3) THEN`
`BEGIN ELSEIF`
   `PRINT "The number is 3.";`
   `number = number +1;`
`END ELSEIF`
`ELSE`
   `PRINT "The number is not 1,`
   `2 or 3.";`
	`SELECT (number)`
	   `CASE 1:`
	   `PRINT "The number is 1.";`
	   `number = number +1;`
	   `BREAK;`
	   `CASE 2:`
	   `PRINT "The number is 2.";`
	   `number = number +1;`
	   `BREAK;`
	   `CASE 3:`
	   `PRINT "The number is 3.";`
	   `number = number +1;`
	   `BREAK;`
	   `DEFAULT:`
	   `PRINT "The number is not`
	   `1, 2 or 3.";`
	`END CASE;`

11.11 Loop Structures

Loops allow a block of instructions to repeat until a condition is met.

11.11.1 FOR Loops

A **FOR** loop uses a counter to execute a set of instructions a given number of times based on the value of that number. For example, if you want to execute a set of instructions 10 times, it would look like the following:

```
FOR count = 1 to 10
BEGIN           #begin FOR loop
    # this could be any instruction or block of instructions
END             #end FOR loop
```

In the above example, *count* is initially set to 1 and every time that the **END** statement is reached, *count* is automatically increased by 1. The computer then returns to the **FOR** statement, checks to see if *count* is between 1 and 10, and if it is, executes the statements in the **FOR** loop again. Otherwise, it skips the instructions in the **FOR** loop and exits the loop without returning again to the **FOR** statement. Thus, the value of *count* is the condition that determines whether or not the loop iterates.

11.11.2 WHILE...DO Loops

WHILE...DO loops are similar to **FOR loops** in that there is a condition that must be true in order for the loop to iterate. In this case, however, the condition is not confined to being an integer and can contain more complex logic. The programmer needs to be cautious in these structures because, in contrast to the **FOR** loops where the variable is automatically increased, the programmer must provide a way for the condition to change or be stuck in an infinite, or un-ending, loop.

The basic structure of a **WHILE...DO** loop is as follows:

```
WHILE (condition) DO
BEGIN WHILE
        STATEMENTS...
END WHILE
```

11.11.3 DO...UNTIL Loops

DO...UNTIL loops are very similar to **WHILE...DO loops**. The only difference is that in a **WHILE...DO** loop the condition must be true for the code within the loop to be executed, whereas in a **DO...UNTIL** loop the code in the loop is executed at least once before the condition is checked. Thus, depending on the condition that is to be satisfied, it is possible that the code in the **WHILE...DO** loop is never executed, and, in the **DO...WHILE** loop, the code will always be executed at least once.

The basic structure of a **DO...UNTIL** loop is as follows:

```
        DO
BEGIN DO
            STATEMENTS...
END DO
        WHILE (condition)
```

11.12 Functions and Procedures

All of the examples thus far have assumed that all program instructions reside in one main section of the program. *Functions* and *procedures* allow the programmer to physically separate the code into smaller manageable sections. The difference between functions and procedures is that a function is written to return a single value while procedures are not. Code that appears in these functions and procedures is often the code that might be repeated in several places in the computer program. While the scope of this chapter does not cover proper use of functions and procedures, it is important to remember that these are a natural extension of the basic skills covered here and are used frequently in good programming practice.

A function call is an expression that passes control and arguments (if any) to a function.

Parameter is an item of information such as name, number that is passed by the user, or any other program to the current program. These will affect the program operation receiving them.

Formal parameter—the identifier used in a function to stand for the value that is passed into the function by a caller. The identifier can be a name of class, variable, method, or a parameter.

Actual parameter—the actual value that is passed into the function by a caller.

Syntax:

```
function1 (list of parameters)
{
Code;
}
```

11.13 Testing and Debugging

11.13.1 Types of Errors

When a computer program does not operate the way the programmer is expecting, the code contains an error or a *bug*. Some errors are simple to identify and repair, while others can be quite challenging to discern. There are two types of errors that can occur in any program: *syntax errors* and *logical errors*.

Syntax errors can be considered grammatical errors and are typically the easiest to locate and fix. Because the compiler cannot translate code that is not in the correct syntax into machine code, the compiler will detect these errors and will not successfully compile until the syntax errors are resolved. Typically, when a syntax error occurs, the compiler responds with an error message that indicates what line in the code contained the error. Because one syntax error can generate several error messages, it is recommended that you start with the first error, repair it, and then compile again. Often, repairing one error will resolve a number of error messages.

Logical errors are usually a bit more challenging to resolve. Depending on the specific error, it is possible that the error will cause the program to crash. Other times, the program completes with incorrect results. These errors tend to be much harder to locate in the program. The programmer may have to check variable values at several points in the program in an effort to narrow down the location of the error or errors. At times, this task can be tedious, but, performed systematically, can effectively identify the source(s) of the error.

It is important to test your program with many types of input. Even when you think that all errors have been resolved, it is important to double check with a wide range of inputs. Be sure to test all possible cases to make sure that your program responds appropriately.

11.14 Summary

In this chapter, we explored programming and its application to data. While data manipulation is the simplest to recreate in the classroom setting, keep in mind that this is not a limitation that exists elsewhere. Inputs and outputs of programming can be vastly different in nature. Even if the program responds to data, the data could be obtained from any sort of sensor or other input device. Likewise, instead of responding by displaying a message, it is possible to respond by performing an action such as lifting a bridge or running a motor.

Computer Programming

The possibilities of programming are endless, and with more and more programmable devices being produced every day, programming is an essential skill for engineers to possess.

Problems

1. Draw a flowchart that represents an algorithm that determines whether you should ride your bike to school or take the bus.
2. Write a program flowchart and the corresponding pseudocode asking the user to enter the number of miles that he/she traveled and the number of gallons of fuel that his/her vehicle consumed. Use this information to calculate and display the user's average mileage per gallon.
3. Using pseudocode, write a program that will ask the user to enter five numbers. The program should then calculate and display the average of those five numbers. If the average is less than 100, prompt the user for five more numbers and calculate the average again until the average is greater than 100.
4. Using pseudocode, write a program that translates the months of the year into a numerical value. January should translate to 1, February should translate to 2, and so on until December that should translate to 12.
5. Using pseudocode, write a program that asks the first user to enter a number between 1 and 100. Then have a second user guess the number that the first user entered. After each guess, the program should tell the user whether he/she is too low or too high until the user guesses the correct number.
6. Adjust the program in problem 5 so that it will also tell the users how many guesses it took until the correct number was found.

Section III

Product Design and Development

12

Product Design and Development

CHAPTER OBJECTIVE AND STUDENT LEARNING OUTCOMES

After completing the chapter, students will be able to

1. Define the design processes and the basic concept of design.
2. Describe the product design factors and techniques for the optimum design process.
3. Explain and analyze the product development processes and distinguish between the product design phase and the product development phase. [ABET outcome c, see Appendix C]
4. Describe various techniques and tools for design improvement and the role of R&D. [ABET outcome k, see Appendix C]

12.1 Introduction

Projects are finite activities that seek to develop new and improved tools, methods, products, and equipment. The goal of this chapter is to give you the tools to develop an efficient design process regardless of the product being developed. This chapter will introduce the important features of design problems and the processes for solving them. These features apply to any type of design problem, whether for mechanical, electrical, software, or construction projects. Subsequent chapters will focus more on mechanical design, but even these can be applied to a broader range of problems.

Before knowing the design process, it is necessary to know the main product design factors. Product design factors focus on the product's function, which is a description of what the object does. Related to the function are the product's form, materials, and manufacturing processes. Of equal importance to form are the materials and manufacturing processes used to produce the product. The choice of form and materials that give the product function affects the manufacturing processes that can be used. The *process* of mechanical design focuses not on the design of any one type of object but on techniques that apply to the design of all types of mechanical objects.

12.2 Overview of the Design Process

For design projects, we require a clear reason for solving a problem, the needs, and a specific measure of satisfactory solution. A very clear set of specifications focuses on the design process, leading to a clear and measurable success. The process of defining the specifications is critical to the success of the project.

Use the specifications to begin conceptual design; selecting a design by brainstorming different ideas and narrowing down to two to three designs are the primary activities of the conceptual design phase. When we generate concepts, the customers' requirements serve as a basis for developing a functional model of the product. Design engineers have to compare the concepts to the requirements. The detailed design ends with a formal agreement between the customer and design team. The goal for the detailed design process is approval of a detailed design that will meet the specifications. During the build and test phase, problems should be dealt with as soon as possible, including updating the detailed designs. The project closure process starts once a customer has accepted the deliverables (Figure 12.1).

12.2.1 Product Discovery

Before the original design or redesign of a product can begin, the need must be established. There are three primary sources for design projects: technology, market, and change. Regardless of the source, a common activity at most organizations is maintaining a list of potential projects. Because companies have limited people and money, they must choose which project to develop. Sometimes, this decision comes before project planning and sometimes it is postponed until later, after planning, product definition, and conceptual design have been done and more is known about each of the options.

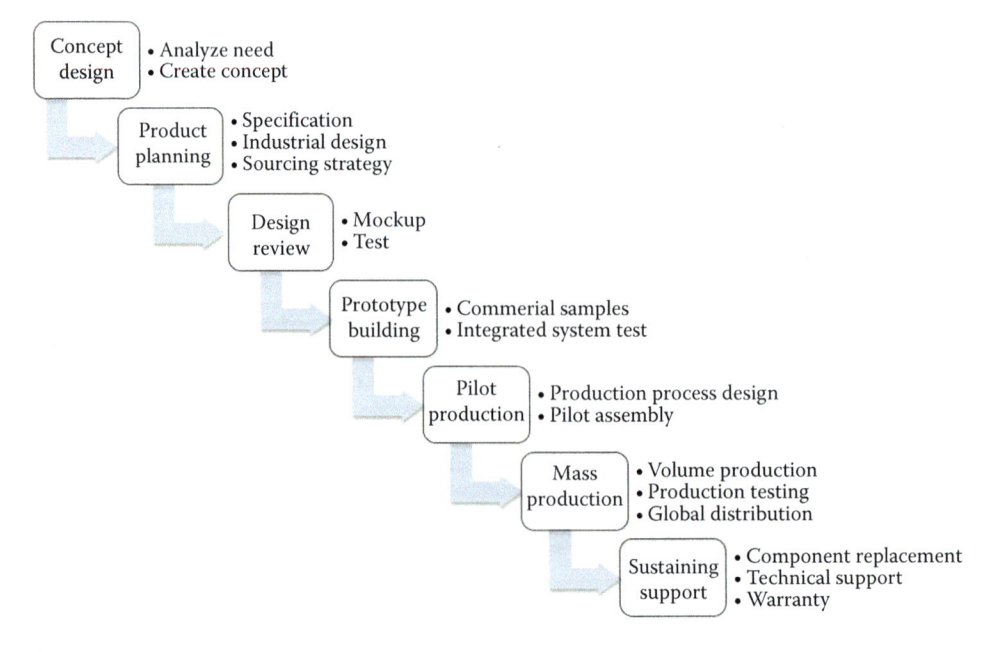

FIGURE 12.1
Product design to production.

12.2.2 Project Planning

The second phase is to plan so that the company's resources of money, people, and equipment can be allocated and accounted for. Planning needs to precede any commitment of resources; however, as with much design activity, this requires speculating about the unknown—and that makes planning for a product similar to an earlier product easier than planning for a totally new one. Planning requires a commitment of people and resources from all parts of the company, so a design team must be formed. Additionally, much planning work goes into developing a schedule and estimating costs. The final goal for this phase is generating a set of necessary tasks and a sequence for them (Figure 12.2).

12.2.3 Product Definition

During the product definition phase, the goal is to understand the problem and lay the foundation for the remainder of the design project. Understanding the problem may appear to be a simple task, but because most design problems are poorly defined, finding the definition can be a major undertaking. In order to define the problem, first customers need to be identified. This activity serves as the basis to generate the customers' requirements. These requirements are then used to evaluate the competition and to generate engineering specifications and measurable behaviors of the product-to-be. That information, later in the design process, will help in determining product quality. Finally, in order to measure the *quality* of the product, we set targets for its performance.

12.2.4 Conceptual Design

Designers use the results of the planning and product definition phases to generate and evaluate concepts for the product or product changes. When we generate concepts, the customers' requirements serve as a basis for developing a functional model of the product. The understanding gained through this functional approach is essential for developing concepts that will eventually lead to a quality product. Concepts must be compared to the requirements developed during product definition. Then, the design team must decide on the optimal design.

12.2.5 Product Development

After concepts have been generated and evaluated, it is time to refine the best of them into actual products. Unfortunately, many design projects begin here, without the benefit of prior specification or concept development. This design approach often leads to poor-quality products and in many cases causes costly changes late in the design process. It cannot be overemphasized: starting a project by developing a product, without concern for the earlier phases, is poor design practice. At the end of the product development phase, the product is released for production. At this time, the technical documentation defining manufacturing, assembly, and quality control instructions must be complete and ready for the purchase, manufacture, and assembly of components.

12.2.6 Product Support

The design engineer's responsibility may not end with release to production. Often there is continued need for manufacturing and assembly support, support for vendors, and help in introducing the product to the customer. Additionally, design engineers are usually involved in the engineering change process. This is the process where changes made to the product, for whatever reason, are managed and documented.

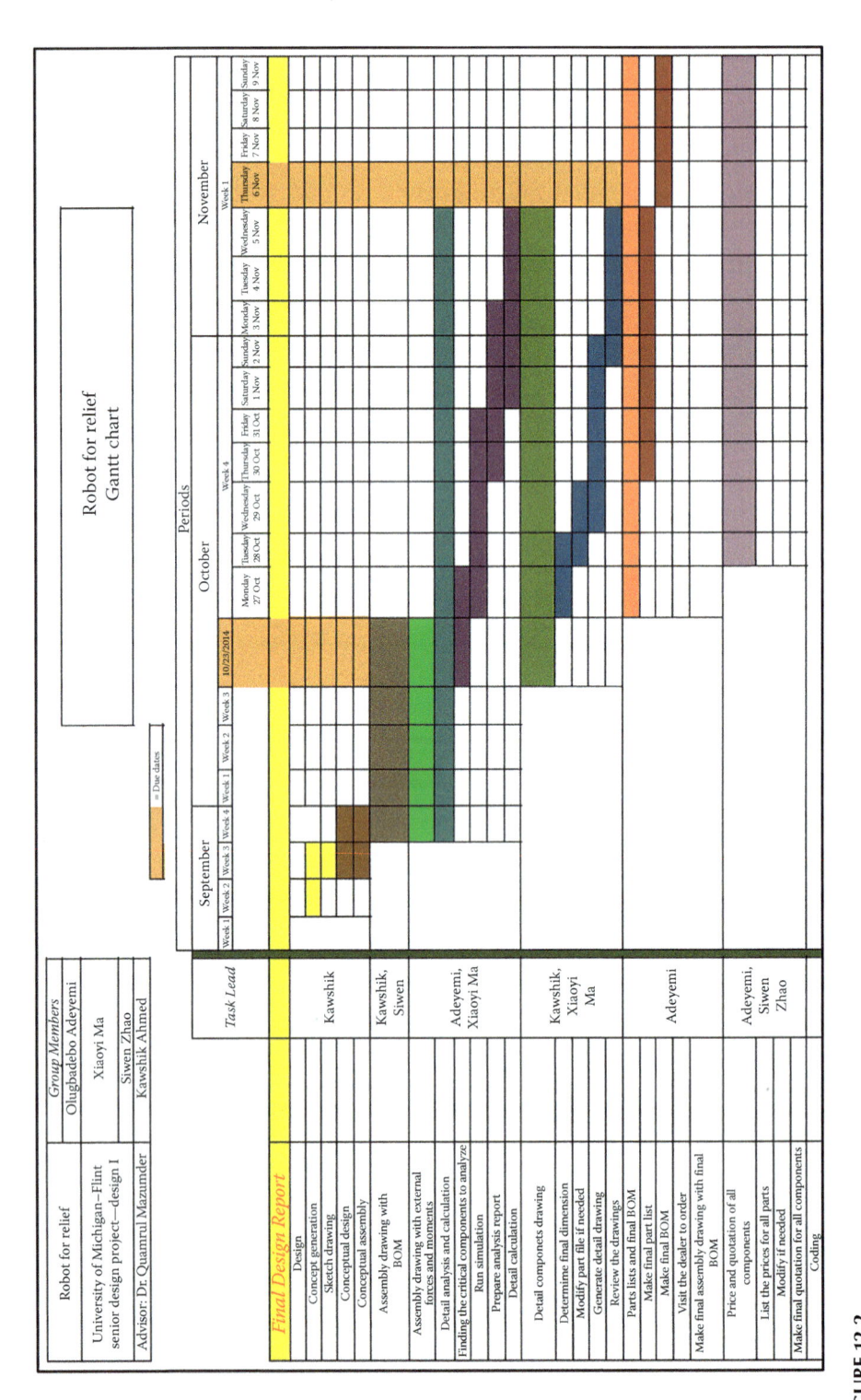

FIGURE 12.2
Gantt chart.

12.3 Definition of the Problem

Designing an industrial product is a multidisciplinary activity as functional, psychological, technological, and economic criteria are all involved. Industrial product designer, acting through these criteria and fulfilling the design function, also acts as a team synthesizer who builds a communication bridge between other professions such as engineering, sociology, and marketing. This formation is because of the demands of the modern world. While specializations have been developed, needs of the modern world, such as airplanes, fast trains, and spaceships, have brought these specializations back together to act in a team toward a common purpose. Organizations may select a number of product development activities as a primary strategic element of competitive advantage in the market. To accelerate the product development activities, corporations develop collaborative program between them and local universities. An example of such collaborations is between industrial organizations and three leading Finnish universities. The aim of this program is to produce professionals (designers, marketers) with a multifaceted view on product development, and with a holistic understanding of the design dimension. This constitution reveals the interdisciplinary approach to both design and business education. One of the concerns is that creating an interdisciplinary approach fails to connect between subdisciplines, fails to reach common understanding, and fails to develop new knowledge and perceptions of design as Nigel Cross states in the proceedings of the Politecnico di Milano Conference (2000: p. 46). Because of dealing with a lot of criteria, the industrial product design field can be stretched to other fields easily, and other fields can be welcomed in the industrial product design field easily, which causes conflicts in developing industrial product design knowledge. Cross stated that the design should be taken as a discipline. Industrial product design is going to be taken as a field of design discipline that accumulates and develops its own design knowledge. Referring to this formation, industrial product design might create and strengthen its place among other overlapping fields and disciplines. Industrial product design, as a field of design discipline, borrows concepts and methods from sciences, arts, engineering, and humanities in order to develop its own knowledge in research and industry contexts. Thinking and acting in this way might strengthen the place of industrial product design while still keeping it as an advantage of the modern world.

12.4 Product Development Process

The narrower definition of products is that they are physical and tangible. A new product in the market is the outcome of varying systematic processes and approaches. It can be assumed that any product in the market or in the hand of customers is masterminded by some group of people; in simpler words, we can say *engineers* or *product development engineers*. Product development is a multistage process in which product design plays the most important role to produce a successful product.

So, product development is a multiphase process in which a set of required activities brings a new product concept to a state of market readiness. The product development process involves the organization and management of people and information they develop in evolution of a product.

To understand the processes of product development, let us discuss the role of a product development engineer. Then we will have a better understanding of the tasks that a product development engineer needs to complete.

The following are major responsibilities for a product development engineer:

- Design and development of new products and processes. This involves complete project responsibility, including concept development, component design, design verification, scheduling, and product cost and reporting.
- Understanding of user needs in order to develop technical product specifications to meet those needs.
- Feasibility studies to assess technology matches to product specifications.
- Proactive participation in cross-functional teams. Collaboration in order to achieve optimum outcomes.
- Prioritization of goals in order to deliver products on time while meeting cost and quality targets.
- Failure analysis and implementation of corrective measures as well as participation in cost and quality improvement programs.

The most general product development flowchart is shown in Figure 12.3.

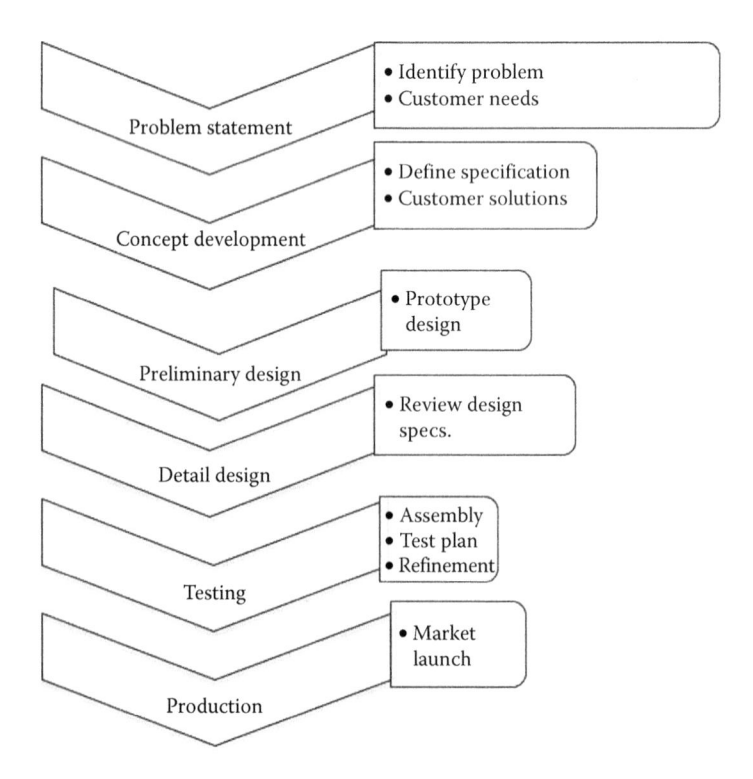

FIGURE 12.3
Product development process.

12.5 Product Design

Product design is the most fundamental step in the product development process. Product design defines a product's characteristics, such as its appearance, the materials it is made of, its dimensions and tolerances, and its performance standards. Because the product development process is very much related to product life cycle (PLC), the processes associated with product design can increase product life. It is very important to generate and formulate a robust design for any product. The designing of a product involves materials, measurements, dimensions, and blueprints. When we think of design, we usually think of car design or computer design and envision engineers working on diagrams. However, product design is much more than that. Product design brings together marketing analysts, art directors, sales forecasters, engineers, finance experts, and other members of a company to think and plan strategically.

Product design is the process of defining all the features and characteristics of just about anything, for example, a Borics Salon or Federal Express. Consumers respond to a product's appearance, color, texture, and performance. In any industry, design and R&D engineers continuously research and brainstorm new ideas for products as well as consider customer needs.

All of its features, summed up, are the product's design. Someone envisioned what this product will look like, taste like, or feel like so that it will appeal to you. This is the purpose of product design.

12.5.1 Steps in Product Design

Step 1: Idea development—All product designs begin with an idea. Someone thinks of a need and a product design that would satisfy it.

Step 2: Product screening—Once an idea is developed, it needs to be evaluated. Often a business comes up with numerous product ideas. At this stage we need to screen the ideas and decide which ones have the greatest chance of succeeding.

Step 3: Preliminary design and testing—This is the stage where preliminary design of the product is made and the company performs market testing and prototype analysis.

Step 4: Final design—This is the last stage, where the final design of the product is made. Next we look at these steps in a little more detail.

- *Idea development*: All product designs begin with an idea. The idea might come from a product engineer or design engineer who spends time with customers and has a sense of what customers want. To remain competitive, companies must be innovative and bring out new products regularly. In some industries, the cycle of new product development is predictable. We see this in the auto industry, where new car models come out every year, or the retail industry, where new fashion is released every season. In other industries, new product releases are less predictable but just as important. The Body Shop, a retailer of plant-based skin care products, periodically comes up with new ideas for their product lines. The timing often depends on the market for a product, and whether sales are declining or continuing to grow.

- *Product screening*: After an idea has been developed it needs to be evaluated to determine its likelihood of success. The company's product screening team evaluates the product design idea according to the needs of the existing customers, new potential customers, and alignment with core value of the organization. During the evaluation phase of new product development, executives from each functional area may explore issues such as the following:

- Operations
  - How to relate production needs and existing resources?
  - Will we need new facilities and equipment?
  - Do we have enough labor skills and material for production to make the product?
- Marketing
  - Potential size of the market for the new product.
  - What needs will develop a market for the product?
  - Long-term product potential.
- Finance
  - Is the financial investment feasible for new product?
  - Evaluate product's commercial potential, cost, and return on investment.

It is very difficult to predict whether a product will be successful due to changes in marketplace, competition, and customers view toward the product. Effective product development also requires all internal and external resources to be available during the development phase. The lead product engineer may possess project managerial skill and experience; however, they may be already committed to a number of other projects. Organizations must continuously generate new product ideas, whether for a new brand of cereal or a new design for a car door. Approximately 80% of ideas do not make it past the screening stage. Management analyzes operations, marketing, and financial factors, and then makes the final decision. Fortunately, we have decision-making tools to help us evaluate new product ideas:

- *Preliminary design and testing*: After passing the screening stage, products undergo preliminary design and testing. This is the time for designers and design engineers. A senior product engineer or development engineer moderates and coordinates the design team. At this stage, design engineers translate general performance specifications into technical specifications. Prototypes are built and tested. There is more than one design engineer working on the same projects and reporting to the product engineer. Design engineers often rely on computer-aided design (CAD), which is a comprehensive tool for designing and simulation. Changes are made based on test results and the process of revising, rebuilding a prototype, and testing continues. Product refinement can be time consuming and there may be a desire on the part of the company to hurry through this phase to rush the product to market. However, rushing creates the risk that all the *bugs* have not been worked out, which can prove very costly.
- *Final design*: Following extensive design testing, the product moves to the final design stage. This is where final product specifications are drawn up. The final specifications are then translated into specific processing instructions to manufacture the product, which include selecting equipment, outlining jobs that need to be performed, identifying specific materials needed and suppliers that will be used, and all the other aspects of organizing the process of product production.

12.5.2 Factors to Consider in Product Design

12.5.2.1 Design for Manufacture

Customer satisfaction is the first priority for the new product design. It is also important to consider how easy or difficult it is to manufacture the product. Otherwise, we might have a great idea that is too difficult or costly to manufacture.

Design for manufacture (DFM) is the method of design for ease of manufacturing the collection of parts that will form the product after assembly. It is the optimization of the manufacturing process. Design for assembly (DFA) is a tool used to select the most cost effective material and process used in production in the early stages of the product design.

DFM is a series of guidelines that we should follow to produce a product easily and profitably. DFM guidelines focus on two issues:

1. *Design simplification* means cutting down the number of parts and features of the product whenever possible. A simpler product is easier to make, costs less, and gives us higher quality.
2. *Design calibration* refers to the use of common and interchangeable parts. By using interchangeable parts, we can make a greater variety of products with fewer inventories, significantly lower cost, and greater flexibility.

12.5.2.2 Product Life Cycle

Another factor in product design is the stage of the life cycle of the product. Most products go through a series of stages of changing product demand called the PLC. There are typically four stages of the PLC: introduction, growth, maturity, and decline (Figure 12.4).

Products in the introductory stage are not well defined and neither is their market. Often all the *bugs* have not been worked out and customers are uncertain about the product. In the growth stage, the product takes hold and both product and market continue to be refined. The third stage is that of maturity, where demand levels off and there are usually

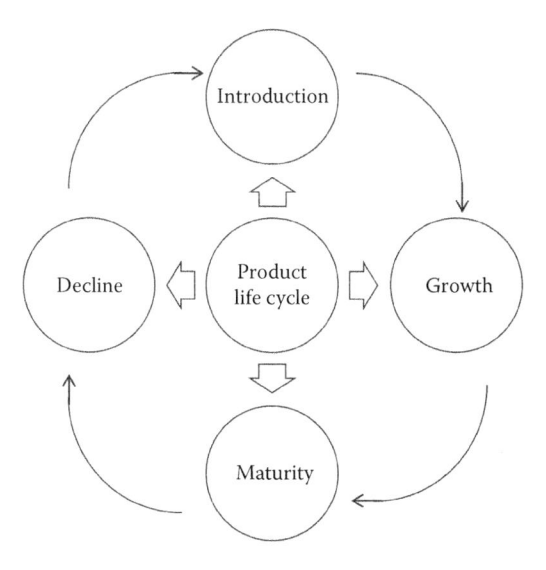

FIGURE 12.4
Product life cycle diagram.

no design changes. The product is predictable at this stage and so is its market. Many products, such as toothpaste, can stay in this stage for many years. Finally, there is a decline in demand, because of the new technology, better product design, or market saturation. The first two stages of the life cycle can collectively be called the early stages of the PLC because the product is still being improved and refined, and the market is still being developed. The last two stages of the life cycle can be referred to as the later stages because here the product and market are both well defined. Understanding the stages of the PLC is important for product design purposes, such as knowing at which stage to focus on design changes. Also, when considering a new product, the expected length of the life cycle is critical in order to estimate future profitability relative to the initial investment. The PLC can be quite short for certain products, as seen in the computer industry. For other products, it can be extremely long, as in the aircraft industry. A few products, such as paper, pencils, nails, milk, sugar, and flour, do not go through a life cycle. However, almost all products do and some may spend a long time in one stage.

12.5.2.3 Concurrent Engineering

Concurrent engineering (CE) is an approach that brings many people together in the early phase of product design in order to simultaneously design the product and the process. This type of approach has been found to achieve a smooth transition from the design stage to actual production in a shorter amount of development time with improved quality results. The old approach to product and process design was to first have the designers of the idea come up with the exact product characteristics. Once their design was complete, they would pass it on to operations who would then design the production process needed to produce the product. This was called the *over-the-wall* approach, because the designers would throw their design *over-the-wall* to operations who then had to decide how to produce the product. There are many problems with the old approach (Figure 12.5).

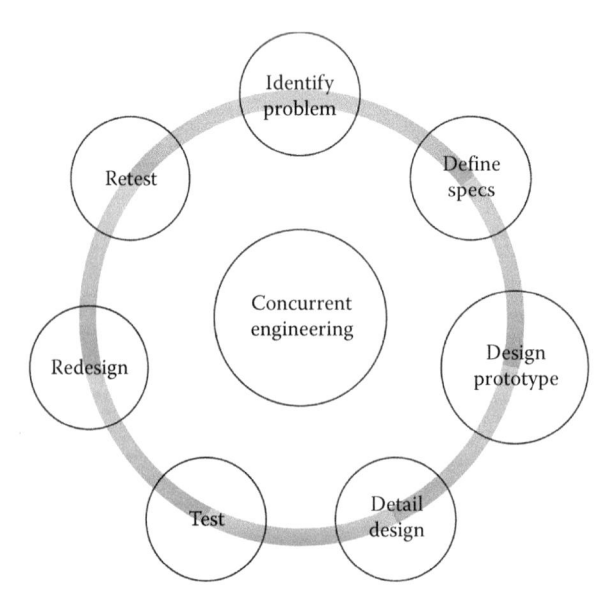

FIGURE 12.5
Concurrent engineering diagram.

First, it is inefficient and costly. For example, there may be certain aspects of the product that are not critical for product success but are costly or difficult to manufacture, such as a dye color that is difficult to achieve. Because manufacturing does not understand which features are not critical, it may develop an unnecessarily expensive production process with costs passed down to the customers. Because the designers do not know the cost of the added feature, they may not have the opportunity to change their design or may do so much later in the process, incurring additional costs. CE allows everyone to work together so these problems do not occur. In today's markets, new product introductions are expected to occur faster than ever. Companies do not have the luxury of enough time to follow a sequential approach and then work the *bugs* out. They may eventually get a great product, but by then the market may not be there! Another problem is that the old approach does not create a team atmosphere, which is important in today's work environment. Rather, it creates an atmosphere where each function views its role separately in a type of *us versus them* mentality. With the old approach, when the designers were finished with the designs, they considered their job done. If there were problems, each group blamed the other. With CE, the team is responsible for designing and getting the product to market. Team members continue working together to resolve problems with the product and improve the process.

12.5.2.4 Remanufacturing

Remanufacturing has been gaining increasing importance as our society becomes more environmentally more conscious. As a concept, it focuses on efforts such as recycling and eliminating waste. Remanufacturing uses components of old products in the production of new ones. In addition to the environmental benefits, there are significant cost benefits because remanufactured products can be half the price of their new counterparts. Remanufacturing has been quite popular in the production of computers, televisions, and automobiles.

12.6 Objectives of Design and Development Techniques

Design techniques form a set of tools that enable product innovation, improved quality, functionality, image, and differentiation, thereby permitting engineers to greatly increase their competitiveness. The main goals of Product Design methodologies are to

1. Help new products meet the specifications related to customers' needs, quality, price, manufacturing, recycling, and so on.
2. Reduce development costs and time necessary for commercialization.
3. Coordinate and schedule the activities involved in the design and development of products within the entire set of activities, taking into account time, tasks, resources, manufacturing, and so on, all in the context of the company.
4. Integrate the above objectives into a development strategy in line with the company's capacities.

12.7 Techniques That Can Be Used in Product Design and Development

Many *technology-ready* techniques and tools are currently in use. The comprehensive and simultaneous conception of the product development process entails specific design and development techniques that permit managing relevant information. These techniques can be classified into two broad groups.

12.7.1 Techniques and Tools for Design Improvement

These provide the company with analytical techniques and tools designed to analyze the product concept in the context of its restrictions. Main techniques of this group are as follows:

12.7.1.1 Concurrent Engineering

CE, also known as concurrent engineering, increases competitiveness by decreasing the lead-time, while improving quality and cost. CE is a system in which various engineering activities in the production and development process are integrated and performed in parallel rather than in sequence. Implementation of concurrent engineering means that multifunctional teams cooperate in the early stages of the product development process to fulfill these objectives. As a result, most modifications of the product will not be made in the production stage (more costs) but in the design stage. The product must be designed taking into account all the requirements necessary for each stage of its life cycle, from design to deactivation. These include functional factors but also aesthetic, ergonomic, easy manufacturing assembly, repair, and recycling. Product oriented organizational strategies establish a specific platform so that those involved in all relevant company functions have the chance to articulate their interest and concerns. Thus, the enterprise establishes cooperation as a basic component of the product development process at an early stage.

12.7.1.2 Quality Function Deployment

This methodology is a means to convert the client's opinions into the specifications of the product at every step of its development. It is helpful to structure and systematize several steps usually carried out in a discovered and incoherent way. To perform quality function deployment (QFD), interdisciplinary teams are formed, bringing together marketing, research and development, process planning, quality assurance, and manufacturing (Figure 12.6). This multidisciplinary approach permits as follows:

- Listening to the voice of the customer
- Improving horizontal and vertical communications
- Setting priorities for product development
- Improving product reliability
- Defining technical goals
- Sequencing individual goals
- Defining areas of cost reduction

Product Design and Development

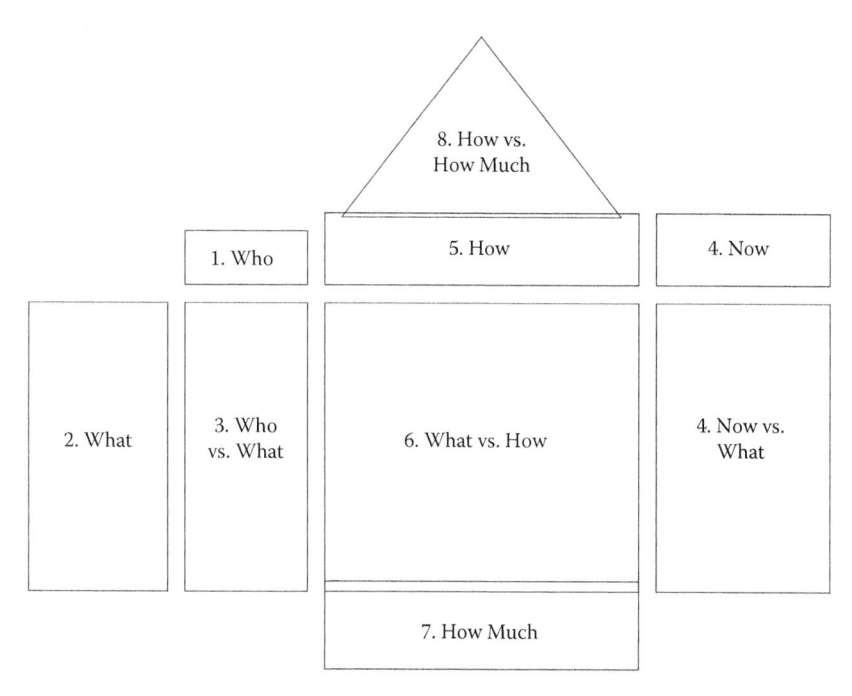

FIGURE 12.6
The house of quality (QFD diagram).

Step 1: Identify the Customers: Who Are They?

Step 2: Determine the Customers' Requirements: What Do the Customers Want?

Step 3: Determine Relative Importance of the Requirements: Who versus What.

Step 4: Identify and Evaluate the Competition: How Satisfied Are the Customers Now?

Step 5: Generate Engineering Specifications: How Will the Customers' Requirements Be Met?

Step 6: Relate Customers' Requirements to Engineering Specifications: How to Measure What?

Step 7: Set Engineering Specification Targets and Importance: How Much Is Good Enough?

Step 8: Identify Relationships between Engineering Specifications: How Are the How's Dependent on Each Other?

12.7.1.3 Design for X

As the process of product development and design is conceived, products must meet a broad series of optimization requirements, generically denominated X. Design of various factors, such as manufacture, assembly, and environment, is defined as DfX and aims to optimize design, manufacture, and support, through the effective feedback of the Xs within the design domain knowledge. In order to incorporate it during the design stages, X in DfX stands for manufacturability, inspectability, recyclability, and so on. These words are made up of two parts: life-cycle business process (x) and performance measures (bility), that is, X=x + bility. For example, *x=total* and *bility=quality* in *design for total quality*; *x=whole life* and *bility=costs* in *design for whole – life costs*; *x=assembly* and *bility=cost* in *design for assembly cost*, and so on. On the other hand, *design* in DfX is interpreted as concurrent design of products and associated processes and systems.

The proliferation and expansion of Xs has led to a string of new terms such as design for manufacturability, design for quality, and design for recyclability. Design for X has been devised as an umbrella for these terms. Most of them are closely related and decisions made on any one of them may affect the other Xs in the final product performance.

DFMA (design for manufacturing and assembling) is the integration of the separate but highly interrelated issues of assembly and manufacturing processes. It aims to help companies make the fullest use of the manufacturing processes that exist, while keeping the number of parts in an assembly to a minimum. First, DFA is conducted, leading to a simplification of the product structure. Then, early cost estimates for the parts are obtained, for both the original design and the new design, in order to make trade-off decisions. During this process the best materials and process for various parts are considered. Once the materials and processes have been selected, a more thorough analysis for DFM can be carried out for the detail design of the parts.

12.7.1.4 Failure Mode and Effects Analysis

Failure mode and effects analysis (FMEA) evaluates, in a systematic and structured way, the effects of failures on customers (Figure 12.7). A list of possible failures, their effects, and causes is drawn up and classified by effects on the client. This evaluation makes it possible to give priority to corrective actions. There are two kinds of FMEA: process (client=final user of next process stage) and design (client=final user). To use this method effectively, FMEA concentrates on selected system components. It is used for the following:

- New development of a product
- Security and problem parts
- Product or process modification
- New operation or other conditions of existing products

FMEA WORKSHEET														
Failure mode and effects analysis														
Name of the part	Passenger car door								Customer	Major automotive company				
Assembly	Passenger car								Original date					
Project engineer	Kawshik Ahmed								Revision date					
Team members														
											Action results			
Product or process description	Potential failure mode	Potential effects of failure	S E V	Potential causes	P R O B	Current controls	D E T	R P N	Recommended actions	Responsibility and target completion date	Action taken	S E V	P R O B	D E T	R P N
Provide access and exit from vehicle	Corroded B-pillar and interior lower panel	Improper fuction of the door	6	Insufficient wall thickness	4	Design reviews	7	168	Perform testing and may increase thickness	Month/Day/yr	Increase material thickness				0
Protect passengers and drivers from weather, noise, and accident		Propagate the rust	4	Wrong material	5		4	80	Review specs	Month/Day/yr	Modify material				0
Surface for appearance items		Insufficient room between	5	Design flaw	6		3	90							0

FIGURE 12.7
Failure mode and effects analysis (FMEA).

12.7.2 Computational Techniques and Tools

These are techniques that support design integration through shared product and process models, and databases. The advantages of effectively using computational techniques and tools are to allow different teams to share information and to manage all data required to proceed on the process. The main techniques of this group are the following:

12.7.2.1 Computer-Aided (CAx) Systems

These computer applications are used in the creation, modification, analysis, and optimization stages of product design. The term CAx means computer-aided (CA) support of the industrial production, where the x stands for different activities within the product development and manufacturing process, such as D for design, Q for quality, and E for engineering. Because 70%–80% of product costs are determined during the development phase, the most important systems in this group of tools are CAD. The competitive pressure in the development of industrial products adds new demands on the development process, which cannot be fulfillment by traditional approaches of CAD technology. Industries that want to keep their market position need to develop strategies that take into account new solutions for information technology. Some characteristic features of current and future trends in CAD development are listed below.

- x–D geometric modeling (2–D, 3–D)
- Employment of process chains and product models
- Parametric and associative design
- Open architecture of hardware and software systems
- User driver software development
- Distributed product development and life cycle engineering

Computer-aided product planning (CAPP) is the meeting of CAD and CAM. This system is used to automate the repetitive functions of process planning. It produces more constant and efficient plans, taking into account the available equipment, updated designs, and the most recent engineering changes. There are two kinds of CAPP systems:

- *Variant*: These systems select one plan from a library of already existing process plans and modify it to adapt it to the specific manufacturing requirements of a new product.
- *Regenerative*: These systems create the production process plan with no reference to any existing plan.

12.7.2.2 Engineering/Product-Based Data Management

These systems are also considered CAx systems. An engineering/product-based data management (E/P BDM) system manages all the information—data and processes—related to the product electronically. It allows two different possibilities: first, creating reports, data transport, images and translation services, files, NC programs, documents, and so on, and secondly, providing interfaces to other systems (CAD, CAM, etc.) or integration with different databases.

12.7.2.3 Knowledge-Based Engineering

Knowledge-based engineering (KBE) is a system that can be programmed to reproduce the decisions that an engineer has to take when producing designs. This system use databases, a knowledge base, and a set of rules are called algorithms, which are able to make decisions using the knowledge contained in the knowledge base. KBE is a step in the development of CAD systems, because it not only uses design information it also includes the rules that are used to create design. KBE systems are also known as expert systems.

KBE systems, such as CAD systems, are used throughout the entire design process, especially in the detail design phases. Moreover, they store engineering information generated during the design process.

12.7.2.4 Finite Element Analysis

The finite element analysis (FEA) method is based on the breaking up of the goal model into a series of finite elements. In other words, the model is divided into numerous small parts that are then used as study units for analysis. The model obtained is a meshwork of elements joined together by common nodes. These elements can be flat (representations of surfaces), or volumetric (representation of solids). They can also be triangular (three nodes), tetragonal (four nodes), parallelepipeds, and so on. The system contains a series of equations that define the behavior of a node, based on the conditions of the nodes contiguous to the first one, and on the surrounding condition, which affect it. The application of FEM permits sidestepping the traditional design cycle based on prototype–testing–modification–prototype.

Figure 12.8 illustrates the FEA output for a C-Channel. Figure 12.8(1) and (2) shows the mesh file with 14,578 nodes and the C-Channel with the area of 30.4 cm^2 where 294.9 Pa pressure is applied, respectively. Figure 12.8(3) shows the stress analysis with maximum stress 1.88e-06 Pa and Figure 12.8(4) shows the maximum deformation might occur at C-Channel, 2.5e-07, which is very small and can be negligible.

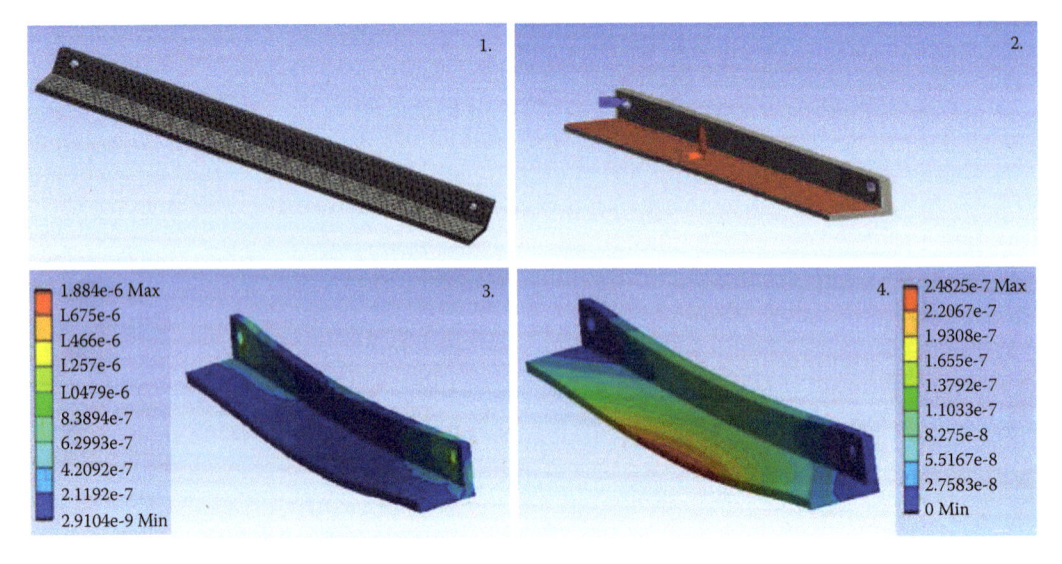

FIGURE 12.8
Finite element analysis (FEA) of C-Channel used in the robotics project.

Product Design and Development

There is no need to physically have the prototype on hand to perform the analyses (stresses, displacements, or deformed model visualizations). The FEM method is a part of process simulation, a technique that has been growing over the past few years, thanks to the ever increasing capability of current day systems to process large quantities of information. The method of process simulation encompasses all stages of manufacturing and operation of a product, from its machining or shaping phases, through assembly (by robot), to its operation and maintenance.

Figure 12.9 shows a mesh file and computational fluid dynamics (CFD) analysis output of an S-bend pipe. CFD is a powerful tool for simulation and analysis of different fluid flow patterns and heat transfer to find real life results without even producing the parts. It shows us how any particular part will react in real life with specified conditions and specifically which part of a component needs more precaution. Figure 12.9 shows a location of maximum erosion of a pipe after performing CFD analysis. It can also tell us the magnitude of this erosion and the external conditions. ANSYS Workbench, AutoCAD CFD, OpenFoam, Creo Parametric, SolidWorks CATIA, and Unigraphics (NX) are some of the commonly used FEA and 3D modeling software packages.

12.7.2.5 Rapid Prototyping

Rapid prototyping (RP) is a generic name for a group of technologies that can translate a CAD model directly into a physical object, without tooling or conventional matching operations. RP requires a CAD solid or surface model, which defines the shape of the object to be built. The electronic representation is then transferred to the RP system, which, using various technologies, transforms this information into a physical object.

Several techniques exist for prototype building. Among the most recent are MIME (material increase manufacturing) techniques: selective laser sintering (SLY), solid group curing (SGC), focused deposition modeling (FDM), and laminated object manufacturing (LOM). RP has had a significant impact on the design process, because designers can obtain a physical object as soon as a CAD object is available. This enables them to make evaluations earlier and more frequently during the design process.

12.7.2.6 3D Printing

3D printing is now a very popular tool in 3D modeling. It is very convenient and useful for printing 3D parts in a few hours. In 3D printing, additive processes are used, in which successive layers of material are laid down under computer control. These objects can be of almost any shape or geometry, and are produced from a 3D model or another electronic data source.

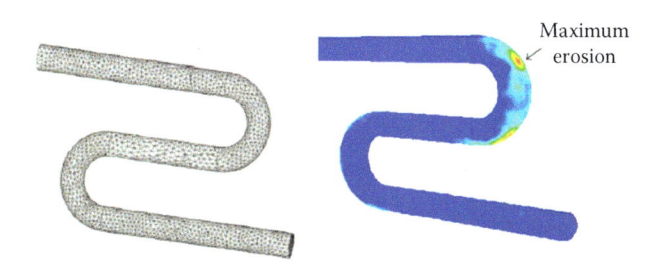

FIGURE 12.9
Computational fluid dynamics (CFD) analysis of an S-shaped pipe.

12.8 Summary

Product design and development is the core activity of most engineering jobs. It is important to develop a good understanding of the design process, factors to be considered during design and design optimization. Project planning is necessary for efficient use of organizational resources such as money, people, and equipment, which must be allocated and accounted for. The four stages of PLC include introduction, growth, maturity, and decline of the product as each product has finite market period and life.

The CE method involves all stakeholders such as customer, design, manufacturing, purchasing, quality, supplier, sales and marketing, and accounting personnel working together in the early phase of product design in order to simultaneously design the product.

The QFD method incorporates the customer's input in the product specification during the product design process. In the QFD process, interdisciplinary teams are formed, with individuals from marketing, research and development, process planning, quality assurance, and manufacturing. FMEA systematically evaluates the effects of failures on the product or service. A list of possible failures and their effects and causes is tabulated to evaluate and prioritize corrective actions. Several computational tools, methods, and techniques are used for design optimization such as FEA and CFD.

Problems

1. What are the major breakdowns of the product design process and at which phase do many design projects begin, without benefit of prior specification or concept development?
2. List the objectives of product design and development techniques.
3. What are the major steps that a product development engineer needs to complete?
4. List all product screening issues.
5. What are DFM and PLC? What are the stages for PLC?
6. What are the advantages of QFD?
7. What is FMEA? Describe how FMEA can help in the product design process.
8. What are the differences between FEA and CFD? How can FEA and CFD help in improving the product design cycle?
9. What is CAPP? What are the types of CAPP?
10. Define the abbreviations: MIME, SLY, SGC, FDM, LOM, E/P BDM.
11. Draw the product development process flowchart with all components.

Bibliography

Cooper, R.G. An investigation into the new product process: Steps, deficiencies, and impact, *Journal of Product Innovation Management*, 3(2) (June 1986), 71–85.

Pawar, K.S. and H. Driva, Performance measurement for product design and development in a manufacturing environment, *International Journal of Production Economics*, 60–61 (1999), 61–68.

Roozenburg, N.F.M. and J. Eekels, *Product Design: Fundamentals and Methods*. John Wiley & Sons, 1995.
Tichem, M. and T. Storm, Designer support of product structuring—Development of a DFX tool within the design coordination framework, *Computers in Industry*, 33 (1997), 155–163.
Ullman, D. *The Mechanical Design Process*, 4th edn. New York: McGraw-Hill, 2008. pp. 85–90.

13
Manufacturing Processes

CHAPTER OBJECTIVE AND STUDENT LEARNING OUTCOMES

After completing this chapter, students will be able to

1. Describe the role of machining processes in manufacturing.
2. Develop a better understanding of machining processes and machines in a workshop.
3. Describe the role of metal forming processes in manufacturing. [ABET outcome k, see Appendix C]
4. Describe the functions of various machines and their operating procedures.

13.1 Introduction to Machines

Human civilization evolved around development of simple machines and systems that can perform task efficiently. Wheel assembly used in carts is one such example that has greatly reduced human effort. Instead of digging deeper into the history of machining in this chapter, we are going to learn about machines and machining in manufacturing. We are talking specifically about metal cutting machines. If you have visited the workshop in your institution, then you might have seen such machines as the lathe machine, milling machine, and hopefully various vertical and horizontal numerical controlled (NC) machines. All these machines use some tools made of harder material to remove material from a relatively softer part or raw materials. Now, the question is: How do we get the part or raw materials? Here comes metal forming processes. These processes are discussed next.

13.2 Metal Forming Processes and Metal Cutting Processes

After extracting metals from ore, we generally shape them in the form of rods, bars, and sheets. These primary shapes are not directly usable in manufacturing components. So we use metal forming processes such as hot forging, cold forging, rolling, drawing, extrusion, and so on. Some of the most used metal forming processes are given next.

13.2.1 Hot Forging

In hot forging, the metal part is heated to its recrystallization temperature. This heated part is then placed on the bed of a forging machine in a bottom die. A hammer with impression of required shape strikes the part. Note that the impression is made on the top die that is attached to the hammer firmly. Because of the high force of the hammer, the heated part is shaped according to the top and bottom die.

13.2.2 Cold Forging

In cold forging, the metal part is either at room temperature or heated to a temperature well below the recrystallization temperature. The rest of the process is the same as for hot forging. Cold forging is also called impression-die forging. Cold forging has better dimension control than hot forging. However, we cannot perform cold forging on hard metals. We generally perform cold forging on metals like aluminum.

13.2.3 Rolling

Rolling is the process in which metal is squeezed between two rotating wheels that have been imprinted with the required shape. Using this process, we can get a continuous rail of the desired shape. We generally use this process to get pipes, tubes, shafts, and so on from raw material.

Most of the time, we get raw material after a forging or casting process. In these metal forming processes, we are not able to get the accurate dimension of the product. So the next step is to perform some machining process.

Almost all machining processes remove material from the object with the help of a tool. These tools can be single point cutting tools or multiple point cutting tools. A single point cutting tool removes material at a single point on the cutting edge; refer to Figure 13.1. A multipoint cutting tool removes material at multiple points of the cutting edge. We will now learn about machines and their related cutting tools.

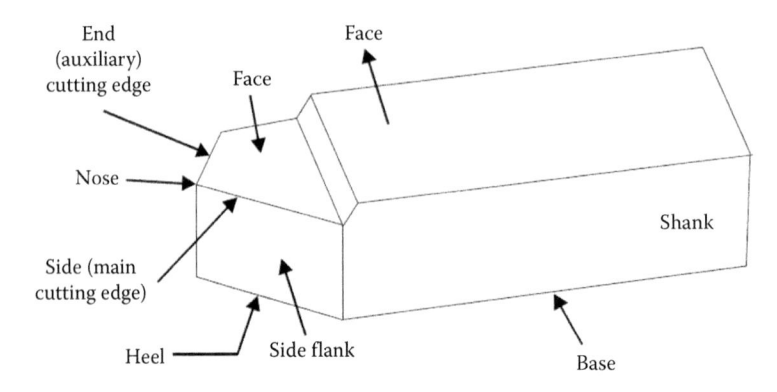

FIGURE 13.1
Single point cutting tool.

Manufacturing Processes

13.3 Machines and Tools

We will now discuss the engine lathe, milling machine, drilling machine, and computerized numerical controlled (CNC) machine.

13.3.1 Engine Lathe and Tools

The lathe is the most general and useful of all machine tools. A lathe is used to produce cylindrical surfaces. This is commonly known as the mother of all other machine tools. Figure 13.2 shows the basic components of the lathe.

A lathe may have *n* number of components, but the basic function of the lathe remains the same. In a lathe, the part is held tightly in a chuck mounted on the head stock. If the part is long, then the tail stock center is used to support the part on other end; refer to Figure 13.3. We rotate the part with the help of a motor or transmission shaft. Then a tool is brought into contact with the rotating part at the desired location. Because the material of the tool is harder than the part, material from the part is removed.

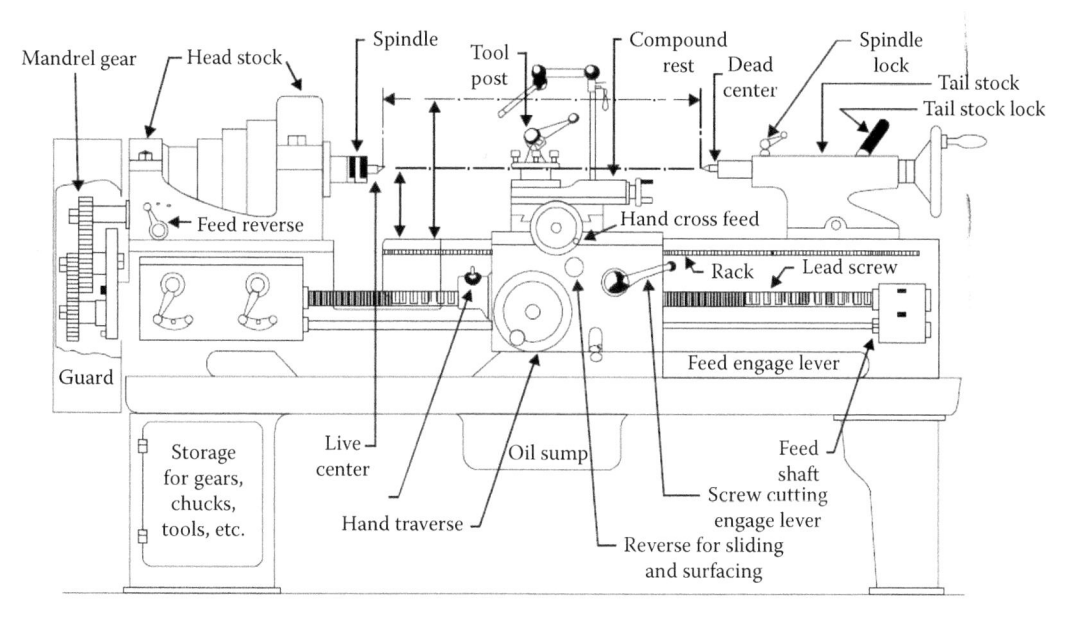

FIGURE 13.2
Engine lathe.

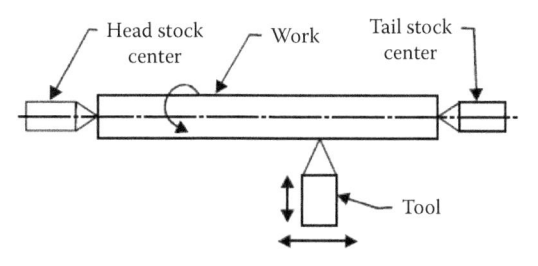

FIGURE 13.3
Working principle of a lathe.

The most basic parts of a lathe involved in the simple task of removing material are discussed next.

13.3.2 Head Stock

The head stock is used to hold the rotating part. The part named chuck is assembled in the head stock to hold the part; refer to Figure 13.4. The part is placed in the jaws of the chuck and then these jaws slide with the help of a key to hold the part firmly in place.

13.3.3 Tool Post

As per the name, it is a post of tools. Tool post is used to hold the tools required for performing operations on a lathe. Refer to Figure 13.5.

13.3.4 Tail Stock

The tail stock is attached to the lathe at the end opposite the spindle (chuck). The dead center of the tail stock is used to hold long workpieces like transmission shafts.

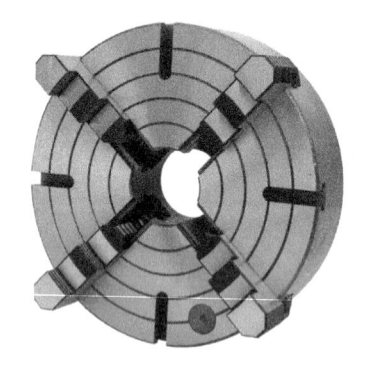

FIGURE 13.4
Chuck for a lathe.

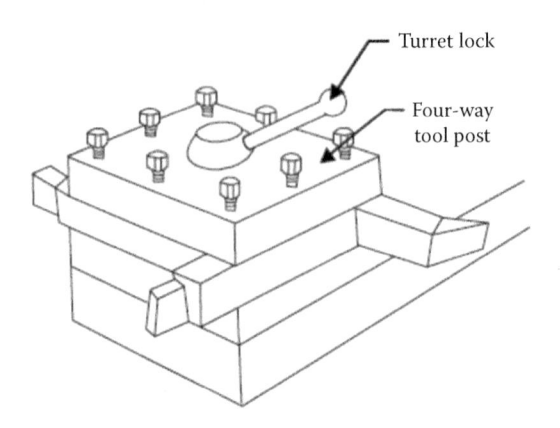

FIGURE 13.5
Tool post.

Sometimes we also attach a drill to the tail stock to facilitate drilling at the centerline of the workpiece.

13.3.5 Carriage Mechanism

A carriage is mounted on the outer guide ways of the lathe bed and moves parallel to the spindle axis. A carriage includes important parts such as an apron, cross-slide, saddle, compound rest, and tool post. The lower part of the carriage is termed the apron. Gears constitute the apron mechanism for adjusting the direction of the feed using clutch mechanism and the split half nut for automatic feed. The cross-slide is basically mounted on the carriage, which generally travels at right angles to the spindle axis. On the cross-slide, a saddle is mounted in which the compound rest is adjusted that can rotate and fix to any desired angle. The compound rest slide is actuated by a screw, which rotates in a nut fixed to the saddle.

The tool post is an important part of the carriage, which fits in a tee-slot in the compound rest and holds the tool holder in place by the tool post screw.

Now we know the main components of the lathe, but in manufacturing, we require the specifications of machines to find those that are suitable for our job. We will now discuss the specifications of a lathe machine.

13.3.6 Lathe Machine Specification

The size of a lathe is generally specified by the following means (Figure 13.6):

1. Swing or maximum diameter that can be rotated over the bed ways
2. Maximum length of the job that can be held between head stock and tail stock centers
3. Bed length, which may include the head stock length also
4. Maximum diameter of the bar that can pass through spindle or collect chuck of capstan lathe

There are n number of operations that can be performed on a lathe with the help of job-specific tools. We will first see the tools used on a lathe and then we will check the operations that can be performed.

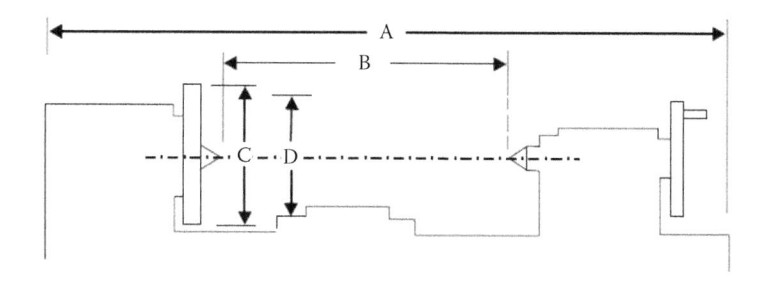

FIGURE 13.6
Specifications of lathe. A, length of bed; B, distance between centers; C, diameter of the work that can be turned; D, diameter of the work that can be turned over the cross slide.

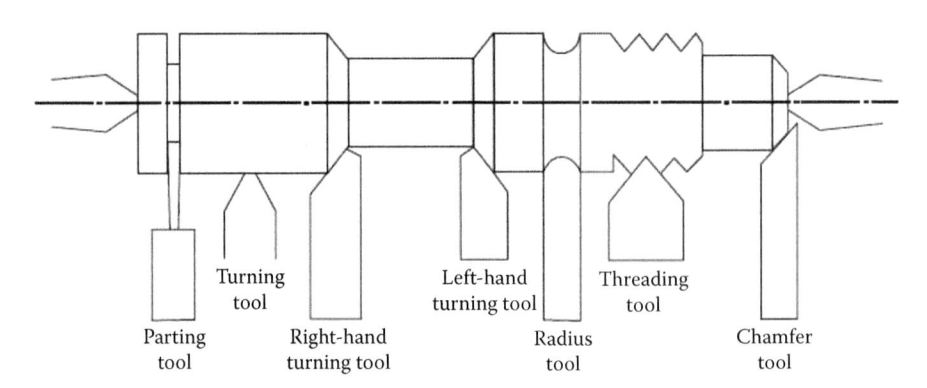

FIGURE 13.7
Tools used on a lathe.

13.3.7 Lathe Machine Tools

Figure 13.7 shows some common tools used on lathes to perform various operations:

Parting tool or slotter: The parting tool is used to create a slot in the part. It is also used to divide the part into two parts.

Turning tool: The turning tool is used to remove material from the walls of the part. In other words, this tool is used to reduce the diameter of the part.

Right-hand/left-hand turning tool: The right-hand/left-hand turning tool is used to create slope in the part. This tool is also used to create chamfers.

Radius tool: The radius tool is used to form radial grooves on the part. We can also use this tool to create fillets in the part.

Threading tool: As the name suggests, this tool is used to form threads in the workpiece.

Chamfering tool: This tool functions in the same way as the right-hand/left-hand turning tool. This tool is specifically used to create chamfers at the outer edges of workpieces.

13.3.8 Lathe Operations

The operations that can be performed using various lathe tools are given next:

1. Straight turning

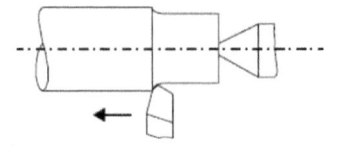

2. Shoulder turning

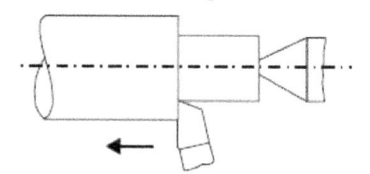

3. Taper turning

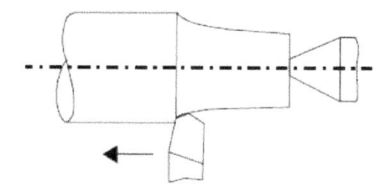

4. Thread cutting

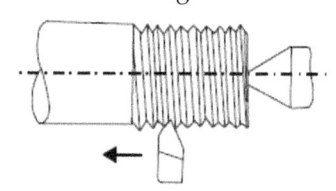

5. Facing

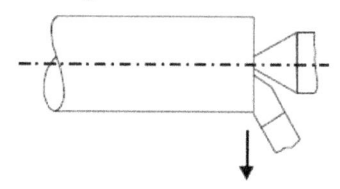

6. Filleting

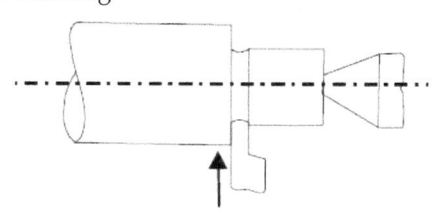

7. Drilling

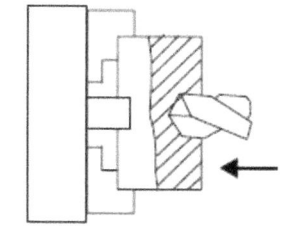

8. Knurling

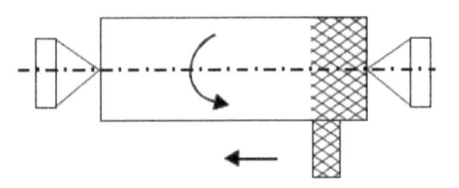

13.3.9 Drilling Machine

As the name suggests, a drilling machine is a machine tool designed for drilling holes in metals. A drill machine also works in a broader scope. Besides drilling round holes, a drilling machine can also be used for counterboring, countersinking, honing, reaming, lapping, sanding, and so on. Before we dig deeper into operations performed by drilling machines, let us understand the construction of a drilling machine.

13.3.9.1 Construction

Figure 13.8 shows a vertical drilling machine generally found in every workshop. In drilling machines, the drill is rotated and fed along its axis of rotation in the stationary workpiece. Various parts of a drill machine are discussed next.

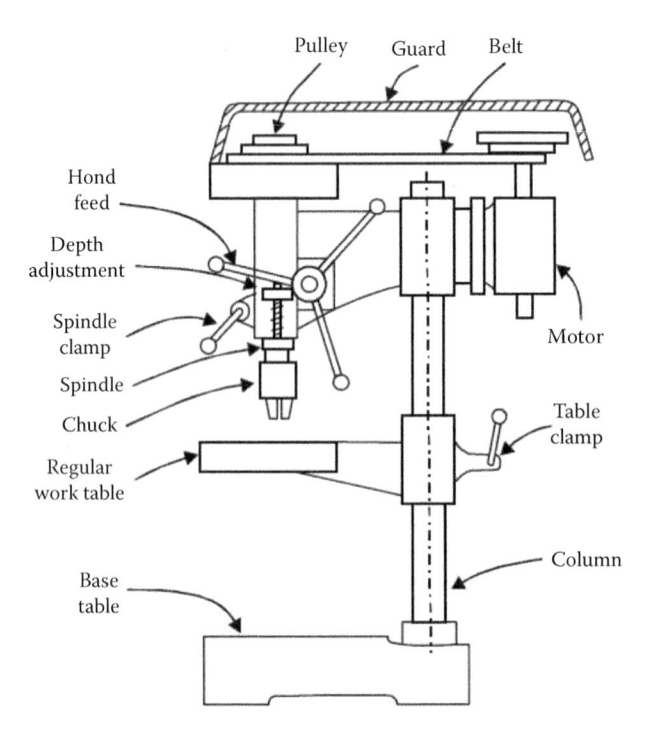

FIGURE 13.8
Drilling machine.

Manufacturing Processes 325

- The head contains an electric motor, V-pulleys, and a V-belt, which transmit rotary motion to the drill spindle at a number of speeds.
- The spindle is made up of alloy steel. It rotates as well as moves up and down in a sleeve. A pinion engages a rack fixed onto the sleeve to provide vertical up and down motion of the spindle and hence the drill so that the same can be fed into the workpiece or withdrawn from it while drilling. Spindle speed or drill speed is changed with the help of V-belt and V-step pulleys. Larger drilling machines have gear boxes for that purpose.
- Drill chuck is held at the end of the drill spindle and in turn holds the drill bit.
- Adjustable workpiece table is supported on the column of the drilling machine. It can be moved both vertically and horizontally. Tables generally have slots so that the vice or the workpiece can be held securely.
- Base table is a heavy casting. It supports the drill press structure. The base supports the column, which in turn supports the table, head, and so on.

13.3.9.2 Types of Drills

A drill is a multipoint rotating cutting tool used to machine holes in the workpiece. It usually consists of two cutting edges set at an angle with the axis. Broadly there are three types of drills (Figure 13.9):

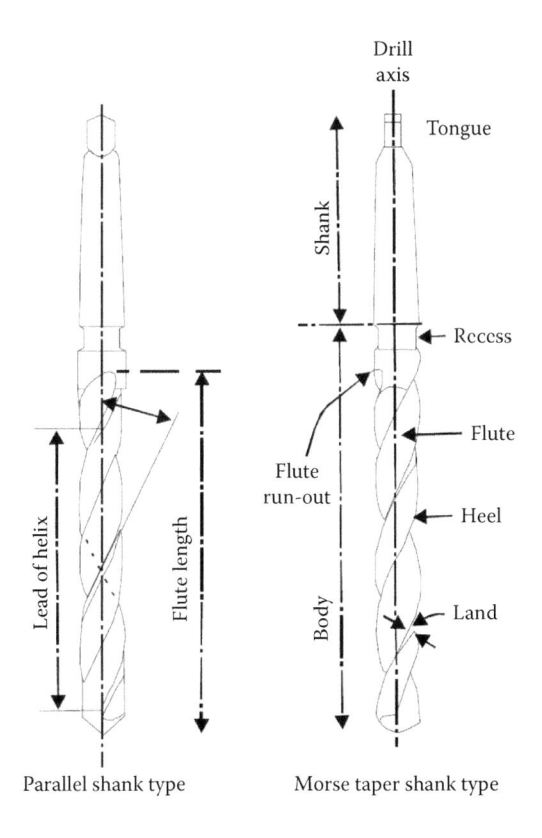

FIGURE 13.9
Drill type.

1. Flat drill
2. Straight-fluted drill
3. Twist drill

Flat drill is usually made from a piece of round steel that is forged to shape and ground to size, then hardened and tempered. The cutting angle is usually 90° and the relief or clearance at the cutting edge is 3°–8°. The disadvantage of this type of drill is that each time the drill is ground the diameter is reduced. Twist drill is the most common type of drill in use today.

13.3.9.3 Operations Performed on a Drill Machine

In addition to a drill tool, other tools can be installed on a drill machine such as taps, reamers, lapping tools, and so on. Various operations that can be performed on a drill machine with the help of various tools are discussed next:

- *Drilling*: Drilling removes solid metal from the job to produce a circular hole (Figure 13.10). Before drilling, the hole is located by drawing two lines at a right angle, and making an indentation at the intersection with a center punch. The indentation serves as a drill point to help the drill get started.
- *Reaming*: This operation sizes and finishes a hole already made by a drill. Reaming is performed by means of a cutting tool called a reamer as shown in Figure 13.11. Reaming operation serves to make the hole smooth, straight, and accurate in diameter.
- *Boring*: Figure 13.12 shows the boring operation where a hole is enlarged by means of adjustable cutting tools with only one cutting edge. A boring tool is employed for this purpose.

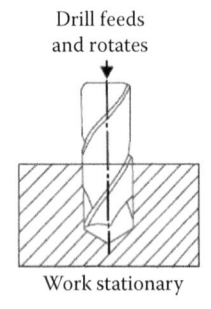

FIGURE 13.10
Drilling operation.

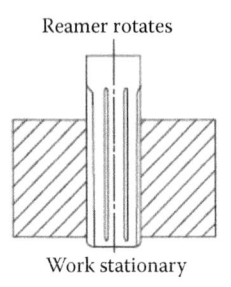

FIGURE 13.11
Reaming operation.

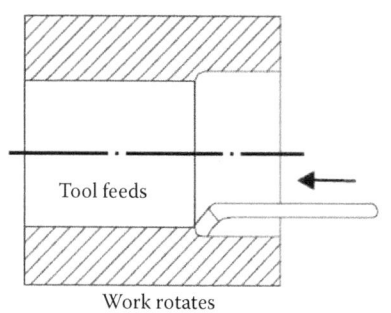

FIGURE 13.12
Boring operation.

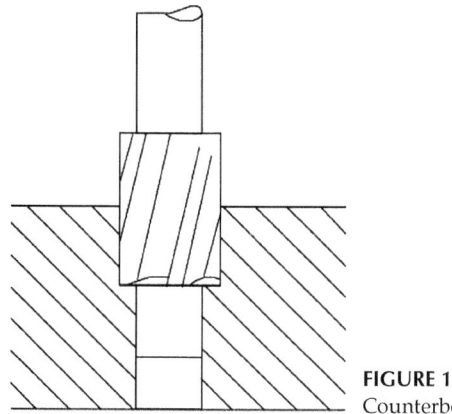

FIGURE 13.13
Counterboring operation.

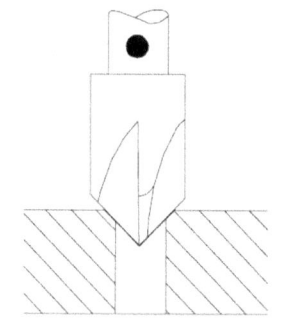

FIGURE 13.14
Countersinking operation.

- *Counterboring*: Counterboring operation is shown in Figure 13.13. It is the operation of enlarging the end of a hole cylindrically, as for the recess for a countersunk rivet. The tool used is known as a counterbore.
- *Countersinking*: In a countersinking operation, shown in Figure 13.14, the end of a hole is made cone-shaped, so it can accept a flat head screw. The cone shape provides a seat for countersunk heads of the screws so that the latter become flush with the main surface of the work.

- *Lapping*: This operation sizes and finishes a hole by removing very small amounts of material using an abrasive. The lapping tool keeps the abrasive material in contact with the sides of the hole that is to be lapped.
- *Spot-Facing*: This operation removes enough material from a flat surface around a hole to accommodate the head of a bolt or a nut. A spot-facing tool is very nearly similar to the counterbore.
- *Tapping*: It is the operation of cutting internal threads by using a tool called a tap. A tap is similar to a bolt with accurate threads cut on it. To perform the tapping operation, a tap is screwed into the hole by hand or by machine.
- *Core drilling*: Core drilling is a primary operation for producing circular holes, deep in the solid metal by means of a revolving tool called a drill. It is performed on radial drilling machines.

13.3.10 Milling Machine

Milling machine is a machine tool that removes metal as the work is fed against a rotating multipoint cutter. The milling cutter rotates at high speed and removes metal quickly with the help of multiple cutting edges. Any metal component that is not cylindrical is most probably machined by a milling machine. This is the reason why a milling machine finds wide application in production work. Milling machines are used for machining flat surfaces, contoured surfaces, surfaces of revolution, external and internal threads, and helical surfaces of various cross-sections.

13.3.10.1 Construction of Milling Machine

There are various types of milling machines, each having a different type of construction. Some common milling machines are given as follows:

1. Column and knee type milling machines
2. Planer milling machine
3. Fixed-bed type milling machine
4. Machining center machines
5. Special types of milling machines

Here, we will examine the construction of horizontal and vertical column and knee type milling machines. Figure 13.15 shows a horizontal column and knee type milling machine, and Figure 13.16 shows a vertical column and knee type milling machine.

13.3.10.2 Milling Cutters

Milling cutters are made in various forms to perform different classes of work. They can be classified as follows:

1. Plain milling cutters
2. Side milling cutters
3. Face milling cutters

Manufacturing Processes

4. Angle milling cutters
5. End milling cutters
6. Fly cutters
7. T-slot milling cutters
8. Formed cutters

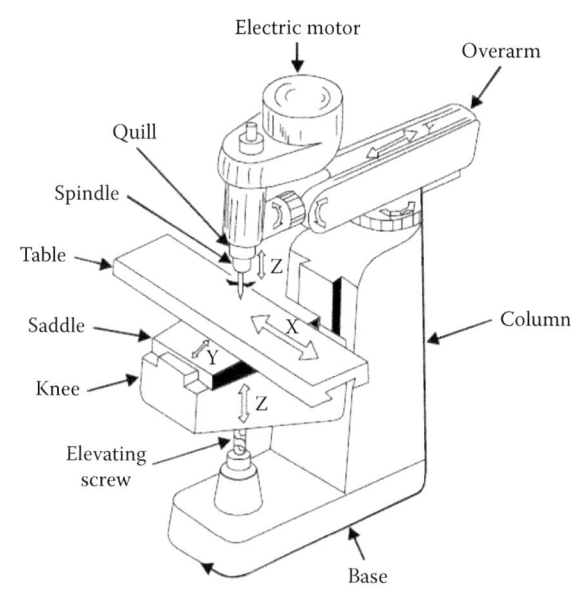

FIGURE 13.15
Horizontal milling machine.

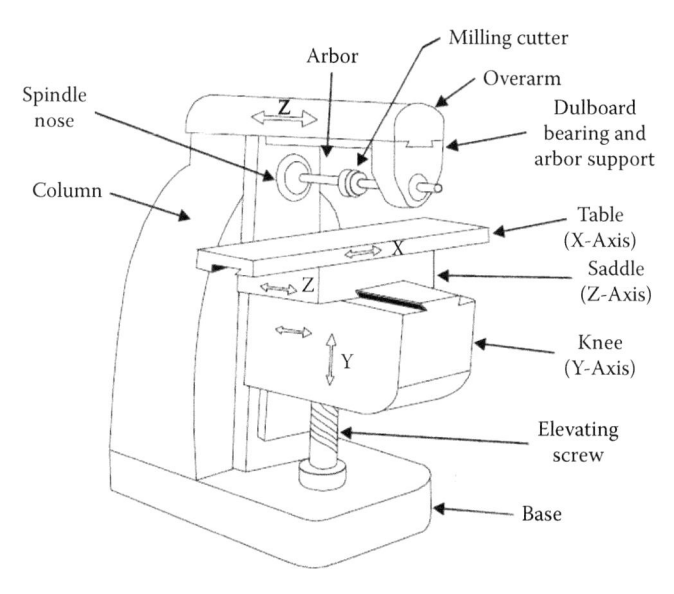

FIGURE 13.16
Vertical milling machine.

13.3.10.3 Operations Performed on Milling Machine

Figure 13.17 gives a brief idea about operations that can be performed on a milling machine.

13.3.10.3.1 Computerized Numerical Controlled Machine

Before we understand the concept of a CNC machine, we need to understand the fundamental concept of numerical control. Numerical control can be defined as a system in

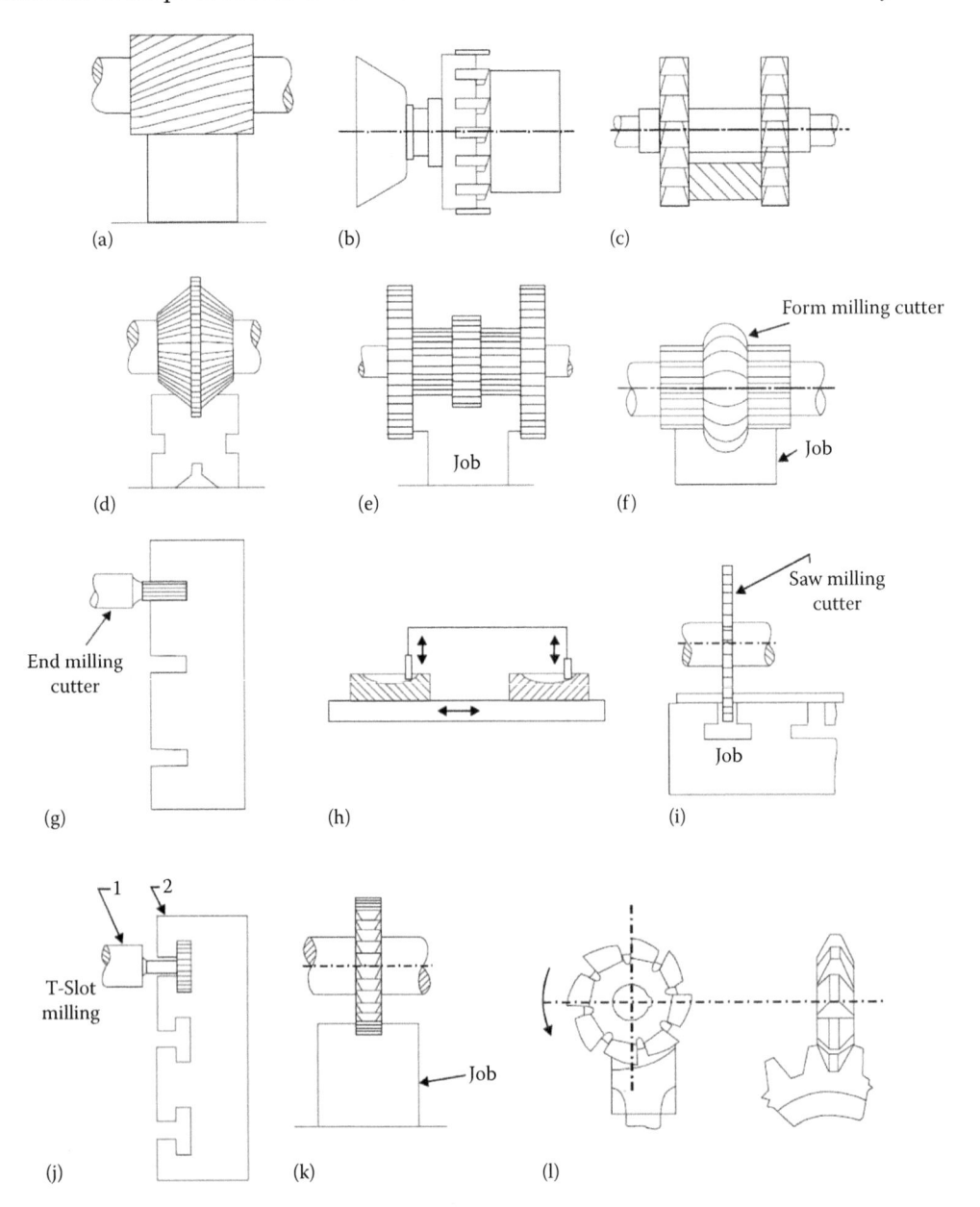

FIGURE 13.17
Milling operations. (a) Plane milling, (b) face milling, (c) side milling, (d) angular milling, (e) gang milling, (f) form milling, (g) end milling, (h) profile milling, (i) saw milling, (j) T-slot milling, (k) key way milling, and (l) gear cutting milling.

which the actions of a machine tool are controlled by recorded information in the form of numerical data. In an NC system, intervention of human beings in the machining process is substituted by coded operating instructions. This coding, otherwise known as a part program, is stored in cards or tapes. Getting the required instructions from these input media, the machine carries out different tasks in sequence.

When the activities of an NC machine are administered by a dedicated computer, it is known as a CNC machine tool. The functions of the machine tool are controlled by the instructions stored as programs in the computer.

13.3.10.3.2 Elements of an NC Machine

The NC system requires the preparation of a manuscript (part program) based on the product drawing, preparation of input media (punched cards and punched tapes), the data entry into the control unit, consequent processing and actuation of the machine tool to produce the desired part (Figure 13.18).

In performing all the above operations, the NC machine may have the following elements:

1. Software
2. Machine control unit
3. Driving devices
4. Manual control unit
5. Machine tool

13.3.10.3.3 Software

Software of an NC system consists of instructions (programs), languages used to write these programs and a variety of input media.

Different types of input media are used to store information and to provide input to various control units of the NC machine. They are

1. Punched cards
2. Punched tapes
3. Magnetic tapes
4. Floppy disks

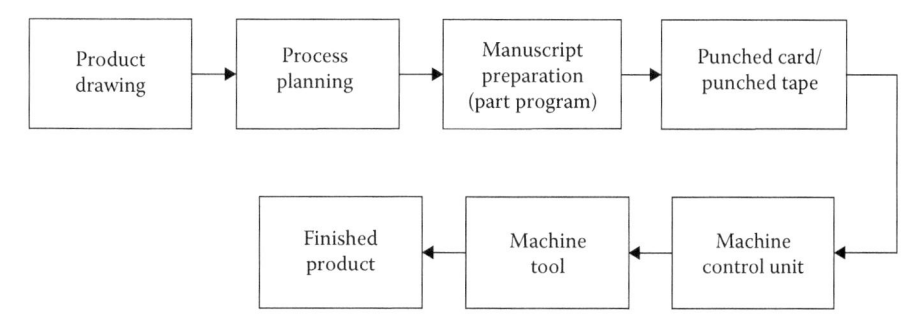

FIGURE 13.18
Layout of an NC machine.

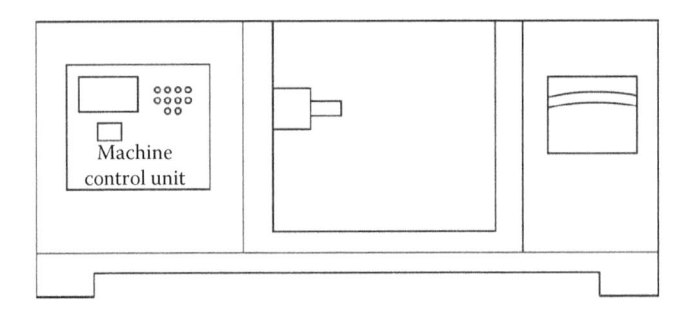

FIGURE 13.19
Machine control unit.

13.3.10.3.4 Machine Control Unit

Machine control unit (MCU) consists of electronic circuits (hardware) that are useful in reading and interpreting the instructions (NC program) fed by means of input media and converted into mechanical actions of the machine tool (Figure 13.19).

13.3.10.3.5 Driving Devices

Driving devices consist of different types of motors and gear trains. They convert the instructions from the MCU into accurate mechanical displacements of the machine tool slides. The motors may be electrical, hydraulic, or pneumatic.

Electrical motors are mainly used as prime movers because of their speed and torque characteristics. AC induction motors are cheap and easy to maintain. For easy and effective speed changes, DC motors are also used.

Hydraulic motors are used in some specific types of CNC machines. Hydraulic motors are driven by oil pumped from a constant delivery hydraulic pump. Hydraulic motors are used where the load is high and a wider range of speed is necessary.

Servomotor, stepper motor, synchros, and resolvers are different types of motors used as drives in NC machines.

13.3.10.3.6 Manual Control Unit

Manual control unit consists of dials and switches to be operated by the operator. It may also have a display unit to provide useful information to the operator. In some machines, the manual control unit may be a part of the MCU (machine control unit).

The operator uses the manual control unit to

1. Switch on and off the machine
2. Load and unload the workpieces
3. Change the tools in certain types of machines

13.3.10.3.7 Machine Tool

This element of the NC machine actually performs the useful work of converting the raw material to finished components. It is designed to perform various machining operations. It consists of a machine table, spindles, cutting tools, work holding devices such as jigs and fixtures, coolant systems, scrap removal systems, and other auxiliary equipment.

Manufacturing Processes

13.3.10.3.8 NC Programs

NC programs are the instructions given to the computer of CNC in the form of numeric codes. If someone has used the logo programming, then programming a CNC machine is very simple. In NC programs, we give instructions to the tool mechanism for required movement in steps.

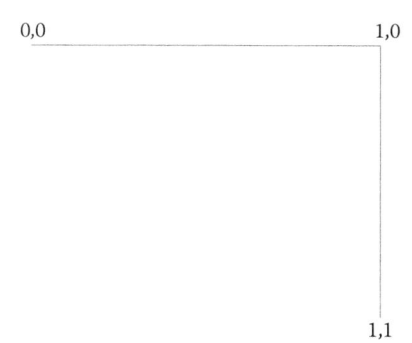

```
0,0 ─────────────────── 1,0
                         │
                         │
                         │
                         │
                        1,1
```

For example, we want the movement of a tool to be as given in the above illustration. Then we will first instruct the MCU to move the tool from 0, 0 to 1, 0 and then to 1, 1. There are some keywords or say codes that tell the machine what to do.

13.3.10.3.9 CNC Codes

Code	Description
G00	Rapid positioning
G01	Linear interpolation
G02	Circular interpolation clockwise (CW)
G03	Circular interpolation counterclockwise (CCW)
G20	Inch input (in.)
G21	Metric input (mm)
G24	Radius programming
G90	Absolute programming
G91	Incremental programming

For the above illustration, we will make three segments of codes as given as follows:

G00 X0.0 Y0.0;

G01 X1.0 Y0.0;

G01 X1.0 Y1.0;

You can find more about the CNC codes in the machine manuals that come with the machine.

13.4 Machine Process Flow

Machining is an intermediate step of manufacturing. There are various processes that have been performed before machining and various processes that will be performed after machining. In Figure 13.20, we have a sample process flow of manufacturing.

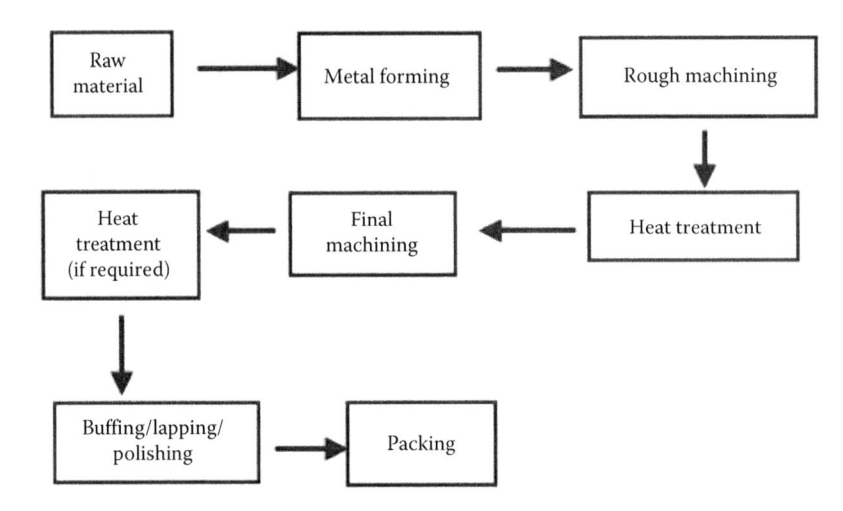

FIGURE 13.20
Sample process flow.

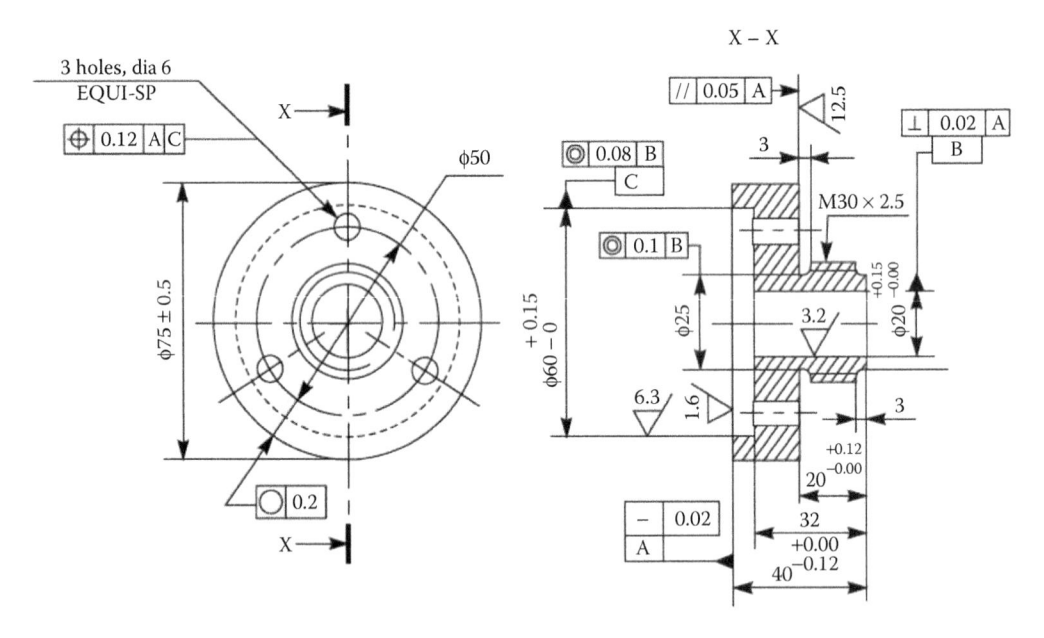

FIGURE 13.21
Machine drawing of a component.

The steps given in this figure are just main categories; there are various subprocesses in each operation. Because we have to concentrate on machining process flow, let us discuss a small machining problem given in the drawing shown in Figure 13.21.

From the drawing, we can understand that the component is a flange with four holes. Generally, for this component, we get forged raw material of approximate dimensions as shown in Figure 13.22.

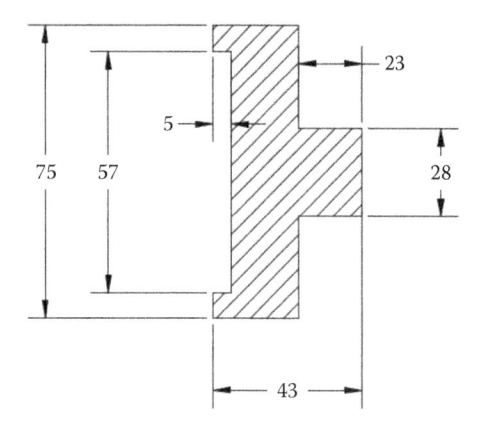

FIGURE 13.22
Forge drawing of a component.

Operations of machining can be written in order as follows:

1. Rough turning and facing (Side 1)
2. Rough turning and facing (Side 2)
3. Finishing (Side 1)
4. Finishing (Side 2)
5. Punch marking for drill
6. Drilling holes
7. Boring the center hole

In industry, each operation will be having a separate engineering drawing with only required dimensions for that operation.

13.5 Welding and Cutting

Welded connections and assemblies represent a very large group of fabricated steel components, and we can only discuss a portion their designs and fabrication here. The welding process itself is complex, involving heat and liquid-metal transfer, chemical reactions, and the gradual formation of the welded joint through liquid-metal deposition and subsequent cooling into the solid state, with attendant metallurgical transformations.

13.5.1 Arc Welding

Arc welding is one of several fusion processes for joining metal. By the generation of intense heat, the juncture of two metal pieces is melted and mixed—directly or more often with intermediate molten filler metal. On cooling and solidification, the resulting welded joint metallurgically bonds the former separate pieces into a continuous structure.

In arc welding, the intense heat needed to melt metal is produced by an electric arc. The arc forms between the workpiece and an electrode that is either manually or mechanically moved along the joint; conversely, the work may be moved under a stationary electrode. The electrode generally is a specially prepared rod or wire that not only conducts electric current and sustains the arc but also melts and supplies filler metal to the joint; this constitutes a consumable electrode. Carbon or tungsten electrodes may be used, in which case, the electrode serves only to conduct electric current and to sustain the arc between the tip and the workpiece, and is not consumed; with these electrodes, any filler metal required is supplied by a rod or wire introduced into the region of the arc and melted there. Filler metal applied separately, rather than via a consumable electrode, does not carry electric current. Most steel welding operations are performed with consumable electrodes.

13.5.2 Welding Process Fundamentals

Heat and filler metal: An AC or DC power source fitted with necessary controls is connected by a work cable to the workpiece and by a *hot* cable to an electrode holder of some type, which in turn is electrically connected to the welding electrode. When the circuit is energized, the flow of electric current through the electrode heats the electrode by virtue of its electric resistance. When the electrode tip is touched to the workpiece and then withdrawn to leave a gap between the electrode and workpiece, the arc jumping the short gap presents a further path of high electric resistance, resulting in the generation of an extremely high temperature in the region of the sustained arc. The temperature reaches about 6500°F, which is more than adequate to melt most metals. The heat of the arc melts both base and filler metals, the latter being supplied via a consumable electrode or separately. The puddle of molten metal produced is called a weld pool, which solidifies.

13.5.2.1 Shielding and Fluxing

High-temperature molten metal in the weld pool will react with oxygen and nitrogen in ambient air. These gases will remain dissolved in the liquid metal, but their solubility significantly decreases as the metal cools and solidifies. The decreased solubility causes the gases to come out of solution. If they are trapped in the metal as it solidifies, then cavities, termed porosities, are left behind. This is always undesirable, but it can be acceptable to a limited degree depending on the specification governing the welding. Smaller amounts of these gases, particularly nitrogen, may remain dissolved in the weld metal, resulting in a drastic reduction in the physical properties of the weld metal. Notch toughness is seriously degraded by nitrogen inclusions. Accordingly, the molten metal must be shielded from harmful atmospheric gas contaminants. This is accomplished by gas shielding or slag shielding or both. Gas shielding is provided either by an external supply of gas, such as carbon dioxide, or by gas generated when the electrode flux heats up. Slag shielding results when the flux ingredients are melted and leave behind a slag to cover the weld pool, to act as a contact barrier between the weld pool and ambient air. At times, both types of shielding are utilized. In addition to its primary purpose to protect the molten metal, the shielding gas will significantly affect arc behavior. The shielding gas may be mixed with small amounts of other gases (as many as three others) to improve arc stability, puddle (weld pool) fluidity, and other welding characteristics.

13.5.2.2 Shielded Metal Arc Welding

The shielded metal arc welding (SMAW) process, commonly known as stick welding, or manual welding, is the most popular and widespread welding process. It is versatile, relatively simple to do, and flexible in application. To those casually acquainted with welding, arc welding usually means shielded-metal arc welding. SMAW is used in the shop and in the field for fabrication, erection, maintenance, and repairs. Because of the relative inefficiency of the process, it is seldom used for fabrication of major structures. SMAW has earned a reputation for providing high-quality welds in a dependable fashion. It is, however, inherently slower and more costly than other methods of welding.

SMAW may utilize either direct current (DC) or alternating current (AC). Generally speaking, DC is used for smaller electrodes, usually less than 3/16 in. diameter. Larger electrodes utilize AC to eliminate undesirable arc blow conditions. Electrodes used with AC must be designed specifically to operate in this mode, in which current changes direction 120 times per second with 60 Hz power. All AC electrodes will operate acceptably on DC. The opposite is not always true.

13.5.2.3 Gas Metal Arc Welding

Gas metal arc welding (GMAW) is a welding process that joins metals by heating them to their melting point with an electric arc. The arc is between a continuous, consumable electrode wire and the metal being welded. The arc is shielded from contaminants in the atmosphere by a shielding gas. GMAW can be performed in three different ways:

- *Semiautomatic Welding*—The equipment controls only the electrode wire feeding. Movement of the welding gun is controlled by hand. This may be called hand-held welding.
- *Machine Welding*—It uses a gun that is connected to a manipulator of some kind (not hand-held). An operator has to constantly set and adjust controls that move the manipulator.
- *Automatic Welding*—It uses equipment that welds without the constant adjustment of controls by a welder or operator. On some equipment, automatic sensing devices control the correct gun alignment in a weld joint.

Basic equipment for a typical GMAW semiautomatic setup:

- Welding power source—It provides welding power.
- Wire feeders (constant speed and voltage sensing)—It controls supply of a wire to the welding gun.
  - *Constant speed feeder*—It is used only with a constant voltage (CV) power source. This type of feeder has a control cable that will connect to the power source. The control cable supplies power to the feeder and allows the capability of remote voltage control with certain power source/feeder combinations. The wire feed speed (WFS) is set on the feeder and will always be constant for a given preset value.
  - *Voltage-sensing feeder*—It can be used with either a CV or constant current (CC)—DC power source. This type of feeder is powered off of the arc voltage and does not have a control cord. When set to CV, the feeder is similar to a constant speed feeder. When set to CC, the WFS depends on the voltage present.

The feeder changes the WFS as the voltage changes. A voltage-sensing feeder does not have the capability of remote voltage control.

- Supply of electrode wire
- Welding gun—It delivers an electrode wire and shielding gas to the weld puddle.
- Shielding gas cylinder—It provides a supply of shielding gas to the arc.

13.5.3 Use of Laser in Manufacturing

13.5.3.1 Welding

Laser welding is recognized as an advanced process to join materials with laser beams of high-power, high-energy density in every industrial field. The power density of a laser beam, which is equivalent to that of an electron beam, is much higher than that of an arc or plasma. Consequently, a deep, narrow keyhole is formed during welding with a high-power laser or electron beam and a deep, narrow penetration weld can be effectively produced. In electron beam welding, the chamber for a vacuum environment and X-ray protection should be used. Arc and plasma welding cannot be employed in a vacuum; however, laser welding can be performed and a sound, deep weld bead can be produced in a similar way to electron beam welding.

The laser welding process can be performed in multiple ways with two different power sources for the beam.

1. The first process involves the YAG laser system. Officially this system is called the ND:YAG (neodymium doped yttrium aluminum garnet) laser system.
    a. With the *YAG* laser system, a crystal and *flash lamps* channel power into a focused beam. The light frequency is in the near-infrared end of the light spectrum at 1064 nm.
    b. The laser beam is sent to the part by fiber optic cables and imparted through a focus head optic assembly. By using the fiber optic for beam delivery, we have much more flexibility in the weld process because of the ability to move the optics in multiple directions.
    c. No cover gas is required to weld using this equipment.
2. The second process we use to channel power is the CO_2 laser. With this laser, the beam is created using gases and delivered to the optics by reflecting off a series of mirrors.
    a. The laser is based on the excitation of three gases (helium, nitrogen, and CO_2).
    b. The gases flow through a glass tube (plasma tube) with a cathode and an anode at opposing ends. DC voltage, AC frequency, or RF frequency is used to excite the gases to create the laser beam.
    c. The light frequency is in the far-infrared end of the light spectrum. The laser beam is sent to the part by reflecting in the mirrors. For this type of welding, a cover gas such as helium is used.

13.5.3.2 Cutting

In laser cutting, the focused laser beam is directed onto the surface of the part to rapidly heat it up, resulting in melting and/or vaporization, depending on the beam intensity and part material. The molten metal and/or vapor is then blown away using assist gas.

Manufacturing Processes

The power intensity required is typically of the order of 10^6–10^7 W/cm^2. Lasers can be used effectively to cut metal plates of thickness up to about 10 cm.

Assist Gas Function:

- Facilitates ejection of molten metal through the backside of the workpiece.
- Protects the lens from the spatter.
- Acts as a heat source resulting in an exothermic reaction that aids in cutting, which may occur in oxygen-assisted cutting of steel.

Overview of Laser Cutting Process

Cutting Types	Flame Cutting	Fusion Cutting	Microjet	Sublimation Cutting
Laser type	Mainly CO$_2$ lasers, solid-state lasers	Mainly CO$_2$ lasers, solid-state lasers	Solid-state lasers, also frequency doubled and tripled	Mainly solid-state lasers, but also CO$_2$ lasers
Material	Steel	Mainly steel, but also fusible plastics	Mainly semiconductors, but also thin metals and plastics	Thin metals, foil, semiconductors, ceramic, wood, plastics
Parameters	Focus diameter, focus position, laser power, cutting speed, nozzle diameter	Focus diameter, focus position, laser power, operating mode, cutting speed, nozzle diameter	Laser power, pulse duration, pulse frequency, nozzle and water jet diameter	Focus diameter, focus position, laser power, operating mode, cutting speed
Focus diameter (mm)	0.1–0.5	0.1–0.5	0.02–0.15	0.05–0.1
Quality criteria	Degree of burr formation, contour accuracy, roughness, perpendicularity of cut edges	Degree of burr formation, contour accuracy, roughness, perpendicularity of cut edges	Contour accuracy, kerf width, perpendicularity of cut edges	Contour accuracy, head effected zone
Machines and systems	Cutting head with lenses or mirrors, flatbed laser cutting machine, 3D system, robots	Cutting head with lenses or mirrors, flatbed laser cutting machine, 3D system, robots	Special lens optics with water chamber, 2D laser machine	Lens optics, scanning optics, 2D laser machine or combined with robots
Applications	Quality cuts in mild steel, mainly metal working and mechanical engineering	Quality cuts in mild steel, mainly metal working and mechanical engineering, medical tech	Precision machining and microprocessing, separation of semiconductors, wafers, stencils, food industry	Precision matching and microprocessing in electronics and medical, industrial textiles, adhesive labels, packaging

13.5.3.3 Surface Treatment/Modification

Bending tools must be able to withstand incredible forces. In deep drawing, the part rubs against the dies. Movement, friction, tension, pressure, dirt, and chemical cause were not only on the tools but also on other machine and system components. To increase the components' resistance to stresses, their surfaces are treated. Alongside the many conventional methods for surface treatment, laser processes also exist. They include laser hardening, remelting, and coating. These techniques increase the stress resistance of the component in different ways. For example, they may increase the component's hardness and toughness, alter its surface, produce compression stresses in the surface, or apply coatings.

Advantages:

- They process an irregular, three-dimensional surface just as easily as they process a flat, regular surface.
- Lasers are capable of treating only selected areas of the surface as needed.

13.5.3.4 Hardening

Laser hardening uses ferrous materials extensively. These include steels and cast iron with a carbon content of more than 0.2%. To harden the part, the laser beam usually warms the outer layer to just under the melting temperature (about 900–1400°C). The high temperature causes the iron atoms to change their position within the metal lattice. Rapid cooling prevents the metal lattice from returning to its original structure, producing martensite.

The hardening depth of the outer layer is typically 0.1–1.5 mm. On some material, it may be 2.5 mm or more. Conventional surface hardening processes include flame hardening and induction hardening. The advantage of laser hardening: part cools itself. Other processes require water to quench the part. The laser also has an edge when irregular, three-dimensional geometries are involved; it can be guided across the part in any pattern desired. Unfortunately, when hardened tracks overlap, the material between the new track and the previous one is reheated. In this tempered zone, the degree of hardening is diminished. On dynamically stressed parts, the location of the tempered zones must be carefully controlled and adjusted as necessary.

13.5.3.5 Remelting

Remelting is the process in which the laser beam not only heats the surface but continuously melts it. Depending on the surface material, remelting can enhance the degree of hardening or increase corrosion resistance. Remelting is chiefly employed on cast iron parts. An example is the camshaft used to operate the engine valves of an automobile. The contact surfaces of the cams are made more wear resistant by remelting and rapid cooling. During processing an extremely hard structure is created in the cast iron.

13.5.3.6 Coating

In the coating process, the laser beam is primarily used to melt a filler material. The filler bonds with the surface of the part, forming a separate layer that is typically between 0.5 mm and several millimeters thick. There are two methods of introducing the filler. The first involves delivering the filler in powder directly to the weld pool through a nozzle, much like direct metal deposition. The other possibility is to apply the filler first to the part surface. In a second step, the laser beam melts the layer so that it bonds with the parent material. Laser coating is especially well suited to selective treatment of part surfaces.

13.5.3.7 Drilling/Piercing

In laser drilling, a short laser pulse with high-power density feeds energy into the part extremely quickly, causing the material to melt and vaporize. The greater the pulse energy, the faster material is melted and vaporized.

Different drilling processes:

- *Single-shot and percussion*: In the simplest case, a single laser pulse with comparatively higher pulse energy is used to produce the hole. This method enables a large number of holes to be created in an extremely short amount of time. In percussion drilling, the hole is produced using multiple short duration, low-energy laser pulses.
- *Trepanning*: It also uses multiple laser pulses to produce the hole. In this process, first a pilot hole is created using percussion drilling. Then the laser enlarges the pilot hole, moving over the part in a series of increasingly larger circles.
- *Helical drilling*: Unlike trepanning, helical drilling does not involve the creation of pilot hole. Right from the start, the laser begins moving in circles over the material as the pulses are delivered, with a large amount of material shooting upward in the process. The laser continues to work its way through the hole in a downward spiral.

Review Questions

1. Write down some of the specifications of the lathe.
2. Draw a single point cutting tool and label different areas.
3. Discuss different types of operations that can be performed by a drill machine.
4. What is a milling machine? What are the different cutters of a milling machine?
5. What is a CNC machine? Discuss some of the elements of the CNC.
6. What is constant speed feeder and voltage-sensing feeder?
7. Discuss laser welding and the major power sources for laser welding.
8. What are the advantages and disadvantages of laser hardening?
9. Why is assist gas important in laser cutting?

Section IV

Engineering Profession

14
Engineering and Society

CHAPTER OBJECTIVE AND STUDENT LEARNING OUTCOMES

After completing this chapter, students will be able to

1. Demonstrate knowledge of engineering and technological advancements and their impact on historical and modern societies. [addresses ABET Criteria 3(h)]
2. Explain how engineering relates to a problem of societal concern. [addresses ABET Criteria 3(h)]
3. Identify important contemporary regional, national, or global problems that involve engineering. [addresses ABET Criteria 3(j)]
4. Propose and discuss ways engineers contribute or might contribute to the solution of specified regional, national, and global problems. [addresses ABET Criteria 3(j)]
5. Describe how engineering and technological developments affect society and the environment. [addresses ABET Criteria 3(j)]

14.1 Social Influence in a Globalized World

With the increasingly symbiotic nature of technology and society, today's engineers cannot ignore the complex interaction between the engineering profession and society. The development and proliferation of technologies affects everyone, regardless of location, culture, gender, or ethnicity. Engineers must be able to create solutions that span in and beyond cultural and societal barriers while remaining keenly aware of the significance of their decisions. In addition to the technical skills traditionally sought, employers are now seeking engineers who also possess a deeper sense of responsibility to the world and have insight, knowledge, and awareness of the significance of their work.

It is all but impossible to thoroughly give an adequate explanation of the interaction between society and engineering in a single book, let alone a single chapter. The examples provided in this chapter are meant to be a brief introduction to some of the interactions between these two entities. Not only are there far more dynamic and complex situations that exist, but also will one find that the relationship between these two entities is continuously transforming. Gaining and obtaining a more thorough outlook should and will be a lifelong journey. Thus, the aim of the following information is to whet your appetite and start you on a transforming journey in which you will be more conscientious to the roles of the engineer and the effects of their designs.

14.2 Evolution of Social Norms and Technology

In every society, there are certain behaviors that are considered acceptable and others that are not. Typically, such behaviors are not governed by law, but rather by an *unspoken* standard. We call these behavioral expectations social norms. These norms vary from country to country and among social classes, groups, and settings. Today's engineers typically design solutions that address the needs of many different social groupings, and an element of their success relies on their ability to be aware of and able to conform to the appropriate social norms.

Social norms can be as simple as flushing the toilet, mowing the lawn or shaking hands when meeting someone. Norms vary from culture to culture. For example, in several Asian cultures, avoiding eye contact while conversing with another person is interpreted as being polite or reverent. In the United States, eye contact is expected and demonstrates a person's interest and engagement in the conversation (htt). Some social norms are also dependent on the setting within a particular culture. For example, dancing that would be considered acceptable at a nightclub might not be considered appropriate at a Sunday morning church service.

Some people believe that social norms somehow develop separately from technology. In reality, however, many social norms are formed in conjunction with technology. This is important for engineers to understand because not only do they create technology that contributes to the development of new social norms, but social norms also dictate the types of future technologies that engineers can produce. Because an engineer's solution is only successful if it is actually employed by those for whom it is intended, creating solutions that go against social norms is futile, regardless of how well the solution addresses the problem.

14.3 Emerging Social Norms in Social Networking

Technology and social norms can be plainly seen evolving concurrently on social networking websites. Specifically, consider the online website, Facebook.com. Similar to the way that email addresses are required to access much of the online world, having a Facebook page is becoming an expectation of being online as well. Facebook.com is a free website where a person, group, or other entity can register and create a profile that will allow them to connect and share with others through posting status updates, messages, pictures, videos, links, and other information.

Because of its increasing popularity worldwide with users who have drastic variations in culture and background, Facebook has become a virtual environment where one can observe new social norms evolving at an incredible pace. In America alone, roughly one in three American adults use Facebook (Shih 2010). With such a large portion of people using Facebook, emerging social norms ripple through society in profound ways that will only become more pronounced.

Technology and social norms can be seen emerging simultaneously in the way that people grieve the death of a friend or family member in the age of social media. Social media provides an environment where sharing information is acceptable and encouraged. Much of what is shared on these sites would not have been appropriate to share so publically 15, 10, or even 5 years ago. In the past, people endured grief from the loss of a loved one privately. Traversing the stages of grief is not pleasurable and can be tremendously emotional. This is typically not an experience that anyone longs to relive once they begin to move past their loss. However, in social media, we frequently see people living out those stages in a very public and permanent manner. Experts claim that this is not a negative change, but, rather, beneficial for two reasons; it is healthy for people to share their feelings, and doing it through social media provides an opportunity for crucial support and response from others. However, the ability to publically share such deep and personal feelings still leaves us with many new questions about acceptable behavior after a loss (Van Der Leun, n.d.).

As part of a Facebook profile, user can upload images or graphics to represent themselves. This image can be changed anytime the user desires. When a friend or family member passes away, some people change their profile picture to an image that includes the deceased person. While some consider this to be an appropriate gesture, people closest to the deceased are sometimes angered and view it to be shallow or mocking that person's memory (Di Donato 2014). If a user does decide to post a profile picture in honor of a deceased friend or family member, the question then becomes, what is the appropriate amount of time to have that person in a profile picture? In reality, most of us understand that everyone grieves differently and at a different pace, but depending on the timing, the simple act of changing a profile picture could be interpreted in many different ways.

In the early phases of Facebook development, dealing with pages of deceased users was not considered high priority. It was not until Facebook.com staff experienced the loss of a staff member that these issues came to the forefront. It then became clear to the developers that even though someone was no longer physically on earth, his or her memory should still remain alive. Simply deleting profiles and pages of those who passed away would not be an acceptable solution, but it was also agreed that these pages should be treated differently from the pages of living people. Because the deceased can no longer accept or reject friend requests, one of the stipulations decided on was that deceased people should not appear in search results and that others should not be able to *friend* someone who was deceased. At the same time, people who were friends with a person at the time of his or her death should be able to keep their friendship and even post messages on that person's page. As they delved deeper into the issue, it was clear to the Facebook staffs that they needed to establish a formal policy, and thus, the Facebook *memorialization* policy was created (Chan 2009).

The establishment of the Facebook *memorialization* policy was a difficult process. Careful consideration was given to both the wants and needs of Facebook users and the desires of family and friends of the deceased that were not Facebook users. In cases where a person or party might have a vested interest in obtaining access to the personal information associated with such an account, compliance with legal policies, guidelines, and precedence also became a consideration. Policy makers quickly realized that these policies varied drastically throughout the world, and even within the same demographical areas. For now, Facebook developers have responded by allowing themselves as much discretional power as possible in their policy. While some consider this a logical choice given the complexity of this situation, others criticize the policy as being insufficient (McCallig 2015).

CULTURAL AWARENESS OR CULTURAL CONVERGENCE

With the world becoming increasingly global and the need for specialization pulling workers together from vastly different cultures, cultural awareness, and acceptance is a hot topic in almost every industry. There are some people, however, who believe that the push for cultural diversity will eventually evolve into cultural oneness. With technology facilitating the spread of ideas and values and making travel feasible to the common person, these people recognize the possibility that, eventually, our values, ideas, and cultural heritage will converge and become the same for everyone, regardless of where we live in the world. For some, this theory may seem a bit extreme. However, it does bring up some interesting questions.

Consider the following:

What are some ways that technology is facilitating the *flattening* of culture?

What are some ways that technology is preserving cultural values and differences?

Most engineers embrace their potential to affect the present and the future, but how might engineers affect the way the past is preserved and perceived?

14.4 Political Influence

Because engineers are constantly developing new technologies and processes, the need for policies and regulation of these new technologies also continuously emerges. Often, the need for such regulation only surfaces after confronted with an issue, like the establishment of the Facebook *memorialization* policy discussed earlier. While the policies and regulation procedures established by the political system significantly impact the development of technological solutions, individuals who are attracted to engineering and technology typically are not interested in politics and sometimes find politics to be frustrating and useless. Engineering and politics, however, should not be viewed as separate entities. As it stands, politicians and policy makers rarely possess the understanding or knowledge to deliberate technologically-related issues based on their technical merits. An engineer who embraces the relationship between engineering and politics can potentially make tremendous impacts on the development of public policy and the direction of future technological endeavors (Augustine 2012).

Technological development has frequently been stagnated by political decisions that experts believe lack technological basis. One such example is the US space program. Budget cuts approved by congress have proven to greatly impede development in this arena despite the technological potential that engineers believe to be on the horizon. Some believe that the space program's current condition would be dramatically different had engineers been more instrumental in voicing the importance of and potential that exists within the aerospace industry. In countries like China and Iran, engineers already play a significant role in the political process and their involvement has been recognized as a significant factor in the rapid technological advancement within their countries (Riley 2008). Such a positive impact of engineer involvement indicates that engineer involvement in political processes could be equally fruitful elsewhere.

USING DRONES

As with many other technologies, drones, also known as unarmed aerial vehicles or quadcopters named for the four engines used for their navigation, find their roots with the military. Initially used for surveillance, they eventually were used as mobile combat units. Together with the advancement of satellite and imagery technology, drones were armed and proved to be excellent at locating and attacking specific targets, reducing civilian casualties, and distancing military personnel from the potential dangers of the battlefield. Interestingly enough, distancing military personnel from those dangers still did not protect drone operators from experiencing the same kind of posttraumatic stress disorder (PTSD) as men on the battlefield. Despite the fact that drone operators could literally be anywhere in the world while controlling drones and executing their mission, the excellent imagery together with consciousness of the true nature of their mission and the ability to endlessly review the videos of past missions renders operators susceptible to PTSD (Sifton 2012).

As more and more nonmilitary uses of drones are identified, some people contest the negative connotation that many associate with the term *drone*. The arming of drones with guns and weapons is only one possible use for these technologically advanced flying machines, but there is tremendous potential for the drones beyond the killing machines that they have become known for. One such use came to light when Amazon.com announced its intentions to implement a drone delivery system for its products. The vision of Amazon's Prime Air delivery service is to deliver packages weighing up to 5 lb to locations located within 10 miles of an Amazon warehouse in as little as 30 min (see Figure 14.1). Representatives at Amazon.com claim that the technology is advanced enough and ready to be implemented, but there is still another set of obstacles. Approval from several governmental agencies must first be obtained. The primary agency that needs to approve the venture is the FAA, aka the Federal Aviation Agency. In March 2015, the FAA granted Amazon the permission to begin testing their drone delivery system with the condition that they adhere

FIGURE 14.1
Amazon.com's prototype delivery drone.

to several restrictions. Of the most prohibiting is the restriction that the drones must always remain in sight of its operator. Amazon fully intends for their drones to be programmed and then operated by computers, but for now, the approval is a step in the right direction for the company (Lavars 2015).

Some people find it frustrating that the technology exists, but political and regulatory issues stand in the way of its implementation. There are, however, numerous concerns that new technologies, such as drones, bring to our attention and new provisions must be made to accommodate these concerns. For every technology, there is a good, intended use and there is also a potential for misuse. If these regulatory bodies and procedures did not exist, there is a potential that a technology could be used to reap mass havoc on our society. Already there is much concern about the invasion of privacy that these drones can invoke. With their advanced surveillance abilities, they have the potential to remain undetected as they listen in on private conversations and obtain photographs and videos from long distances. Where should the lines be drawn to protect people from invasion of privacy? Other concerns exist with safety as drones are becoming accessible to members of the general public. Do they have the potential to interfere and possibly cause the malfunctioning of larger airplanes if not operated within the regulation? These are all questions and concerns that politicians and engineers must work together to answer and address. Thus, for engineers, it is not just about creating solutions; it is also about working within their means to create an environment where the technologies can be safely assimilated.

14.5 Learning from Tragedy: The Fukushima Disaster

On March 11, 2011, Japan was hit by a devastating tsunami. The largest waves were as high as 40 ft and going over 100 mph. In the aftermath, 20,000 people were dead or missing and survivors were left to deal with mass destruction and loss all around them. In all the chaos, most people had no idea that there was a much greater threat looming inside the nuclear power plant in Fukushima. Figure 14.2 shows the location of Fukushima nuclear plant.

The tsunami had penetrated the seawall and flooded the Fukushima Daiichi power plant. Nuclear reactors are designed to automatically shut down in such cases, but, even after shutting down, the radioactive rods in the reactor's core will continue to generate intense heat and require cooling to keep them stable. At Fukushima, the backup generators designed to provide the coolant to the reactors were located in the basement and destroyed from the flooding. Workers at Fukushima knew that it was only a matter of time before a complete meltdown occurred.

Over the next few days, workers at the power plant worked furiously to gain control of the situation. In an attempt to release pressure from the reactors and avoid a much larger disaster, Japan's prime minister authorized the venting of radioactive gases into the air, and areas surrounding the power plant were evacuated. Workers exposed themselves to dangerous levels of radiation as they worked to vent the reactors, and eventually they were successful.

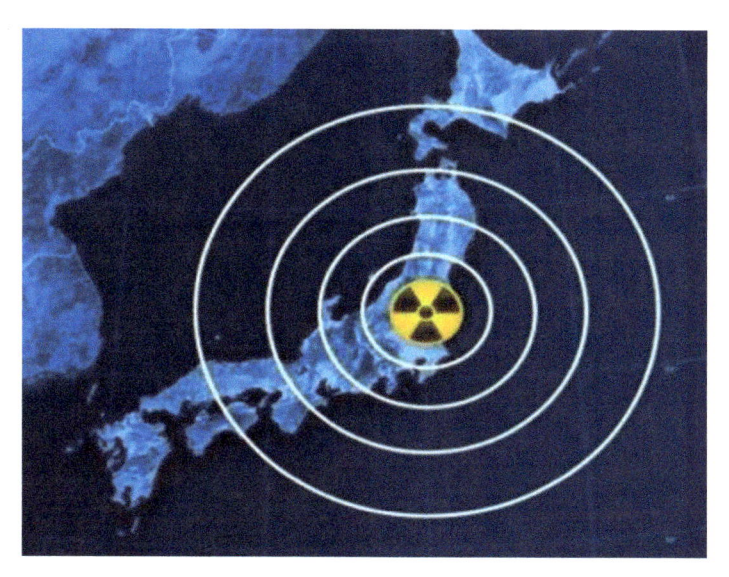

FIGURE 14.2
The Fukushima nuclear plant location in Japan.

Still the reactor cores needed to be cooled. Before workers could get water into the cores, two more explosions occurred, leaking more radioactive gases into the air. Thankfully, the explosions were in the roofs of the reactor buildings and the cores remained intact. Eventually, all but a skeleton crew of men were sent home while the remaining workers continued to battle on the front lines. Later, these remaining workers became known as the Fukushima 50 (Edge 2011).

With the help of the military and firefighters, workers were able to get water into to cool the nuclear cores. There were no fatalities at the nuclear power plant during the days after the tsunami, but more than 100 people were exposed to high radiation levels, increasing their risk of developing cancer in the future. The worst-case scenario would have meant the evacuation of the area within a 120–190 mile radius from the power plant, but because of the workers' selfless actions, the evacuation zone was kept to a 12-mile radius (Edge 2011) (Figure 14.3).

In the aftermath of these events, the workers known as the Fukushima 50 face many struggles. While their name suggest that there were 50 men in this group, the Fukushima 50 actually comprised hundreds of men who are now dealing with depression, symptoms of PTSD, and more surprisingly, hatred and intolerance from the Japanese people. While some cultures would consider these men to be heroes, the Japanese people hold the workers from the nuclear power plant responsible for the disaster. In Japan's culture, much of a person's identity is associated with the place where they work, and Japanese people often introduce themselves by the company name first followed by their name. Anyone who worked or continues to work for TEPCO, the operating company of the power plant, is viewed as personally responsible for the disaster. Instead of receiving praise for their actions in the days following the tsunami, these men are forced to hide in shadow for fear of the backlash that might be inflicted if people knew of their affiliation with TEPCO (Wingfield-Hayes 2013) (see Figure 14.4).

352   Introduction to Engineering

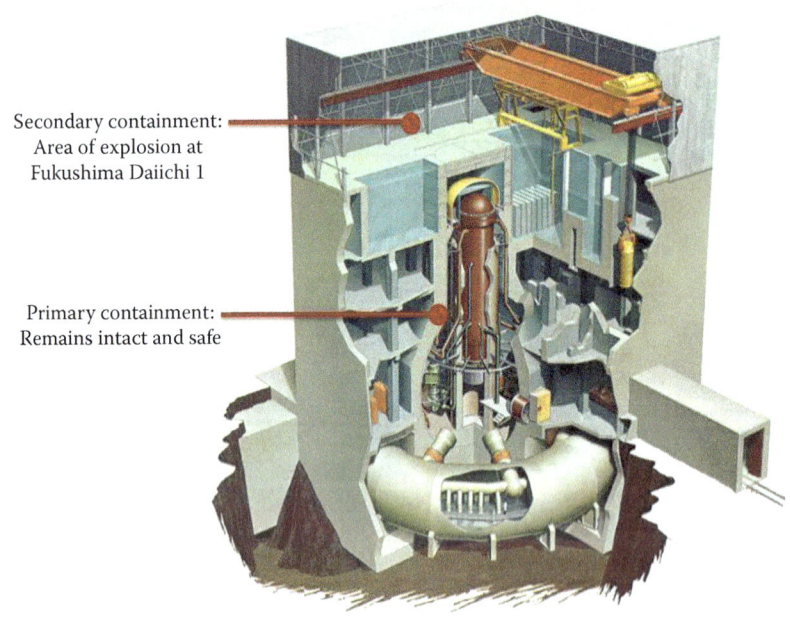

Secondary containment:
Area of explosion at
Fukushima Daiichi 1

Primary containment:
Remains intact and safe

Boiling water reactor design

FIGURE 14.3
Diagram of a reactor at the Fukushima Daiichi power plant. (Courtesy of Brook, *Fukushima Nuclear Accident—A Simple and Accurate Explanation, Brave New Climate.* Retrieved from: http://bravenewclimate.com/2011/03/13/shima-simple-explanation/, 2011.)

FIGURE 14.4
Hydrogen explosion at Fukushima Daiichi.

14.6 Lessons from Fukushima

The lessons from the tragedy at the Fukushima power plant are numerous and far reaching. People from all over the world were affected by the disaster but reacted in vastly different ways. For some, it fueled mistrust and apprehension of nuclear power. While many people embrace alternative energy resources, general public opinion about the feasibility of nuclear power as one of these alternatives has been greatly damaged since the Fukushima disaster.

Before Fukushima, Germany derived 23% of its energy from nuclear power, but after the tragedy, the country began phasing out nuclear power production completely (After Fukushima: Global Opinion on Energy Policy 2012). Scientists and nuclear power experts view this as an unnecessary and rash decision derived from mass hysteria and fear and not a scientific fact. Because public opinion is a significant driving force behind policy establishment, supporters of nuclear power now face even stronger opposition against furthering nuclear power technologies.

14.7 Pursuit of Social Understanding

The decisions that engineers make are significant not only to their employers or current projects but also to society as a whole. Engineers have made it possible for us to do what was once thought impossible and live in ways that were once only dreams of the few. On the other hand, engineers have also created our most destructive weapons and placed the power to obliterate the world in the palms of our hands. Technologies on all points of the spectrum have impacted the way that people live. As engineers, embracing the importance of our decisions and contributions is a responsibility that should not be ignored.

Problems

1. In your own words, explain the importance of having an understanding of societal norms to the engineering profession. [addresses ABET Criteria 3(h)]

2. Consider the development of technology over the past 50 years. In general, do you believe that these developments have made the world a better or worse place to live? Are people's lives happier or unhappier as a result of modern technology? How has the amount of an individual's leisure time been effected as a result of this technology? [addresses ABET Criteria 3(h)]

3. One capability that has resulted from recent technological developments is the ability for people to manage most, if not all, of their finances virtually. There is little to no need for physical money. Few people get written paychecks anymore; instead, paychecks are automatically deposited into bank accounts. Bill payments are automatically withdrawn from bank accounts—no need to keep track of all those due dates or worry about sending creditors a physical check or other form of payment. Almost everything can be paid for using debit and/or credit cards. Some people even tithe (i.e., give money to their church) and give their children their allowance by virtually transferring money.

a. Nonetheless, money management continues to be a significant challenge among American households. How and to what extent have new technological developments affected our concept of money? Do we need to see physical cash for us to visualize our budgets? Or is it possible for us to completely understand our finances virtually? If adults struggle with these concepts, is it possible to teach future generations the true *value* of money? [addresses ABET Criteria 3(h)]

b. There are a wealth of tools now available to assist with financial management. Are these tools helping enough to justify *virtual* banking? As a whole, do you believe that we are more or less capable of successfully managing our money than people 50 years ago? 20 years ago? Do you think that there may eventually be a day when physical cash no longer exists? [addresses ABET Criteria 3(h)]

4. Great advancements have been made in the autonomous vehicle arena over the past few years, and it is projected that the first fully autonomous vehicle will hit the market in the near future.

   a. Automakers claim that these vehicles will be the solution for many societal issues. List and describe some of the benefits that proponents of this technology claim will be produced. [addresses ABET Criteria 3(j)]

   b. What are some of the primary concerns with this technology? Do you feel those concerns are reasonable? [addresses ABET Criteria 3(j)]

   c. Consider the regulations, laws, and policies that govern user-controlled vehicles and predict some issues that might require new policies and regulations with the introduction of autonomous vehicles. Describe the differences between old and new policies. [addresses ABET Criteria 3(j)]

   d. Many people express concerns about trusting this technology. Assuming that autonomous vehicles prove to perform reliably, what challenges might exist in changing the opinions of those who mistrust automation on the road? [addresses ABET Criteria 3(j)]

5. Another force that should be considered in the interaction of engineering and society is the role of religion. The Amish religion is a fascinating example that should be considered for its societal contribution in the United States. Research the fundamentals of this religion and explore the significance of their view of modern technology. Then answer the following questions:

   a. On what basis do the Amish reject electricity and other forms of modern technology? How does this affect their daily life and their sense of community? [addresses ABET Criteria 3(h)]

   b. How do the Amish shape their society through technological decision making? [addresses ABET Criteria 3(h)]

   c. How has technology affected their society and beliefs? [addresses ABET Criteria 3(h)]

6. As of 2014, approximately 6000 languages were spoken in the world. Culture is primarily transferred from generation to generation, and language is the most significant tool for transmitting culture. However, in a world that is becoming increasingly global, languages are disappearing at an alarming rate. As a result, many struggle to embrace their native culture and customs. As languages rapidly fade from existence, we are also experiencing a loss of culture.

a. What are the potential impacts on society that might result from the loss of a language? [addresses ABET Criteria 3(j)]
b. What are some things that can be done to address the issue of language survival in cyberspace? [addresses ABET Criteria 3(j)]
c. Overall, is this a significant problem? [addresses ABET Criteria 3(j)]

7. Investigate some of the school shootings that have occurred in the United States over the past decade. Many people believe that violent material on TV, the Web, and in computer games is responsible for the recent increase in school shootings. To what extent do you believe these influences are responsible for these incidents? How might this issue be addressed while still preserving the fundamental rights established by the First Amendment? [addresses ABET Criteria 3(j)]

8. Veterinarians implant computer chips into pets and farm animals to help locate them if they get lost. Some people would like to do a similar thing for their children. What are some of the benefits of doing this? What privacy issues might result? Do the benefits outweigh the risks? [addresses ABET Criteria 3(j)]

9. Consider the technology available for image editing. Rarely, if ever, is there an image that appears in magazines, webpages, and so on that has not been enhanced in some way. We can fix blemishes, make people appear skinnier, and change almost anything with just a few clicks of the mouse. In general, how and to what extent is this affecting our conception of beauty and self-worth? Are children today growing up with a different perception of beauty than we did? In a few paragraphs, explain your view on the usage of this type of technology and the impact that such use has had on society. [addresses ABET Criteria 3(j)]

References

After Fukushima: Global Opinion on Energy Policy. (March 2012). Ipsos Social Research Institute. Retrieved from http://www.ipsos.com/public-affairs/after-fukushima-global-opinion-energy-policy/.

Augustine, N. (March 28, 2012). *Engineers: Our Government Needs You*. Retrieved December 29, 2014, from Forbes.com: http://www.forbes.com/sites/ieeeinsights/2012/03/28/engineers-our-government-needs-you/.

Brook, B. (March 13, 2011). *Fukushima Nuclear Accident—A Simple and Accurate Explanation*, Brave New Climate. Retrieved from: http://bravenewclimate.com/2011/03/13/fukushima-simple-explanation/.

Chan, K. H. (October 26, 2009). *Memories of Friends Departed Endure on Facebook*. Retrieved February 05, 2015, from Facebook.com: https://www.facebook.com/notes/facebook/memories-of-friends-departed-endure-on-facebook/163091042130.

Di Donato, J. (March 4, 2014). *Grief in the Time of Facebook*. Retrieved from HuffingtonPost.com: http://www.huffingtonpost.com/jill-di-donato/grief-in-the-time-of-facebook_b_4893588.html.

Edge, D. (Director). (2011). *Inside Japan's Nuclear Meltdown*. Motion Picture.

Lavars, N. (April 12, 2015). *Amazon to Begin Testing New Delivery Drones in the US*. Retrieved from GizMag.com: http://www.gizmag.com/amazon-new-delivery-drones-us-faa-approval/36957/.

McCallig, D. (March 6, 2015). *Facebook After Death: An Evolving Policy in a Social Network*. Retrieved January 30, 2015, from http://ijlit.oxfordjournals.org/content/early/2013/09/25/ijlit.eat012.full.

Riley, S. (April 2, 2008). *Engineering 'Mindset' Doesn't Include Politics*. Retrieved January 24, 2015, from EETimes: http://www.eetimes.com/document.asp?doc_id=1168269.

Shih, C. (September 9, 2010). *The New Social Norms*. Retrieved from http://www.informit.com/articles/article.aspx?p=1617519.

Sifton, J. (February 7, 2012). *A Brief History of Drones*. Retrieved March 12, 2015, from TheNation.com: http://www.thenation.com/article/166124/brief-history-drones.

Van Der Leun, J. (n.d.). *Using Facebook to Grieve*. Retrieved from Coping with Loss and Grief: http://www.coping-with-loss-and-grief.com/grieffacebook.html.

Wingfield-Hayes, R. (January 1, 2013). *Why Japan's 'Fukushima 50' Remain Unknown*. Retrieved September 15, 2014, from BBC News: http://www.bbc.com/news/world-asia-20707753.

15
Engineering Ethics

CHAPTER OBJECTIVES AND STUDENT LEARNING OUTCOMES

After completing this chapter, students will be able to

1. Define the concept of ethics and professional behavior.
2. Demonstrate the need for ethical standards and their ability in ethical decision making. [ABET outcome f, see Appendix C]
3. Describe academic integrity policy and professional engineers' code of ethics.
4. Emphasize the importance of ethical behavior in classrooms and professionalism for engineers. [ABET outcome f, see Appendix C]

15.1 Introduction

Ethics is a branch of philosophy that investigates between right and wrong actions performed by individuals, groups or society. Ethics defines concepts of good, evil, right, wrong, virtue, justice, and crime, most of which relate to human morality. Engineering ethics is (1) the study of moral issues and decisions confronting individuals and organizations involved in engineering and (2) the study of related questions about moral conduct, character, ideals and relationships of people and organizations involved in technological development.

Martin and Schinzinger
Ethics in Engineering

As engineers are required to make decisions that may be critical to the well-being of others, it is essential that they have a good understanding of ethics and make ethical decisions.

To prepare students with high moral and ethical standards, engineering educators may make students aware of ethics in education. Although case studies and examples of ethical dilemmas faced by engineers in the workplace are important, students can gain valuable understanding of ethics by using standards that may be applicable to them.

15.2 Ethics in Engineering Education

The first step in learning ethics is to apply ethical standards and expect ethical behavior in the classroom, homework, assignments, and exams. Both instructor and student can affect the ethical climate of the classroom by asserting student learning as central to all activities. Awareness of the relationship between ethical behavior and student learning and encouraging engagement in the learning environment may discourage unethical behavior.

Most institutions have a written academic integrity policy and academic standard committee, which students may not be familiar with. The instructor can include the policy in the syllabus and provide copies to students as a reading assignment. A brief survey or quiz may assess the level of understanding of the policy and encourage students to develop better understanding of ethical practices. Students lean toward academic dishonesty when instructors use test banks, assign homework problems for which solutions may be available, do not engage students in the learning process, and do not assign challenging problems. The classroom or department policy toward ethical expectations from students has great influences on how other students behave as do ethical practices by peer students. For example, if a student copies homework from another student or a solution manual, the consequences of such behavior will either discourage or encourage other students' ethical behavior. It is also important to realize that ethical standards should be enforced by other faculty in the department and across the institution.

Academic integrity policies or institutional honor codes can be effective deterrents to academic dishonesty if they are implemented properly. The classroom environment, institutional climate, and how ethically student peers behave greatly influence students' perception of ethics. If students see others cheating or involved in plagiarism without consequences, then they will more likely be involved in similar activities. However, a university cannot create an academically ethical climate merely instating a policy or honor code. The institution must also educate students about the policies and the consequences for academic dishonesty. Such information should be clearly stated in classroom syllabi as well as discussed and reinforced during lecture.

Student motivation is a key factor in developing high ethical standards, so it will never be enough to focus on punishing dishonesty. While unethical behavior is discouraged, institutions and instructors must actively encourage ethical behavior. Ethics are not about avoiding doing wrong but about doing right. Instructors must emphasize the importance of ethical behavior to learning materials in the course. Students can motivate each other toward higher ethical standards when classroom discussion reinforces the relationship between ethical behavior and achievement of learning objectives. Instructors should remind students about their views toward unethical behavior from students and how dishonesty conflicts with their self-images as honest people. Most institutions have an academic standards committee to deal with confrontational situations between student and instructor.

To proactively prevent academic dishonesty and encourage higher ethical standards in classroom, a number of steps can be used by the instructor. These can prevent possible cheating and convey the message that the instructor is serious about assessment of learning through examination. Vigilance is also important because if students observe others cheating successfully, then they are more likely to cheat too (Keith-Spiegel, Tabachnick,

Whitley, & Washburn, 1998). The following teaching tips can be effective during classroom and examination as presented in American Psychological Association, Psychology Teacher Network, May 2013 (Prohaska, 2013):

1. Space desks and students as far apart as possible during an examination.
2. Instruct students to turn off phones, pagers, and other electronic devices and store them out of sight.
3. Before the exam begins, tell students they cannot leave the room during the exam.
4. If using multiple choice questions, use several forms of the exam so that questions and/or answer orders are scrambled. Hollinger and Lanza-Kaduce (2009) found that students reported multiple versions of an exam as the most effective technique for preventing cheating.
5. Try not to use the same examination questions from semester to semester.
6. Do not allow students to keep or photograph graded exams.
7. Actively proctor examinations (i.e., periodically move or at least look around). Do not leave the room unattended or become obviously engrossed in another activity such as reading a newspaper or journal, speaking on a phone, or grading completed exams.
8. Online quizzes or examinations present special difficulties (Young, 2010). Brothen and Wambach (2001) suggested applying time limits to each question and randomly selecting questions from a much larger pool of questions.

Plagiarism is increasingly becoming a serious problem in classrooms as students have access to online resources and solution manuals for most courses. There are also online services that solve problems, write papers, and complete assignments for students. Instructors are overwhelmed with the new techniques used by students and always look for ways to prevent plagiarism and encourage students to do their own assignments. A number of strategies can be used by instructors to help prevent plagiarism as described below:

1. Develop assignments and homework that make plagiarism difficult. For example, assignment should require integration of materials discussed in classroom and textbook rather than summarizing or describing a concept. Also require a higher level of analysis and synthesis of the topic so that students must understand the material to complete the assignment. Another example is to develop homework problems and test problems instead of using problems for which solutions are available to students.
2. Provide instructions about plagiarism in the classroom and how to avoid it. It helps students explicitly learn about proper citation techniques and allows instructors to monitor students' understanding about plagiarism so that it can be avoided.
3. Careful consideration should be made on the level of difficulty and relevance of the assignment to the materials covered in the classroom. Students' are more likely to plagiarize an assignment that is beyond their comprehension than one that they can complete with a reasonable level of difficulty.

15.3 Ethics in the Engineering Profession

Engineering profession requires high moral and ethical standards for practicing engineers. Professional societies and organizations develop their own code of conduct to assure that their members comply with them. National Society of Professional Engineers (NSPE) has developed a comprehensive code of ethics as stated below.

15.3.1 NSPE Code of Ethics for Engineers

15.3.1.1 Preamble

Engineering is an important and learned profession. As members of this profession, engineers are expected to exhibit the highest standards of honesty and integrity. Engineering has a direct and vital impact on the quality of life for all people. Accordingly, the services provided by engineers require honesty, impartiality, fairness, and equity, and must be dedicated to the protection of the public health, safety, and welfare. Engineers must perform under a standard of professional behavior that requires adherence to the highest principles of ethical conduct.

15.3.1.1.1 Fundamental Canons

Engineers, in the fulfillment of their professional duties, shall

- Hold paramount the safety, health, and welfare of the public
- Perform services only in areas of their competence
- Issue public statements only in an objective and truthful manner
- Act for each employer or client as faithful agents or trustees
- Avoid deceptive acts
- Conduct themselves honorably, responsibly, ethically, and lawfully so as to enhance the honor, reputation, and usefulness of the profession

15.3.1.1.2 Rules of Practice
- Engineers shall hold paramount the safety, health, and welfare of the public. If engineers' judgment is overruled under circumstances that endanger life or property, they shall notify their employer or client and such other authority as may be appropriate.
- Engineers shall approve only those engineering documents that are in conformity with applicable standards.
- Engineers shall not reveal facts, data, or information without the prior consent of the client or employer except as authorized or required by law or this Code.
- Engineers shall not permit the use of their name or associate in business ventures with any person or firm that they believe is engaged in fraudulent or dishonest enterprise.
- Engineers shall not aid or abet the unlawful practice of engineering by a person or firm.

- Engineers having knowledge of any alleged violation of this Code shall report thereon to appropriate professional bodies and, when relevant, also to public authorities, and cooperate with the proper authorities in furnishing such information or assistance as may be required.
- Engineers shall perform services only in the areas of their competence.
- Engineers shall undertake assignments only when qualified by education or experience in the specific technical fields involved.
- Engineers shall not affix their signatures to any plans or documents dealing with subject matter in which they lack competence, nor to any plan or document not prepared under their direction and control.
- Engineers may accept assignments and assume responsibility for coordination of an entire project and sign and seal the engineering documents for the entire project, provided that each technical segment is signed and sealed only by the qualified engineers who prepared the segment.
- Engineers shall issue public statements only in an objective and truthful manner.
- Engineers shall be objective and truthful in professional reports, statements, or testimony. They shall include all relevant and pertinent information in such reports, statements, or testimony, which should bear the date indicating when it was current.
- Engineers may express publicly technical opinions that are founded upon knowledge of the facts and competence in the subject matter.
- Engineers shall issue no statements, criticisms, or arguments on technical matters that are inspired or paid for by interested parties, unless they have prefaced their comments by explicitly identifying the interested parties on whose behalf they are speaking, and by revealing the existence of any interest the engineers may have in the matters.
- Engineers shall act for each employer or client as faithful agents or trustees.
- Engineers shall disclose all known or potential conflicts of interest that could influence or appear to influence their judgment or the quality of their services.
- Engineers shall not accept compensation, financial or otherwise, from more than one party for services on the same project, or for services pertaining to the same project, unless the circumstances are fully disclosed and agreed to by all interested parties.
- Engineers shall not solicit or accept financial or other valuable consideration, directly or indirectly, from outside agents in connection with the work for which they are responsible.
- Engineers in public service as members, advisors, or employees of a governmental or quasigovernmental body or department shall not participate in decisions with respect to services solicited or provided by them or their organizations in private or public engineering practice.
- Engineers shall not solicit or accept a contract from a governmental body on which a principal or officer of their organization serves as a member.
- Engineers shall avoid deceptive acts.
- Engineers shall not falsify their qualifications or permit misrepresentation of their or their associates' qualifications. They shall not misrepresent or exaggerate their

responsibility in or for the subject matter of prior assignments. Brochures or other presentations incident to the solicitation of employment shall not misrepresent pertinent facts concerning employers, employees, associates, joint ventures, or past accomplishments.
- Engineers shall not offer, give, solicit, or receive, either directly or indirectly, any contribution to influence the award of a contract by public authority, or which may be reasonably construed by the public as having the effect or intent of influencing the awarding of a contract. They shall not offer any gift or other valuable consideration in order to secure work. They shall not pay a commission, percentage, or brokerage fee in order to secure work, except to a bona fide employee or bona fide established commercial or marketing agencies retained by them.

15.3.1.1.3 Professional Obligations
- Engineers shall be guided in all their relations by the highest standards of honesty and integrity.
- Engineers shall acknowledge their errors and shall not distort or alter the facts.
- Engineers shall advise their clients or employers when they believe a project will not be successful.
- Engineers shall not accept outside employment to the detriment of their regular work or interest. Before accepting any outside engineering employment, they will notify their employers.
- Engineers shall not attempt to attract an engineer from another employer by false or misleading pretenses.
- Engineers shall not promote their own interest at the expense of the dignity and integrity of the profession.
- Engineers shall at all times strive to serve the public interest.
- Engineers are encouraged to participate in civic affairs; career guidance for youths; and work for the advancement of the safety, health, and well-being of their community.
- Engineers shall not complete, sign, or seal plans and/or specifications that are not in conformity with applicable engineering standards. If the client or employer insists on such unprofessional conduct, they shall notify the proper authorities and withdraw from further service on the project.
- Engineers are encouraged to extend public knowledge and appreciation of engineering and its achievements.
- Engineers are encouraged to adhere to the principles of sustainable development in order to protect the environment for future generations.
- Engineers shall avoid all conduct or practices that deceive the public.
- Engineers shall avoid the use of statements containing a material misrepresentation of fact or omitting a material fact. Consistent with the foregoing, engineers may advertise for recruitment of personnel. Consistent with the foregoing, engineers may prepare articles for the lay or technical press, but such articles shall not imply credit to the author for work performed by others.

- Engineers shall not disclose, without consent, confidential information concerning the business affairs or technical processes of any present or former client or employer, or public body on which they serve.
- Engineers shall not, without the consent of all interested parties, promote or arrange for new employment or practice in connection with a specific project for which the engineer has gained particular and specialized knowledge.
- Engineers shall not, without the consent of all interested parties, participate in or represent an adversary interest in connection with a specific project or proceeding in which the engineer has gained particular specialized knowledge on behalf of a former client or employer.
- Engineers shall not be influenced in their professional duties by conflicting interests.
- Engineers shall not accept financial or other considerations, including free engineering designs, from material or equipment suppliers for specifying their product.
- Engineers shall not accept commissions or allowances, directly or indirectly, from contractors or other parties dealing with clients or employers of the engineer in connection with work for which the engineer is responsible.
- Engineers shall not attempt to obtain employment or advancement or professional engagements by untruthfully criticizing other engineers, or by other improper or questionable methods.
- Engineers shall not request, propose, or accept a commission on a contingent basis under circumstances in which their judgment may be compromised.
- Engineers in salaried positions shall accept part-time engineering work only to the extent consistent with policies of the employer and in accordance with ethical considerations.
- Engineers shall not, without consent, use equipment, supplies, laboratory, or office facilities of an employer to carry on outside private practice.
- Engineers shall not attempt to injure, maliciously or falsely, directly or indirectly, the professional reputation, prospects, practice, or employment of other engineers. Engineers who believe others are guilty of unethical or illegal practice shall present such information to the proper authority for action.
- Engineers in private practice shall not review the work of another engineer for the same client, except with the knowledge of such engineer, or unless the connection of such engineer with the work has been terminated.
- Engineers in governmental, industrial, or educational employ are entitled to review and evaluate the work of other engineers when so required by their employment duties.
- Engineers in sales or industrial employ are entitled to make engineering comparisons of represented products with products of other suppliers.
- Engineers shall accept personal responsibility for their professional activities, provided, however, those engineers may seek indemnification for services arising out of their practice for other than gross negligence, where the engineer's interests cannot otherwise be protected.
- Engineers shall conform with state registration laws in the practice of engineering.

- Engineers shall not use association with a non-engineer, a corporation, or partnership as a *cloak* for unethical acts.
- Engineers shall give credit for engineering work to those to whom credit is due, and will recognize the proprietary interests of others.
- Engineers shall, whenever possible, name the person or persons who may be individually responsible for designs, inventions, writings, or other accomplishments.
- Engineers using designs supplied by a client recognize that the designs remain the property of the client and may not be duplicated by the engineer for others without express permission.
- Engineers, before undertaking work for others in connection with which the engineer may make improvements, plans, designs, inventions, or other records that may justify copyrights or patents, should enter into a positive agreement regarding ownership.
- Engineers' designs, data, records, and notes referring exclusively to an employer's work are the employer's property. The employer should indemnify the engineer for use of the information for any purpose other than the original purpose.
- Engineers shall continue their professional development throughout their careers and should keep current in their specialty fields by engaging in professional practice, participating in continuing education courses, reading in the technical literature, and attending professional meetings and seminars.

Sustainable development is the challenge of meeting human needs for natural resources, industrial products, energy, food, transportation, shelter, and effective waste management while conserving and protecting environmental quality and the natural resource base essential for future development. As revised on July 2007:

> By order of the United States District Court for the District of Columbia, former Section 11(c) of the NSPE Code of Ethics prohibiting competitive bidding, and all policy statements, opinions, rulings, or other guidelines interpreting its scope, have been rescinded as unlawfully interfering with the legal right of engineers, protected under the antitrust laws, to provide price information to prospective clients; accordingly, nothing contained in the NSPE Code of Ethics, policy statements, opinions, rulings, or other guidelines prohibits the submission of price quotations or competitive bids for engineering services at any time or in any amount.

Statement by NSPE Executive Committee

In order to correct misunderstandings that have been indicated in some instances since the issuance of the Supreme Court decision and the entry of the Final Judgment, it is noted that in its decision of April 25, 1978, the Supreme Court of the United States declared "The Sherman Act does not require competitive bidding."

It is further noted that, as made clear in the Supreme Court decision:

- Engineers and firms may individually refuse to bid for engineering services.
- Clients are not required to seek bids for engineering services.
- Federal, state, and local laws governing procedures to procure engineering services are not affected, and remain in full force and effect.
- State societies and local chapters are free to actively and aggressively seek legislation for professional selection and negotiation procedures by public agencies.

- State registration board rules of professional conduct, including rules prohibiting competitive bidding for engineering services, are not affected and remain in full force and effect. State registration boards with authority to adopt rules of professional conduct may adopt rules governing procedures to obtain engineering services.
- As noted by the Supreme Court, "nothing in the judgment prevents NSPE and its members from attempting to influence governmental action . . ."

NOTE: In regard to the question of application of the Code to corporations vis-à-vis real persons, business form or type should not negate nor influence conformance of individuals to the Code. The Code deals with professional services, which services must be performed by real persons. Real persons in turn establish and implement policies within business structures. The Code is clearly written to apply to the Engineer, and it is incumbent on members of NSPE to endeavor to live up to its provisions. This applies to all pertinent sections of the Code.

The National Society of Professional Engineers believes that employed engineers have an obligation to their profession, their employer, and the public to make known ethical concerns in the workplace. NSPE believes that in order to establish a healthy professional working environment, it is vital for the following conditions to exist in the workplace:

An atmosphere of trust between the employer and the employee

An empowering environment where employees feel secure in raising and seeking the resolution of sensitive issues

An absence of fear of employer retribution against employees for raising and seeking resolution of sensitive issues

NSPE believes that employees should raise and seek resolution of issues in a professional manner, and that employers should respond in a way that permits timely and effective resolution of those issues without damaging the reputation of the employee or the employer.

15.3.2 The Path to Resolution of an Ethical Dilemma in Employment

The consequences of an employee making a misstep in notifying his or her employer of an ethical and/or safety situation within the company is currently a concern to many employed engineers. The fear of losing one's job and the consequences that obviously flow from that loss are very real. The possibility of litigation and its effect on one's personal and professional life can be overwhelming. As the engineer moves up and establishes credibility within a company, the consequences of a misstep may be lessened but are still of serious concern.

Most companies understand these concerns and have developed opportunities for an employee to raise issues without fear or apprehension. Bringing these issues to the attention of management has become mandatory in many companies. In fact, in some companies, an employee may be terminated for not reporting a professional concern or problem to management. At least one company states in its company policy that an employee will be protected from retribution, whether or not the employee's report was accurate, as long as the report was made in good faith.

The path to resolution of professional concerns is not nearly as clear in companies that have not developed policies and procedures on reporting unethical, illegal, or safety matters. However, this fact alone does not necessarily release the licensed engineer or engineer of their responsibility to themselves, their company, or to the public to address the issue in a professional manner.

In fact, a licensed professional engineer is generally required by state laws and regulations to act for the protection of the public health and safety. While in most states engineers in industry are exempt from licensing laws for work done for their industrial employers, the engineer and the engineer's company may still be liable for unethical, unsafe, or illegal activities.

The 1991 Federal Sentencing Guidelines also offer compelling reasons for companies to develop improved corporate ethics programs. Moreover, corporate fines and the damage to a company's reputation and bottom line for ethical lapses can be staggering. Lynn Sharp Paine, an associate professor at the Harvard Business School who specializes in management ethics and who authored "Managing for Organizational Integrity," published in the *Harvard Business Review*, has identified the serious consequences to an organization for management that does not require ethical behavior of its employees.

15.3.3 Seeking Advice and Reporting Concerns and Violations

NSPE believes that the following three principles form a foundation for employees facing an ethical dilemma:

> When faced with an unethical business conduct or a legal question, the employee has a right and an obligation to seek advice and guidance as necessary to resolve the employee's concern or question.
>
> The employee has an obligation and a responsibility to promptly bring to the company's attention any actions, situations, or conditions that the employee believes are or may be violations of the company guidelines or the law.
>
> As a licensed engineer, the employee has an obligation to protect the life, health, and property of the public.

The representatives of the three companies that helped to develop this document all consider unethical behavior by a company employee to be very damaging to their company's reputation. All three companies encourage employees to bring improper situations to the attention of supervisory management.

The first action an engineer faced with an ethical dilemma should take is to obtain a copy of his or her company policy, read it thoroughly, and proceed in a manner consistent with the company policy. If action is required, the engineer should then go to the appropriate individual within the department for advice on further action.

If there is no company policy, or if the policy does not address the particular concern, the employee should get advice from a person in the company that the employee trusts. That person should be in a position to assist the employee in the proper handling of the situation. Some possibilities, as appropriate, are

> Any member of supervisory management Legal Department Corporate Compliance Administrator Personnel
>
> Corporate Security
>
> Ethics Hot Line (1-800-888-XXXX) if available

In the event that there is no company policy or no one within the company with which the employee is comfortable discussing the matter, the employee should turn to a trusted and respected professional in the community for advice in properly handling the situation.

NSPE is an organization consisting of licensed professional engineers and engineering graduates from Accreditation Board for Engineering and Technology/Engineering Accreditation Commission programs from all engineering disciplines. For that reason, NSPE is uniquely suited to assist engineers in a wide range of ethical matters. Resources available through NSPE include

- Licensed engineers (members) who can lend their experience to the situation
- The NSPE Board of Ethical Review, which consists of NSPE members knowledgeable in ethical matters who may review cases and could provide an opinion or an interpretation

An additional resource for licensed engineers is their state engineering licensure board staff, who may be a valuable source in providing information on how to handle the situation appropriately and ethically. State engineering licensure board staff also may provide the licensed employee with advice on how a situation would be addressed under the board's jurisdiction.

If all other actions fail, the employee may need to consult a private attorney. The attorney may advise the employee on a variety of matters (e.g., what legal options are available, what legal obligations the employee may have, whether the employee should resign or remain on the job, etc.). A private attorney has an obligation to act and advise solely on the employee's (client's) behalf and will maintain the employee's (client's) confidentiality.

15.3.4 Other Questions to Consider before the Employee Takes Action

- Is my action in compliance with all applicable local, national, and international laws?
- Is my action in keeping with the values of the company I am employed by?
- Is my action honest and fair in every respect?
- Will my action be viewed positively if it becomes known to my supervisor, coworkers, friends, or subordinates?
- Will my action reflect positively on my company and me if it is disclosed in the newspaper or other media?
- Is my action in compliance with company policy, procedures, or principles?

If the employee answers *yes* or *true* to all questions and the employee follows company procedure in making known the concern, the employee is acting ethically and is probably in compliance with the company's policy. If the employee answers *no* to any of the questions, or if the employee determines that *it just does not feel right* to take action, the employee should seek further advice. The advice should come from a supervisor the employee trusts or another resource in the company if it is available. The employee should only seek assistance outside of the company when all available resources within the company have been exhausted.

NOTE: Although this document uses the term *Company* to refer to the employer, the document may also apply to other public or private employers, including government and educational institutions.

References

Brothen, T., & Wambach, C. (2001). Effective student use of computerized quizzes. *Teaching of Psychology, 23*, 292–294.

Hollinger, R. C., & Lanza-Kaduce, L. (2009). Academic dishonesty and the perceived effectiveness of countermeasures: An empirical survey of cheating at a major public university. *National Association of Student Personnel Administrators Journal, 46*, 587–602.

Keith-Spiegel, P., Tabachnick, B. G., Whitley, B. E., Jr., & Washburn, J. (1998). Why professors ignore cheating: Opinions of a national sample of psychology instructors. *Ethics & Behavior, 8*, 215–227.

NSPE Code of Ethics, National Society of Professional Engineers, Alexandria, VA. July 2007, Publication #1102. http://www.nspe.org/sites/default/files/resources/pdfs/Ethics/CodeofEthics/Code-2007-July.pdf.

Prohaska, V., Encouraging students' ethical behavior, American Psychological Association, Psychology Teacher Network, May 2013.

Young, J. R. (2010, April 2). High-tech cheating abounds, and professors bear some blame. *Chronicle of Higher Education, 56*(29), A1.

16
Communication and Teamwork

CHAPTER OBJECTIVE AND STUDENT LEARNING OUTCOMES

After completing this chapter, students will be able to:

1. Define the purpose and list attributes of effective communication. [addresses ABET Criteria 3(g)]
2. Outline and select appropriate material to include in oral presentations depending on analysis of audience and purpose. [addresses ABET Criteria 3(g)]
3. Communicate information, concepts, and ideas effectively in writing. [addresses ABET Criteria 3(g)]
4. Orally communicate information, concepts, and ideas effectively. [addresses ABET Criteria 3(g)]
5. Select and use appropriate professional graphics in written and oral presentations. [addresses ABET Criteria 3(g)]
6. Acquire and evaluate credibility of information from a variety of sources. [addresses ABET Criteria 3(g)]
7. Identify specific behaviors and skills that support team effectiveness. [addresses ABET Criteria 3(d)]
8. Describe what one must do to participate fully in team projects. [addresses ABET Criteria 3(d)]

16.1 Essential *Soft* Skills in a Globalized Workforce

Teamwork and communication are both essential skills for engineers. These skills are typically considered *soft skills* and viewed by engineers to be of minor importance compared to technical skills. Yet, in practice, these skills have proven to be equally important to an engineer's success as the prioritized technical skills. These soft skills are grouped together in this chapter only because of their nature and overlap, but this should not diminish their individual importance to the engineering profession.

Two of the largest complaints from companies that hire first-year engineers are that they lack the ability to communicate with others effectively and successfully function as a productive team member. Job candidates often possess all of the desired technical skills for a given position, but their inability to communicate and/or work with others is a cause for great concern. Candidates who do possess these skills, however, clearly stand out from other applicants. Possessing such skills will not only be a benefit when seeking a job but also in numerous other endeavors throughout your lifetime (Schawbel 2015).

16.2 Communication

Communication skills are essential to our existence and act as the foundation for professional and personal success. Engineers especially need to use these skills with colleagues, customers, and other stakeholders so that they can understand the importance of engineering work. However, communication skills, often classified as *soft skills*, are often undervalued by engineers; engineers fail to realize that they cannot be fully effective in their jobs if they are inadequate communicators (White 2013).

Effective communication is essential not only for the diffusion and sharing of ideas but also for gaining the confidence and respect of professionals in the field. Inadequate and ineffective communication skills reflect poorly on engineers and undermine their credibility. Regardless of how knowledgeable an engineer is with scientific and technical details, an inability to communicate ideas and concepts may cast doubt on his or her abilities. An inability to gain the confidence of this audience and adequately relay ideas to others may also translate into missed opportunities as those who communicate more effectively gain support and backing over those with lesser communication skills.

Communication skills cannot be developed overnight. In fact, good communication skills take years to develop and require continuous practice. Those who struggle with communicating with others often tend to avoid situations in which they are forced to utilize these skills. While this is an understandable response, it is important to embrace any weaknesses and actively develop strategies for learning how to act effectively and efficiently as communicators. The sooner students start their endeavors to improve in this arena, the earlier they will reap the benefits.

Examples provided in Box 16.1 and Figures 16.1 and 16.2 show how the nanotechnology topic was introduced with examples.

BOX 16.1 NANOTECHNOLOGY

Nanotechnology is an emerging field that engineers are applying in multiple engineering disciplines. *Nanotechnology* describes the manipulation of molecules and atoms at a molecular level and encompasses anything that occurs on the *nanoscale*. The *nanoscale* is defined as anything between 1 and 100 nanometers. A *nanometer* is one-billionth of a meter. To give you an idea of exactly how small we are talking about, a nanometer is 1/25,000th the width of the average human hair (Strickland and Bosnor 2007).

Although we have long been aware of nanosized entities, the ability to see, understand, and control these objects is recent. Many people are unaware of the fact that nanotechnology is already used in the manufacturing of everyday products. Applications for nanotechnology have already been discovered in aspects of medicine, materials science, bioengineering, computing, energy, environment, aerospace, agriculture, and manufacturing, to name a few, and new applications of nanotechnology are continuously being discovered.

Communication and Teamwork

FIGURE 16.1
In 1989, IBM demonstrated their nanotechnology ability to manipulate atoms when IBM spelled out its initials with xenon atoms on a nickel surface. (Courtesy of IBM, *IBM's 100 Icons of Progress*, Image. March 30, 2015, http://www-03.ibm.com/ibm/history/ibm100/us/en/icons/microscope/breakthroughs/, 2011.)

FIGURE 16.2
Examples of consumer products manufactured using nanotechnology. (1) *Curad® Silver Bandages* contain silver nanoparticles in the bandage pad to inhibit bacterial growth; (2) *L'Oreal RevitaLift™* contains nano-sized nutrients that let light in and give a translucent appearance; (3) *Head® Nano Titanium Tennis Racquets* are made by weaving nanoparticles with titanium to create a racquet with superior stiffness; (4) *Behr® Premium Plus Kitchen & Bath Paint* uses nanotechnology additives to make the finish more durable; (5) *Eagle One® NanoWax* uses nanosized carnauba particles to better conceal scratches and increase durability and shine; (6) *Pilkington Activ™ Self Cleaning Glass* uses a nanoscale transparent coating to spread water into sheets that carry dirt with it as the water runs off the surface; (7) *CDs and DVDs* use nanotech concepts to increase their data; (8) *Blue Lizard® Australian Sunscreen* contains nanometer-sized particles of titanium dioxide and zinc oxide that allow the sunscreen to be transparent on the skin. (Courtesy of National Nanotechnology Infrastructure Network, *Exploring Nanotechnology Through Consumer Products*, Image. June 1, 2015, http://www.nnin.org/education-training/k-12-teachers/nanotechnology-curriculum-materials/exploring-nanotechnology, 2006.)

Public opinion has long been a matter of concern for engineers because public opinion has great say on the speed and direction of technological development. Public opinion becomes especially concerning with innovative, emerging technologies like nanotechnology. Policy development, support of research initiatives, government and private funding allocations, and innovation adoption are all influenced by public opinion. Positive public attitudes have the potential to greatly facilitate innovation while negative public attitudes have the potential to inhibit or even halt development in these areas. For engineers and scientists, understanding how public opinion is formed could be extremely beneficial (National Science Board 2010).

In 2007, a survey was implemented to better understand the public's knowledge and perception about the risks and benefits of nanotechnology. Of the approximately 1,850 people that participated in this initiative, 1,500 of the participants' responses were used to evaluate existing public opinion of nanotechology. Eighty-one percent of the participants indicated they had no knowledge or very little knowledge about nanotechnology, yet, 89% of the participants indicated that they already possessed an opinion on whether nanotech's benefits outweighed its risks. From this, we can conclude that the public perception of risk is not necessarily based on factual knowledge (Braman et al. 2007).

Emotional responses can have tremendous influence in forming public opinion of technologies. Emotion becomes especially significant in the absence of knowledge or familiarity with a technology. Nanotechnology concepts, in particular, are extremely complicated and hard to understand. When it comes to the unknown, fear is a common emotional response, and the public's negative attitude toward nanotechnology can be attributed, at least in part, to fear. Initiatives to better educate the public about nanotechnology concepts can help alleviate irrational fears by providing more factual, scientific evidence supporting the benefits of nanotechnology. This communicating complex nanotechnology concepts in ways that can be understood by members of the general public is essential.

When it comes to knowledge about nanotechnology concepts, there is no better source than nanotechnology engineers and scientists. Even engineers and scientists in other fields rarely have an adequate understanding of how nanotechnologies work, so it is unlikely that anyone outside of the nanotechnology field would be able to provide adequate education on these concepts. Translating such complicated concepts for members of the general public will prove to be one of the biggest challenges in promoting nanotech awareness among the general public. Because the future direction of nanotechnology hinges greatly on public perception and these educational efforts, this is an excellent example of a situation where it is not only beneficial but also essential for engineers to possess superb communication skills.

16.2.1 Oral Communication

For an engineer, oral communication can range from casual progress reports presented to small groups to formal presentations aimed at large audience. Regardless of the type of presentation, the following are some general guidelines to keep in mind when preparing for any presentation.

16.2.1.1 Know Your Audience

The purpose of any presentation is to relay information to your audience. It is essential to be aware of who your audience is so that you can present the information they are expecting to hear. The role and motivation of your audience will determine the information that is useful to them. For example, when presenting a progress report to members of upper management, technical details may be irrelevant. Upper management is likely to be more concerned with whether the project is developing as expected, on time and within budget. Boring them with elaborate technical details that they may not even understand can only serve to frustrate your audience. Your goal is to keep your audience engaged with relevant information for the duration of your presentation.

Some audience are harder to assess than others. The larger and more diverse your audience, the more dimensions need to be considered and accommodated. In general, however, you will always want to consider several basic attributes of your intended audience. You may begin your *audience analysis* by gathering information on *who will be in your audience* and *the approximate number of audience members*. Next, you will want to know *what your audience is expecting to learn* from your presentation. It is also important to consider *your audience's general knowledge* on the presentation subject; providing details that are common knowledge among your audience may give the impression that you are talking down to them while providing too little information will cause confusion and prevent you from communicating your intended message. If you are speaking to a global or international audience, you will also want to consider the *culture of your audience*. We frequently hear of instances where people have inadvertently offended people of a different culture simply because they did not realize that behavior deemed to be acceptable in their native culture was offensive or improper among people of their audience's culture (Stack and Freifeld 2013). Another valuable aspect that should be considered is *your audience's view* on your presentation topic and/or of you in general. If they have any preconceived opinions, concerns, or biases about you or your presentation topic, you may want to consider the roots of negative perceptions and attempt to address and ease any fears that may prevent your audience from accepting the information that you are presenting. Of course, there are times when addressing these fears would be futile or not appropriate within the scope of a particular presentation, but awareness of any preexisting biases should still be taken into consideration and handled sensitively during the presentation (Executive Communications Group 2003).

Depending on your individual situation, there could be any number of additional aspects that would help you to better understand your audience, and the more you know about your audience, the better presentation that you can prepare. Tailoring your presentation to your intended audience serves to ease your stress and also acts as a demonstration of respect to your audience members. You are essentially relaying to your audience that you value their time while simultaneously reinforcing their view of your competence and professionalism.

16.2.1.2 Know Your Content

The better you understand what you are presenting, the more comfortable you will be when making your presentation. Ideally, presenters will have ample knowledge of their particular presentation topic, but this is not always the case. For example, there may be a time when an engineer is asked to present information on a project on which she is collaborating with many other engineers. While that engineer may have ample knowledge of specific aspects of that project, she may have only a general idea of the current status of other components of the project. If the audience is expecting information that requires a more detailed knowledge of other project components, the presenter may not actually possess knowledge of the content she needs. If at all possible, the presenter should defer to the team member that has the most relevant knowledge for the intended audience. Otherwise, the presenter should make special efforts to gain the necessary knowledge prior to the presentation. A strong grasp of the material being presented greatly increases the presenter's confidence and comfort level and is a key contributor to a successful presentation.

16.2.1.3 Presentation Software

If you have a projector available for your presentation, you may choose to take advantage of presentation software as a visual aid. Presentation software, such as PowerPoint or Prezi, allows the presenter to graphically organize his presentations and include visuals to help keep the audience engaged. Most commonly, presentation software allows the presenter to add information or graphics to a series of *slides* in order to reinforce the information being relayed.

One of the most common mistakes that presenters make with presentation software is filling their slides with long, wordy text and reading directly to the audience from their slides. While the slides may serve to keep the presenter on task, they are meant to enhance the information that the presenter is relaying, not replace it. Instead, bullets with keywords or short phrases should be on the slides, along with relevant images or other carefully selected multimedia whenever possible (University of Texas at Austin 2014). Presenters then expand on whatever short keywords or phrases are on the slide.

The following are some other simple suggestions to keep in mind when using the presentation software to enhance your presentations:

- Start your presentation with an overview of what you will be covering. This introduction will help clear up misconceptions in audience expectations and help ensure that your audience will receive your intended message.
- Avoid distracting colors, fonts, and media. Remember that the presentation software is meant to accompany the information that the presenter is attempting to relay, and is not meant to overshadow the actual presenter.
- Avoid reading to your audience as much as possible. It is boring and some consider it condescending.
- Use large text so that everyone in the room will be able to read the text on your slide. For presentations to larger groups, it may be necessary to have even larger text than you might use when presenting to smaller groups.
- Be sure to include appropriate references for any information obtained from outside sources including images and other multimedia.
- Conclude your presentation with a summary of what you have discussed.
- Practice your presentation with your digital slide presentation and other visual aids.
- Always try to arrive early to the presentation location to verify equipment functionality and load your slide presentation.
- Save your presentation in multiple places. Nothing is worse than thinking you are prepared and then realizing that the file you uploaded to your digital Dropbox or saved on your flash drive cannot be opened. Saving your presentation in multiple accessible locations can help avoid such a situation.
- Always have a backup plan. At times, the worst does happen and you may have equipment issues and/or not be able to access your slide presentation as intended. Still, the presentation must go on—with or without your slide presentation. Having a printed copy of your slide presentation and/or your presentation outline may still provide you the guide that you need to successfully deliver your presentation despite the unfortunate circumstances.

Adhering to the above suggestions, the presentation software can produce excellent accompaniments to any presentation, and are recommended to be utilized in all feasible situations (University of Texas at Austin 2014).

16.2.1.4 Nonverbal Messages

Many presenters fail to consider the role that body language and other forms of nonverbal communication play in the delivery of a presentation, but things such as eye contact, posture, facial expression, and gestures can greatly impact the effectiveness of the presenter's intended message. In fact, research has shown that nonverbal forms of communication can be just as important as the words spoken when relaying a message.

When you make a presentation, your audience forms an opinion of you and your message based not only on what they hear but also on what they see. Body language can speak to the presenter's confidence, credibility, honestly, enthusiasm, and sincerity. When a presenter displays body language that his or her audience interprets as an indication of calmness, confidence, and sincerity, the audience is far more likely to judge the presentation on the merits of the information presented. Positive nonverbal communication can greatly enhance the presenter's ability to connect with his or her audience. On the other hand, when a presenter appears uncomfortable, nervous, or uneasy, the audience forms a negative opinion of the presenter and the presentation content even before any information is verbally transmitted.

Most of our nonverbal communication is done subconsciously, so you may not be aware of an unusual habit you have when public speaking. In order to thoroughly evaluate your own use of nonverbal communication, you can record video of yourself practicing. Ideally, you could watch video of yourself delivering a presentation to a live audience. On viewing such a video, presenters often find themselves surprised at the nonverbal messages they have been sending. The following are behaviors that presenters should consider when evaluating their own nonverbal behaviors:

- *General appearance*: The manner in which the presenter dresses and grooms makes a significant impact on the audience's opinion of that presenter. Be sure to dress appropriately and professionally when giving any presentation.
- *Eye contact*: When delivering a presentation, it is important to make eye contact with members of the audience. This helps to enhance the connection between the presenter and his or her audience and is interpreted as a significant indicator of the presenter's confidence and comfort level. Throughout the presentation, the presenter's eyes should scan the audience, engage with individual audience members for a short moment, and then slowly continue on.
- *Facial expressions*: Throughout a presentation, facial expressions should be appropriate to the message that the presenter is trying to relay. Never should a presenter maintain a single expression throughout a presentation, but rather, his or her expressions should vary to increase impact and relay appropriate emotion.
- *Gestures*: Presenters should use meaningful and deliberate gestures throughout their presentation. Executed correctly, gestures can be a great asset to enhancing the impact of a message. Be careful not to cross your arms or legs during a presentation. Additionally, when presenting to international audience, special care should be taken to avoid using gestures that are already associated with a different meaning within the audience's culture.

- *Body movement: Body movement* refers to the physically changing position or location during a presentation, like when walking around. Remaining in a single location, especially behind a lectern, is generally interpreted as a sign of insecurity, but when presenters freely move about the room, they relay a message of confidence and comfort. Of course, too much movement can become a distraction and can also be interpreted as a sign of nervousness and insecurity. Movement that develops into a consistent pacing, swaying, or rocking is an example of excessive movement that should be avoided. The presenter's movement should not be constant, but, rather, done with purpose and followed by moments of stillness.

The first step in correcting or improving nonverbal messages is becoming conscious of one's own behaviors. When presenters are keenly aware of the impact of their own body language, they can learn to adapt their nonverbal signals to reinforce the message that they are presenting (Feloni 2014).

16.2.1.5 Rehearsing Presentations

After planning, organizing, and preparing your presentation, supplementary materials and presentation aids, it is highly recommended that you practice your presentation before actual delivery. It is better to do this out loud instead of rehearsing in your mind alone. Forcing yourself to verbalize your presentation will help you identify places in your presentation where better explanation and/or connections are needed. Pronunciation issues also become apparent when verbalizing aloud. When reviewing in your mind, your brain often fills in for any gaps and fails to identify the weak areas that verbalization can reveal (Witt 2012).

If at all possible, also try to practice your presentation standing and walking around like you might be when delivering your presentation. Recording voice or full video of your presentation is also useful in critiquing your readiness. However, if these additional suggestions are not feasible, rehearsing mentally and verbally will still provide an adequate amount of preparation. Remember that any and all efforts to practice prior to delivery will increase your confidence and improve the quality of your presentation (Witt 2012).

16.2.1.6 Presenting with Groups

There are many times throughout your education and career where you will find yourself presenting with one or more colleagues or team members. While we discuss some of the basic attributes of teamwork later in this chapter, there are a few things that you should keep in mind regarding presentations as well.

The first mistake that teams tend to make is allowing teammates with stronger communication skills to dominate the presentation. Especially in situations where all presenters are presumed to have made equal contributions, having dominant presenters can give the appearance that other teammates did not actually contribute in the capacity that they should have. It can also negatively reflect on the dominant presenter, who may not be viewed as a team player. Neither interpretation reflects positively on the team as a whole.

Team presenters also struggle to know what they should do when they are not presenting. Often, you will observe team members looking out into the audience or staring into the distance while a teammate speaks. Remember that everyone should be focused on the presenter, and that includes fellow presenters. Focusing your attention on the active presenter reinforces to the audience that their attention should be focused there as well.

Presenting in teams requires more practice than being the sole presenter. Without proper practice, important transitions and cues can be awkward or even missed, but rehearsing together as a group can be instrumental in ironing out unexpected kinks.

16.2.1.7 Concluding Remarks on Oral Communication

Oral communication skills do not come naturally to most people and are developed over time and with repeated practice. Some individuals struggle more than others to develop these skills, yet the ability to communicate effectively through oral communication is essential to an engineer's professional success. In any communication, always keep in mind that having people dedicate time to listening to us is a privilege that should not be taken for granted, and we then have the responsibility to relay our message in the most effective, engaging, and impactful way possible.

16.2.2 Written Communication

Throughout his or her career, an engineer may be asked to produce written work in various capacities. Proposals, reports, letters, email, and lab reports are just a few types of written communication that the engineer might utilize. Learning to write using a concise and clear writing style is extremely important for an engineer, and just like oral communication, it is a skill that must be continuously developed over time.

As with oral communication skills, knowledge of the intended *audience* must be a primary consideration in developing written communication. Engineers rarely write for audience comprised of fellow engineers with similar levels of expertise. Instead, engineers frequently produce written work intended for customers, members of management, or other stakeholders that do not possess the same technical background. Having an understanding of who will be reading your document enables you to select appropriate content that will meet your audience's expectations. Audience analysis also facilitates appropriate language choice and determines when special care should be taken to explain highly technical jargon in terms that can be understood by a particular audience.

Not only should content be customized for a specific audience, but the expected formatting and report elements can vary depending on the audience. Companies, for example, often require their own writing formats, and it is important you conform to the adopted formats when submitting written works within that company. When working with multiple companies in the same industry, expectations will likely differ. Sometimes the formats may be similar to what you have used in educational environments, and other times, the formats could vary drastically from what you are accustomed to. During your educational experience, you will likely find that the format expectations vary significantly from instructor to instructor and course to course, as well. Thus, before preparing any formal written works, it is essential that you clarify that both your content and format conform to your audience's expectations.

16.2.2.1 Letters and Memos

As an engineer, you may need to communicate in writing with other companies, the public, management, and/or colleagues. All written communications should be concise, clear, and accurate. *Letters and memoranda*, commonly called *memos*, are common types of written correspondence used in companies today. In general, *memos* are written to people within your place of work, whereas *letters* are written to people outside your place of work. Memos are often less formal and less concerned with structure and formatting, whereas letters are carefully proofread to avoid even the smallest mistake and usually printed on high-quality paper, rather than copy paper. Typically memos are used to quickly deliver important information, while letters provide more detail, support, and justification in order to make a sale or answer a question. While letters and memos can be similar in content, they are written in different formats. Most companies have specific policies for correspondence that establish company-specific formatting expectations. It is important to become familiar with the requirements of each document type for your employing organization (Blinn College-Bryan Writing Center 2008).

Both memos and letters include headers that give information about the author, the intended recipient, subject, date, and the names of other people who may see the memo. For both memos and letters, people who are mentioned in a memo or are directly affected by the memo should receive a copy. The writer's company name, logo, address, and other contact information are found at the top of the page of a letter, but this information is not required in memos (Blinn College-Bryan Writing Center 2008).

A memo begins by stating its purpose in the first sentence followed by any necessary information. A typical letter, on the other hand, should state their purpose in the first paragraph but not necessarily the first sentence. Only the most necessary detail and explanation is included in a memo, whereas letters usually contain more supporting information. In letters, the message is summarized in the last paragraph, but memos do not include a summary. Instead, memos often conclude with a call to action that details any responses or actions expected from the recipient. While omitted in memos, letters often end with a conclusion thanking the recipient and/or requesting that the recipient contact the writer followed by a formal closing such as *Sincerely,* or *Yours Truly* and the writer's name (Blinn College-Bryan Writing Center 2008).

Example 16.1: Business Memo Example

ABC Consulting
12000 Fenton Rd
Fenton, MI 48844
(800) 700-10100

Memo

To: All Staff

From: Jackson Hutton, Director of Human Resources

Date: June 1, 2015

Subject: New Vacation Policy Effective July 1

This memo serves to inform employees about a change in ABC Consulting's vacation policy.

Notice of your intention to take a vacation must now be given no less than 7 days prior to your vacation. This will enable us adequate time to obtain temporary help, if necessary, and to schedule vacations in a manner that is not disruptive to regular company operations.

This vacation policy change will take effect July 1, 2015. If you have any questions, please don't hesitate to call me.

> **Example 16.2: Business Letter Example**
>
> May 15, 2015
>
> Dr. Laura Sutton
> 303 East Keaskey Street
> Flint, MI 48502
>
> Dear Dr. Sutton,
>
> I would like to take this opportunity to express my heartfelt thanks to you for your valuable contributions to our recent conference, "Engaging in Engineering." Judging from the comments of those who attended, the conference was very successful, and the majority of the credit is owed to you and the others who gave such interesting presentations.
>
> Your skill in delivering your presentation on "Controversies in Genetic Engineering" was very much appreciated by those representing all sides of that extremely sensitive topic. We have received numerous post-conference requests for the paper you delivered on "Eugenics and the Ethics of Designing a Deaf Baby." I was not alone in my fascination and engagement as we explored these issues from the viewpoint of deaf parents.
>
> On both a professional and a personal level, I really appreciated the time that the two of us were able to spend together during conference down times. The personal experiences that you shared were truly captivating. I certainly learned a lot about some of the unique aspects of genetic engineering and the challenges that you confront in your line of work.
>
> Again, thanks so much for your participation in our conference. I have no doubt that it would not have been the success that it was without your presence. Your enthusiasm is contagious, and we hope that you will consider presenting again in the future. Thank you again for your contribution.
>
> Please keep in touch, and drop in and visit us whenever you are in the area.
>
> Sincerely,
> Melinda Donaldson
> Chairman of Events

16.2.2.2 Email and Electronic Communications

Email, or *electronic mail,* is a very useful communication tool used in most work environments. Email communication contains the same basic components as other correspondence mediums such as the letters and memos previously discussed. Emails differ from letters and memos because they can be used for communication both internal and external to the company. Many companies have found that email has been an efficient and effective replacement for traditional hard copy correspondence, especially internal communications, and has allowed for a significant reduction in the need for telephone communication.

Most people receive large volumes of emails every day. Managing these messages can sometimes be tedious and challenging. Because email messages can be so easily composed and sent, extreme care should be taken in what is said. It is important to remember that spelling and grammar is just as important in emails as it is in other forms of correspondence. Many people have a natural tendency to quickly type messages and/or responses and click send before adequately proofreading. It is important to fight this natural tendency and always take an extra moment to review the message for typos, grammatical errors, and clarity before sending. Additionally, remember that not all email recipients will interpret messages in the same way. Be sure to watch for anything that might be misconstrued within your message.

Appropriate email communication in the workplace should adhere to standards and etiquette that reflect your professionalism. Never use abbreviations or *text talk* such as "u" instead of "you" or BTW instead of "by the way." Using jargon of this type gives the recipients the impression that you are lazy and can be a language barrier for people unfamiliar with these language trends. Similarly, the use of smiley faces or other emoticons are also deemed inappropriate in professional emails (Office Angels 2013).

Workplace email should never be used for personal reasons. Instead, maintain separate email accounts for professional and personal usage. Employees are not entitled to privacy in their workplace email, and most companies monitor employee email for relevant and appropriate use. Never say anything negative or offensive in a work email. Avoid writing anything in an email that you would not say in person. Once sent, all email becomes part of the corporate and/or public record (Bussing 2011).

The following are some additional tips to keep in mind when communicating via email:

- Always include a short, meaningful subject. A good subject helps recipients with searching, filing, and filtering.
- Always use proper salutations and closings in your messages.
- When including attachments, be sure to assign meaningful file names to your attachments and always refer to the attachments within your message text.
- Keep emails as short as possible. A general rule of thumb is to limit messages to the amount of text that can be viewed on the readers' screen without having to scroll.
- For the first email in a conversation, be sure to use your full name and provide contact information in your signature block. Some companies have established guidelines for the format and content in the signature block. If your company does not have an established guideline, conform your signature block to signature blocks that other employees include.
- Only send messages to people or groups that need to receive the email. When responding to messages sent to email groups, only reply to all recipients when absolutely necessary.
- If you want a recipient to perform an action after receiving your message, make your expectations clear. For example, if you need a response before a certain date and time, directly state that. If you need action on multiple items, list them individually.

16.2.2.3 Reports

Technical reports are often required in business and industry. All technical reports should adhere to the following principles of technical writing:

- *Clear*—*Clear* technical communication means that information is presented in a simple and effective manner so that readers easily understands its meaning. Technical documents that lack *clarity* can lead to confusion and costly mistakes. Word choice is of the utmost importance and requires ingenuity and attention to detail. Avoid vague and ambiguous terms and include sufficient detail.
- *Correct*—*Correctness* encompasses both the use of proper spelling and grammar and the presentation of current and accurate information. Grammar and punctuation rules bring order and dignity to our language and make it possible for others to understand what has been written. Effective technical communication cannot be achieved successfully unless rules of grammar and punctuation are consistently adhered to.
- *Concise*—*Conciseness* means to write in a direct, precise manner and to relay information using the fewest words possible. Avoid redundancies and wordy expressions, and rambling.
- *Complete*—To achieve *completeness*, all necessary information should be provided for the reader to understand a message or topic. This includes relevant background information, process or methodology descriptions, results, conclusions, and recommendations for future action.

The aforementioned writing principles are commonly referred to as the *4 Cs of Technical Writing* (Altmann and Hallesky 2011).

Progress reports, lab reports, proposals, project completion reports, and evaluation reports are examples of technical reports that an engineer may be asked to compose. Technical reports should be in paragraph form with indented first lines with an average paragraph's size of 6–10 sentences. Technical documents should be written in third person. Each type of report has a different purpose and requires different structural elements to properly meet the objective. At a minimum, all technical reports include an *introduction*, a *body*, and a *conclusion*. The following section defines and explains what is expected in these sections:

16.2.2.3.1 Fundamental Report Elements

The *introduction* explains what the paper is about and why it is important. It defines the parameters and scope of the remainder of the paper, and informs the readers of what they should expect. An introduction always contains a thesis statement at the end of the section. It is possible for the thesis statement to be comprised of more than one sentence, but they should all be located together at the end of the introduction. Because the thesis statement is a declaration of what you want your readers to do, it is essential that the body of the paragraph supports the thesis and stays on track.

The *body* of the report contains all background and supporting information for your paper. It is often divided into subsections for easier navigation. Common subsections include Methodology, Procedure, Data, Calculations, and Analysis. The body contents should be organized in a manner so that information flows logically.

Lastly, the *conclusion* presents your professional opinion as to the meaning of the data as well as your analyses, and if appropriate, includes suggestions for future actions. Be sure to discuss any options or limitations that you encountered.

16.2.2.4 Citations and Bibliographies

Whether you are making a presentation or writing a report, it is always necessary to give proper credit to sources of content, data, images, and so on that you reference in your endeavors. Not only is it ethical to give credit where it is due, but it also helps to strengthen the validity of your work by providing your audience a way to verifying the credibility of others' works that your work is based on. Failure to give adequate credit to your sources is considered to be *plagiarism* and can have severe repercussions if not avoided.

Any content considered to be common knowledge does not require citation. In general, common knowledge can include widely known facts, dates, or data such as common sense observations, folklore, and legends. However, there is a gray area surrounding the idea of common knowledge, and there is no guarantee that everyone will agree whether something should be considered common knowledge or not. Direct quotes from anyone's writings should always be cited, regardless of whether or not the information presented is considered to be common knowledge. If there is any question as to whether a piece of information is considered common knowledge, it is best to cite this information. There can never be too many citations, but a lack of adequate citation can have severe repercussions. Therefore, it is best to err on the side of caution and cite whenever there is any concern.

It is important that you use the same style of references throughout your paper. MLA and APA are two very popular ways of formatting your sources, but many others exist as well. A list of references should be prepared and attached to the end of all types of writing in a *works cited* or *bibliography* page. In written works that have appendixes, reference pages are usually found after the text and before the appendixes. In the body of the report, abbreviated forms of the references appear typically in parenthesis, brackets, or subscripts after information from that source has been quoted or paraphrased.

16.2.2.5 Source Reliability

The 21st century may be the information age, but with a surplus information comes the challenge of sorting credible sources from unreliable and inaccurate ones. Anyone can post information on the World Wide Web. It becomes even more difficult when the information being researched is controversial and involves a number of different stakeholders who all have varying viewpoints. In many cases, finding reasonably unbiased sources is challenging if not impossible, and often, you may find yourself sifting through an overwhelming amount of information from a wide spectrum of viewpoints in order to develop your own conclusions.

That being said, the following are a few tips to help you determine whether or not an Internet source can be considered credible:

- Scholarly journals and articles are written by experts in a particular field and usually contain information on the most recent research, findings, and news in that field. These sources have undergone some sort of peer-review process and are typically published by a professional society or journal publication or by a university or educational institution. In general, these are considered to be credible sources and are preferred over other Internet sources.
- Look for information on the authors of the source. Try to determine their qualifications and/or any organizational affiliations. If there is no author listed, check to see if it is hosted by a reputable organization or group. If no author or host information can be found, you will want to look for a different source.
- Check out the home page of the source you are examining. From the home page, you should be able to find more information about the author and/or organization behind your source information and/or locate an "About Us" link that might give you the information you are looking for. Determine whether the website was developed by an individual or an organization and if the identified entity is reputable. If you cannot find a link to the home page, find the first single backslash ("/")

in the URL address, delete everything from that point to the end of the address, and navigate to the resulting address.
- Check to see the publication date of the source. In many fields, changes occur rapidly and older material may no longer be valid. Always try to find the most current sources, and, as a general rule of thumb, try not to cite anything written more than 5 years ago.
- If your source contains facts or statistics, look for citations or other documentation that indicates where the information was obtained from. Often times, there will be a list of references at the end of the page. If the source does not reveal where they obtained their information, you should question the validity of this information.
- Avoid using any sources from a wiki web site. On wiki websites, anyone modify information regardless of whether they are the initial authors or not. There is no way to verify authorship of information on these sites. While links to valuable, reputable information might be found on these pages, they should not be used as a direct source.
- When using websites, look at the ads on the page. Excessive ads are a bad sign. Types of ads can also indicate the professionalism of a website.

Selecting credible sources is a testament to your own credibility. If your work is based on unreliable or inaccurate information, then your work is just as unreliable and inaccurate. As an engineering professional, you are tasked with the ethical responsibility to take reasonable measures in order to ensure that your sources are credible.

The importance of grammar and correct spelling is provided in Box 16.2 and Figure 16.3.

BOX 16.2 UTILIZING WORD PROCESSING TOOLS FOR GRAMMAR AND SPELLING CORRECTIONS

Many word processing programs are equipped with excellent spelling, grammar, and language tools that can be extremely beneficial during the writing process and their capabilities are consistently being improved upon. While using such tools is highly recommended, understand the limitations of their capabilities. Spellcheck is never a replacement for a read through.

Consider the following scenario: I once had a student who chose to do his final research paper on prosthetic limbs for a university technology class that I was teaching. All in all, it was an excellently written paper. That is, except for one thing. Throughout the entire paper, the word *limb* was spelled L-I-M-E, and, thus, his paper was about prosthetic *LIMES* instead of prosthetic *LIMBS*. I can only imagine that this individual probably left the letter *b* off the end of the word *limb* when writing his paper. When his word processing software identified the misspelling and recommended changing the word to *lime*, he most likely applied the changes to all instances of the word *lim* throughout his paper. Despite quality of paper's content, it was very difficult to take it seriously with the image of limes in my head.

This is an excellent example of how overreliance on these tools can go wrong. Credibility in the workplace is extremely important, and simple mistakes such as the one described above can greatly undermine your efforts. Keep in mind that just because the software makes a recommendation does not mean that you should blindly accept its suggestion. Be sure to read the suggestions carefully and always perform your own manual proofreading in addition to using these tools.

FIGURE 16.3
Changing one single letter can be the difference between talking about prosthetic limbs, pictured left, to limes, pictured right. (Courtesy of Ottobock, *Lower Limb Prosthetics Overview*, http://www.ottobock.ca/prosthetics/lower-limb-prosthetics/solution-overview/, 2015; Jughandle Fat Farm, *Limes—Why Are There No Seeds?* http://jughandlesfatfarm.com/wp-content/uploads/2011/12/limes-2.jpg, 2011.)

16.2.2.5.1 Common Optional and Special Report Sections and Elements

In the next section, we will describe a general format for a short report that can be adapted to meet the needs of other types of reports.

Most organizations have their own expected writing formats, and it is unlikely that you will use the exact format that you used at your university in your professional engineering practice. Regardless of the format adopted by a school or engineering organization, it is important you conform to the adopted format of your company or institution. The following are descriptions of a few additional sections that a company may require to be included in a report:

- *Abstracts* are the most commonly required additional elements in technical writing. These are required so often that there is some debate as to whether it should be classified as a fundamental element or an additional element. Abstracts summarize the whole report in one concise paragraph. It provides enough information for the readers to decide whether they want to read the full report. Abstracts establish the reason for the paper, concisely describe the content and scope of the report, highlight key points, and briefly summarize the results. Abstracts are placed before the introduction of a report, and should never contain text that has been copied directly from another location in the report (Hicks 2013).

- A *Title page* is a page that contains the report title, author(s), and other relevant publication information. A professor or company will often expect a title page. Typically, the citation style that you are using will provide format guidelines for a title page. APA is one citation style that provides formatting guidelines; however, MLA does not require a title page and provides no guidelines. When formatting guidelines are unavailable, check with your professor or organization for their formatting expectations.

- For longer publications, a *Table of Contents* can greatly assist readers in easily finding the information that they are looking for. A *Table of Contents* includes all the headings and subheadings in a report and lists the page number where each of these begins. A good *Table of Contents* uses indentation to distinguish headings from subheadings and aligns these with the appropriate page numbers.
- *Appendixes* are typically used to present material that supports information in the report. Complex background or history information, detailed charts and tables, supporting documents, calculations, and raw data are all things that might be included in an appendix. Appendixes should be lettered and arranged in the order in which they are referred to within the report text (Unilearning Project 2000).

Example 16.3: Table of Contents Sample

Table of Contents

1.0	Project Definition	5
	1.1 Background	7
	1.2 Project Description	9
	1.3 Importance	10
	1.4 Scope	11
	1.5 Life Cycle with Due Dates	11
	1.6 Interdependencies and Conflicts	13
	1.7 Uniqueness	14
	1.8 Types of Resources	15
2.0	Conflict Management	16
	2.1 Types of Conflict	16
	2.2 Appropriate Conflict Interventions	18
3.0	Project Program	19
4.0	Project Team Skills	22
5.0	Work Breakdown Structure (WBS)	23
	5.1 Resource Identification	25
	5.2 Project Phases and Sub Tasks	29
	5.3 Resource and Task Assignments	30
	5.4 Estimate Task Duration	30
	5.5 Inflexible Start or Finish Date Constraints	30
6.0	Budget	33
7.0	Resource Management	35
8.0	Project Status Report	37
	8.1 Accomplishments	38
	8.2 Upcoming Tasks	39
	8.3 Issues	40
9.0	Earned Value Analysis	43
10.0	Resource Analysis	44
11.0	Resource Deployment and Control Plan	45
12.0	Suggestions for Improvement	47
13.0	Final Report and Project History Reflection	48
14.0	References	55
Appendix A: Accompanying Documents		57

16.2.2.5.2 Headings and Subheadings

In general, short reports, essays, reflections, and other shorter writings do not require the use of headings and subheadings unless they are specifically requested. In other works, headings and subheadings are intended to help your readers find the information that they are looking for. Headings divide a paper into sections of significant importance. The following are some additional suggestions to keep in mind when using headings and subheadings in papers:

- Keep your headings and subheadings short. Complete sentences should never be used as a heading or subheading.
- Do not overuse headings and subheadings. These are meant to highlight significant sections, but not every paragraph needs a heading.
- Use the same style of headings throughout your paper. Make subheadings smaller than headings.
- There should never be only one subheading. If the topic cannot be subdivided into at least two topics, then subheadings are unnecessary.

Lastly, remember that papers that use headings and subheadings should still employ good transitions between sections. In most cases, it is a good idea to write your paper first and insert headings and subheadings after the paper is completed. Because they are intended to facilitate your readers' navigation of your paper, they should not replace key sentences, and the readers should be able to read your paper and gain the same benefit from the information with or without the use of headings and subheadings (Hatalla 2015).

16.2.2.6 Figures, Images, and Tables

Images, figures, and tables are often used in writings. These elements are intended to enhance the written content but not replace it. Special care and consideration should be taken to select the most appropriate, effective, and relevant elements. Images should appear in the text after the paragraph in which they are introduced, but figures and tables should be presented before the paragraph in which they are introduced.

Every image, figure, and table should be labeled and a number assigned according to the order in which they appear in the text. For example, the first table will be called Table 1 and the first figure called Figure 1. Graphics should always have a descriptive caption indicating what it is showing. It is not enough to include a figure along with the essay or report. If an element is significant enough to be included in the work, then it deserves some sort of discussion. As a writer, you can briefly explain the significance of the element or more thoroughly examine the content of the element, depending on individual context. When discussing a graphic, specifically name the label when referencing a particular figure, table or image from anywhere in the document.

16.2.3 Other Technical Communication Mediums

We have already discussed some types of technical communication in earlier chapters of this textbook. CAD and computer programming are two types of technical communication that engineers of all disciplines should be familiar with. It is extremely uncommon for engineers to manufacture their own designs. Instead, engineers typically communicate their designs so that manufacturers or machinists can create the physical product. Both reading and writing in technical mediums allow engineers to relay important specifications quickly and easily to each other without interference from language or cultural barriers. Throughout their educational and professional careers, engineers will encounter numerous technical communication mediums. The valuable role of these communication tools in the engineering field should not be minimized.

16.2.4 Communicating by Listening

Listening is often overlooked when discussing communication skills; however, communication is multifaceted and requires the ability to both give and receive information. Like oral and written communication skills, listening is a hard skill to master and requires a conscious effort to improve. Furthermore, true listening is often sacrificed in a world where multitasking is a way of life. Instead, we often find ourselves listening for key pieces of information instead of considering the entire conversation and tend to overlook subtle cues in an interaction.

To help ensure that you are engaged in active listening, make a conscientious effort to limit distractions. If you are reading an email, you cannot be listening. You can turn your chair away from your computer screen or temporarily turn off your monitor. Consider taking notes during meetings. It is also wise to paraphrase what you believe you heard. Try not to parrot the speaker, but in your own words, describe what you understand to have happened, especially any calls to action or corrections that have been discussed. This gives both parties an immediate chance to clarify and can save confusion and headache later on.

The things that matter when you are speaking still matter when you are listening. Consider your body language, your nonverbal cues, and eye contact. Respect the person you are speaking with by actively listening him or her. Avoid any activity that may make it appear that you are not interested in the conversation, such as fidgeting, or playing with a cellphone.

Good listening is essential in building relationships, solving problems, and identifying potential opportunities. It is a demonstration of respect and appreciation to those that we are listening to. It takes time and conscious presence, but is well worth the efforts that we invest.

16.3 Teamwork Essentials

In the working world, it is rare to find an engineer that works on a project by himself or herself. Most often, engineers work in teams consisting of members with different but complementing specialties and disciplines. A *team* can be defined as a group of two or more people working together to achieve a common goal (What is a team? definition and meaning). The size and duration of a particular team can vary drastically, yet there are some consistent characteristics that exist in all effective team situations regardless of these differences.

16.3.1 Fundamental Characteristics of Effective Teams and Team Members

Because the first half of this chapter is dedicated to communication skills, it should come as no surprise that communication is essential for effective teams. An effective team must have communication among all group members. Communication proliferates into all aspects of teamwork, and any weakness in communication can potentially cripple a team's chance of success. Of utmost importance is facilitating an environment with *open communication* where everyone feels free to share ideas and feels respected by other members of the team. Problems can be identified much earlier in environments of open communication, which can only benefit engineers and their companies. The sooner a problem is identified, the earlier it can be addressed so that potentially devastating long-term effects can be avoided. Team members also share status reports and other vital information about their project frequently at meetings and through other announcement mediums.

To be successful, team members must all understand and embrace a *clear purpose*. A clear purpose clarifies what the team is attempting to produce or achieve, and also defines the parameters of success for the project. Often, trouble within ineffective teams can be traced back to a lack of a clear, agreed upon purpose or goal. As an example, when students in engineering courses are asked to participate in team projects, they assume that everyone on their team has the same ultimate purpose and motivation—to earn the best possible grade in the course. It may even be possible that team members initially have the same goal in mind, but as the semester progresses and demands on students' time increase, students find that they have to prioritize their efforts. Because students often have course loads or personal situations, one student might not place as much value on the course as another student might. Because their goals were never explicitly clarified, individual team members' definitions of "best possible grade" diverge, and these teams frequently find themselves frustrated and struggling. The team's probability for success would have been dramatically higher had they established a clear goal early in their partnership.

Team members also must *understand and accept their individual roles* within the team. Members of a team are accountable to each other for completing their work on schedule and adhering to the rules and procedures of the team. Everyone must be willing to take initiative in order to ensure the team's ultimate success. Team members should also make special efforts to encourage each other and show respect and considerations of others' ideas and feelings in the team.

16.3.2 Leadership in Teams

Typically, there is one team member who takes the role of a leader for a particular team and manages all team endeavors. This leader is usually referred to as the project manager and is tasked with establishing and facilitating the adherence to timelines and budgets. For this person to successfully fulfill his or her duties, he or she must openly communicate with all team members. Team members must be comfortable and willing to approach their project manager with any difficulties, obstacles, or concerns. Additionally, team members have the responsibility to initiate this communication as soon as they become aware of any issues, so as to allow the project manager time to initiate a solution before the problem escalates to the point where project success is compromised.

It is important to note that leadership also takes on less official forms within a group. A good team member is equipped with some unique skills and knowledge that can assist the team in successfully achieving their goal. When that expertise is needed, that particular team member then steps into a leadership role as he or she leads the team through the successful application of his or her specialized skills. At various stages and times within the project timeline, team members frequently find themselves stepping in and out of leadership roles within the team.

16.3.3 Embracing Diversity

Diversity can be defined and viewed in several ways. In general, diversity refers to all of the ways in which people or groups of people are different. Typically, team members are selected for their unique skills and qualifications that complement other members in the group. The nature of this practice ensures that team members possess a diversity of work skills. *Workplace diversity* is one of the greatest strengths of an organization. Embracing diversity in the workplace means creating an environment that values and supports the contributions of people with differences.

It is uncommon for a team member to have a significant influence, if any at all, on whom will be assigned to their team. Many times, team members start out as complete strangers and are tasked with building relationships with their colleagues from scratch. Often, teams are assembled with members from drastically different cultural backgrounds and ethnicities. Some people have much stronger feelings about these dimensions of diversity than they do about other dimensions. Some people possess strong prejudices and biases within these dimensions that greatly inhibit the establishment of effective team relationships and render the team as a whole ineffective. Many companies implement specialized trainings within their company to help reduce the existence of such prejudices and promote acceptance of others.

References

Altmann, E. J. and Hallesky, G. J. (2011). *Technical Writing That Works*. Bloomington, IN: Authorhouse.

Blinn College-Bryan Writing Center. (October, 2008). Writing memos, *Blinn College: Writing Center*, Document. May 15, 2015, www.des.ucdavis.edu/faculty/handy/ESP171/Writing_Memos.pdf.

Braman, D., et al. (March 2007). *Affect, Values, and Nanotechnology Risk: An Experimental Investigation, GW Law Faculty Publications & Other Works*, Paper., http://scholarship.law.gwu.edu/cgi/viewcontent.cgi?article=1276&context=faculty_publications.

Bussing, H. (October 4, 2011). http://www.hrexaminer.com/employee-privacy-what-can-employers-monitor/, Web: March 20, 2015, http://www.hrexaminer.com/employee-privacy-what-can-employers-monitor/.

Executive Communications Group. (2003). *You Talkin' To Me?: Know Your Audience and What It Takes To Persuade, Inspire and Motivate Them*, Web: May 28, 2015, http://totalcommunicator.com/vol1_4/audience_article.html.

Feloni, R. (September 5, 2014). *The 10 Worst Body Language Mistakes People Make While Giving Presentations*, Web: May 29, 2015, http://www.businessinsider.com/worst-body-language-presentation-mistakes-2014-9?op=1.

Hatalla, M. (January 2015). *Headings & Subheadings*, Web: June 2, 2015, http://www.sophia.org/tutorials/headings-subheadings.

Hicks, F. (October 13, 2013). *How to Write your Introduction, Abstract and Summary*, Web: May 30, 2015, https://thesistips.wordpress.com/2012/03/25/how-to-write-your-introduction-abstract-and-summary/.

IBM. (February 21, 2011). *IBM's 100 Icons of Progress*, Image. March 30, 2015, http://www-03.ibm.com/ibm/history/ibm100/us/en/icons/microscope/breakthroughs/.

Jughandle Fat Farm. (December 4, 2011). *Limes—Why Are There No Seeds?* Image, http://jughandlesfatfarm.com/wp-content/uploads/2011/12/limes-2.jpg.

National Nanotechnology Infrastructure Network. (2006). *Exploring Nanotechnology Through Consumer Products*, Image. June 1, 2015, http://www.nnin.org/education-training/k-12-teachers/nanotechnology-curriculum-materials/exploring-nanotechnology.

National Science Board. (2010). *Science and Engineering Indicators 2010*, Arlington, VA: National Science Foundation. Web: http://www.nsf.gov/statistics/seind10/c7/c7h.htm.

Office Angels. (2013). *Don't LOL: Could Using Text Speak at Work Be Harming Your Career?*. Web: June 2, 2015, http://www.office-angels.com/help-and-advice/news-and-opinion/news/dont-lol.aspx.

Ottobock. (2015). *Lower Limb Prosthetics Overview*. Web: http://www.ottobock.ca/prosthetics/lower-limb-prosthetics/solution-overview/.

Schawbel, D. (May 14, 2012). *Millennial Branding and Experience Inc. Study Reveals an Employment Gap between Employers and Students*, Web: May 12, 2015, http://millennialbranding.com/2012/millennial-branding-student-employment-gap-study/.

Stack, L. and Freifeld, L. (June 13, 2013). *Supercompetent Speaking: Tailoring Your Presentation to Your Audience*. Web: May 25, 2015, http://www.trainingmag.com/content/supercompetent-speaking-tailoring-your-presentation-your-audience.

Strickland, J. and Bosnor, K. (October 25, 2007). *How Nanotechnology Works*, Web: June 30, 2015, http://science.howstuffworks.com/nanotechnology.htm.

Unilearning Project. (2000). *Report Writing: Using Appendices*. Web: March 30, 2015, http://unilearning.uow.edu.au/report/1i.html.

University of Texas at Austin. (August 14, 2014). *Presentation in the Classroom (Slide Decks)*. Web: June 3, 2015, https://www.lib.utexas.edu/services/instruction/tips/tt/tt_present.html.

What Is a Team? Definition and Meaning. n.d., http://www.businessdictionary.com/definition/team.html.

White, M. C. (November 10, 2013). *The Real Reason New College Grads Can't Get Hired*. Web: March 15, 2015, http://business.time.com/2013/11/10/the-real-reason-new-college-grads-cant-get-hired/.

Witt, C. (October 19, 2012). *Practice a Speech or Presentation Out Loud*. Web: May 27, 2015, http://christopherwitt.com/practice-a-speech-or-presentation-out-loud/.

Appendix A: Unit Conversion

	Unit Conversion	
Multiply	By	To Obtain
Atmospheres	1.0133	bars
Atmospheres	29.921	in of Hg
Atmospheres	14.696	lbf/in^2
Btu per hour	0.21601	ft lbf/s
Btu per hour	3.9275×10^{-4}	hp
Btu per hour	0.29288	W
Centimeters	3.2808×10^{-2}	ft
Centimeters	0.39370	in
Cubic centimeters	6.1024×10^{-2}	in^3
Cubic centimeters	2.6417×10^{-4}	gal (US liquid)
Cubic centimeters	0.001	L
Cubic centimeters	3.3814×10^{-2}	oz (US liquid)
Cubic yards	0.76455	m^3
Dynes	1×10^{-5}	N
Feet	30.48	cm
Feet per second	1.0973	km/h
Feet per second squared	0.3048	m/s^2
Foot pounds-force	1.2859×10^{-3}	Btu
Foot pounds-force	5.0505×10^{-7}	hp h
Foot pounds-force	1.3558	J
Foot pounds-force	3.7662×10^{-7}	kWh
Foot pounds-force	1.3558	N m
Gallons (US liquid)	0.13368	ft^3
Gallons (US liquid)	231	in^3
Gallons (US liquid)	3.7854	L
Gallons (US liquid)	128	oz (US liquid)
Horsepower	2.5461×10^3	Btu/h
Horsepower	550	ft lbf/s
Horsepower	0.7457	kW
Joules	9.4782×10^{-4}	Btu
Joules	0.73756	ft lbf
Joules	2.7778×10^{-7}	kWh
Joules	3.7251×10^{-7}	hp h
Kilometer	3.2808×10^3	ft
Kilometer	0.62137	mi
Kilometer per hour	54.681	ft/min
Kilometer per hour	0.91134	ft/s
Kilowatt hours	3.4144×10^3	Btu
Kilowatt hours	2.6552×10^6	ft lbf

(Continued)

Unit Conversion

Multiply	By	To Obtain
Liters	0.26417	gal (US liquid)
Liters	61.024	in^3
Liters per second	2.1189	ft^3/min
Miles	5.28×10^3	ft
Miles	1.6093	km
Miles per hour	1.4667	ft/s
Miles per hour	1.6093	km/h
Newtons	0.22481	lbf
Pascals	9.8692×10^{-6}	atm
Pascals	1.4504×10^{-4}	lbf/in^3
Pounds-force	4.4482	N
Pounds-mass	0.45359	kg
Pounds-mass	3.1081×10^{-2}	slugs
Pounds-force per square foot	4.7254×10^{-3}	atm
Pounds-force per square foot	47.88	Pa
Pounds-force per square inch	6.8948×10^{-2}	bars
Pounds-force per square inch	2.036	in of Hg
Pounds-mass per cubic foot	1.6018×10^{-2}	g/cm^3
Radians	57.296	°
Radians	0.15915	r (revolutions)
Slugs	14.594	kg
Slugs	32.174	lbm
Watts	3.4144	Btu/h
Watts	44.254	ft lbf/min
Watts	1.3410×10^{-3}	hp
Watts	1	J/s

Appendix B: Rubrics and Key Performance Indicators

Learning Outcome a: An Ability to Apply Knowledge of Mathematics, Science, and Engineering

Performance Indicator	0-Missing 1-Emerging	2-Developing 3-Practicing	4-Maturing 5-Mastering
1. Application of mathematical and scientific principles to engineering problems.	Does not understand the connection between the mathematical models and scientific principles to solve problems related to engineering.	Chooses a mathematical model and scientific principle but has difficulties in model development.	Combines mathematical and/or scientific principles to solve problems relevant to engineering.
2. Engineering interpretation of mathematics and scientific theories and terms.	Theories and terms are not interpreted or interpreted incorrectly.	Shows nearly complete understanding of calculus and/or linear algebra but difficulties in interpretation of theory to engineering problems.	Shows appropriate engineering interpretation of mathematical and scientific terms.
3. Ability to perform mathematical calculation.	Calculations not performed or performed incorrectly and did not obtain result.	Minor errors in calculation and obtain an incorrect solution.	Executes calculations correctly with the appropriate unit and evaluate the solution.

Learning Outcome b: An Ability to Design and Conduct Experiments, Analyze, and Interpret Data

Performance Indicator	0-Missing 1-Emerging	2-Developing 3-Practicing	4-Maturing 5-Mastering
1. Conduct an experiment in a logical way based on related theory by following safety rules.	Unsafe practice with no plan of attaining experimental goals, even random and incomplete experiment.	Develop a simplistic experiment plan without recognizing the entire scope theory, thus leading to some more guidance.	Formulate experimental plan to attain stated objectives in observing safety rules and collecting and documenting data carefully by following the logical experimental procedure.
2. Ability to analyze and interpret collected data using theory.	Not following experimental procedure with poorly documented data with making frequent disorganization and risky behavior.	Experimental procedure most often followed, but additional oversight is needed due to making errors in simple math calculations.	Translate theory into practice by analyzing collected data with respect to measurement error.
3. Ability to operate instrumentation and testing equipment.	No ability to select proper equipment and operate it.	Need some guidance for how to select proper equipment and operate it.	Select proper test and measurement tools with the ability to operate those pieces of equipment in a safe way.

Learning Outcome c: An Ability to Design a System, Component, or Process

Performance Indicator	0-Missing 1-Emerging	2-Developing 3-Practicing	4-Maturing 5-Mastering
1. Set up a proper design strategy and develop solutions.	No design strategy and cannot design process.	Use guidance for design strategy to develop best solutions with some neglecting of several key aspects.	Develop design plan, timetable, suggestive approaches, and several potential solutions.
2. Using computer tools and engineering resources to find solutions.	Unable to use computer tools and engineering resources.	Minimal or incorrect use of computer tools and engineering resources.	Know how to use computational tools effectively for all the design procedure.
3. Support design procedure with documentation and references.	Design is done incompletely without proper equations and references.	Design is done, but procedures and equations are not documented.	Properly record all the related documentation and references.

Learning Outcome d: An Ability to Function on Multidisciplinary Teams

Performance Indicator	0-Missing 1-Emerging	2-Developing 3-Practicing	4-Maturing 5-Mastering
1. Recognizes participants' multidisciplinary backgrounds and roles in a team and fulfills appropriate roles to assure team success.	Does not understand the role of himself/herself in a team and fail to execute the tasks pertaining to the role to assure team success.	Understands teammates' background in a team and is able to execute the tasks pertaining to the role to assure team success.	Best utilize the strengths of multidisciplinary roles and efficiently execute the tasks pertaining to the role to assure team success.
2. Integrates input from all team members and makes decisions in relation to objective criteria.	Not be able to summarize the inputs from all team members and fail to make decisions to objective criteria based on the consensus.	Be able to summarize the inputs from all team members but unable to make decisions to objective criteria based on the consensus.	Be able to integrate input from all team members and make decisions to objective criteria based on the collective consensus.
3. Improves communications among teammates and asks for feedback and uses.	Not be able to develop feedback mechanism for improving communications and fail to improve communications during the class.	Be able to provide feedback on teammates' communication but unable to demonstrate improvements during the class.	Be able to develop feedback-based continual improvement on communications among teammates during the class.

Appendix B: Rubrics and Key Performance Indicators

Learning Outcome e: An Ability to Identify, Formulate, and Solve Engineering Problems

Performance Indicator	0-Missing	1-Emerging	2-Developing	3-Practicing	4-Maturing	5-Mastering
1. Application of engineering principles and scientific concepts to solve engineering problems.		Does not use any resources to solve engineering problems.		Uses limited resources to solve engineering problems.		Uses appropriate resources that can relate theoretical concepts to practical engineering problem solving.
2. Ability to formulate strategy for solving engineering problems.		Has no coherent strategies for problem solving.		Has some strategies for problem solving, but does not apply them consistently.		Formulates excellent strategies for solving problems.
3. Ability to identify engineering problems.		Made no connection between theory and practical problem solving.		Made connection between concepts and practical problems but with incorrect solutions.		Made excellent correlation between concepts and practical problems and developed appropriate solution techniques.

Engineering Professional Skills Assessment (EPSA) Rubric Mechanical Engineering

University of Michigan–Flint, USA

Rater's Name:_____ Date:_____ Student Work:_____

NOTE: The engineering professional skills that comprise this rubric are taken directly from the ABET Engineering Criterion 3, Student Outcomes. Each dimension of the EPSA Rubric comprises one ABET student outcome, an EPSA definition of the outcome, and the outcome's performance indicators. Thus, "ABET skill 3f" can also be read as "ABET criterion 3 student outcome 3f" with three performance indicators: stakeholder perspective, problem identification, and ethical considerations.

SCORING PROTOCOL

1. Skim the scenario students used for the discussion.
2. Quickly read the discussion, marking passages where a given skill is exhibited. A given passage may exhibit more than one skill simultaneously.
3. During a second read, highlight passages that provide strong evidence (either positive or negative) related to the skills.
4. Read the skill definition. Assign scores for each of the performance indicators.
5. In the comment boxes, provide line numbers and a short phrase, such as 3f = lines 109–112: tradeoff of wall height/plant safety versus costs; lines 828–836: risk analysis. Be sure to refer back to the skill definition.
6. Update your initial scores should the data provide evidence for a score change.
7. Ultimately assign one score for the skill. Use whole numbers; no increments.

GENERAL DISCUSSION RULES

1. Assess what is transcribed. Do not "read between the lines" (e.g., do not make assumptions about what the group should know given what is transcribed).
2. When conflicted on assigning a score, refer to adjacent score description boxes to determine whether a higher or lower score within the description box is more appropriate.
3. Assign the higher score associated with a box only when evidence for all performance criteria is present.
4. Weigh all performance indicators within a category equally in assigning the overall score.
5. Read the skill definition after scoring to check the score for accuracy.
6. When averaging scores for the performance indicators, round down. For example, 2.6 would be 2 not 3.

SCORING TIPS

1. Supply line numbers and/or student numbers for reference in the comment box.
2. Strive to complete transcript review and scoring within 45–60 min.

Source: EPSA Rubric May 2012 © 2012—ABET; Washington State University; University of Idaho; Norwich University; NSF DUE#: 1008755

ABET Skill 3f. Understanding of Professional and Ethical Responsibility

Rater Score for Skill _____

DEFINITION: Students clearly frame the problem(s) raised in the scenario and begin the process of resolution. Students recognize relevant stakeholders and their perspectives. Students identify related ethical considerations (e.g., health and safety, fair use of funds, risk, schedule, and doing "what is right" for all involved).

	0-Missing	1-Emerging	2-Developing	3-Practicing	4-Maturing	5-Mastering
Stakeholder Perspective	Students do not identify stakeholders.	Students identify few stakeholders vaguely stating their positions or misrepresenting their positions.		Students consider perspectives of major stakeholders and convey these with reasonable accuracy.		Students thoughtfully consider perspectives of all relevant stakeholders and articulate these with great clarity, accuracy, and empathy.
Problem Identification	Students do not identify the problem(s) in the scenario.	Students begin to frame the problem, but have difficulty separating primary and secondary problems. If approaches to address the problem are advocated, they are quite general and may be naïve.		Students are generally successful in distinguishing primary and secondary problems. There is evidence that they have begun to formulate credible approaches to address the problems.		Students convincingly frame the problem and parse it into subproblems. They suggest detailed and viable approaches to resolve the problems.
Ethical Considerations	Students do not give any attention to ethical consideration.	Students give passing attention to related ethical considerations.		Students are sensitive to some relevant ethical considerations and discuss them in the context of the problem(s).		Students clearly articulate relevant ethical considerations and address these in discussing approaches to resolve the problem(s).

Comments

SCORING RULES
1. Review the general decision rules and scoring tips on the first page.
2. *Stakeholder perspective*: To score at the 3–4 level, explanations of stakeholder positions are required.
3. *Problem identification*: To score at the 3–4 level, linkages between issues and problems must be made.
4. *Problem identification*: To score at the 3–4 level, primary problems must be identified and justified.
5. *Ethical considerations*: To score at the 3–4 level, linkages between ethical considerations and stakeholder interests must be made.

Source: EPSA Rubric May 2012 © 2012—ABET; Washington State University; University of Idaho; Norwich University; NSF DUE#: 1008755

Appendix B: Rubrics and Key Performance Indicators 399

ABET Skill 3g. Ability to Communicate Effectively

Rater Score for Skill _____

DEFINITION: Students work together to address the problems raised in the scenario by acknowledging and building on each other's ideas to come to consensus. Students invite and encourage participation of all discussion participants. Note: The ABET communication outcome can include several forms of communication, such as written and oral presentation. This definition focuses on group discussion skills.

	0-Missing	1-Emerging	2-Developing	3-Practicing	4-Maturing	5-Mastering
Group Interaction	Students do not interact as a group.	Students pose individual opinion, without considering other students' ideas.		Students try to balance everyone's input and build on/clarify each other's ideas.		Students clearly encourage participation from all group members, generate ideas together, and actively help each other clarify ideas.
Problem Identification	There is no evidence of group self-regulation.	Some students may monopolize or become argumentative. There may be some tentative, but ineffective, attempts at reaching consensus.		Students attempt to reach consensus, but have some difficulty in developing ways that equitably consider multiple perspectives.		Students clearly work together to reach a consensus to clearly frame the problem and develop appropriate, concrete ways to resolve the problem.

Comments

SCORING RULES

1. Review the general decision rules and scoring tips on the first page.
2. Consider frequency of utterances among teammates.
3. Consider level of individual engagement (as measured by the length and depth of utterances).
4. Trace discussion threads backward and forward to understand conversation flow.
5. *Group interaction*: To score at a 2 level, fewer than 75% of students participate in the discussion. To score at a 3 level, at least 75% of students in the group should be giving input, attempting to build on and/or clarify other's ideas.
6. *Group self-regulation*: To score at the level, one or two students attempt to regulate group discussion without success or lasting impact. To score at a 3 level, at least 75% of the students should participate in attempting consensus.
7. Give greater weight for building on ideas of others.
8. Give greater weight for successful attempts to achieve consensus/closure related to the performance task.

Source: EPSA Rubric May 2012 © 2012—ABET; Washington State University; University of Idaho; Norwich University; NSF DUE#: 1008755

ABET Skill 3h. Broad Understanding of the Impact of Engineering Solution in Global Economic, Environmental, and Cultural/Social Contexts

Rater Score for Skill _____

DEFINITION: Students consider how their ways to address the problem impact relevant global, economic, environmental, and cultural/social contexts.

Global: Students relate the issue or proposed approaches to larger global issues (such as globalization and world politics).

Economic: Students relate the issue or proposed approaches to trade and business concerns (such as project costs).

Environmental: Students relate the issue or proposed approaches to local, national, or global environmental issues (such as ozone depletion).

Cultural/societal: Students relate the issue or proposed approaches to the needs of local, national, or ethnic groups affected by the issue.

	0-Missing	1-Emerging	2-Developing	3-Practicing	4-Maturing	5-Mastering
Impact/Context	Students do not consider the impacts of the solutions.	Students give cursory consideration to how the ways to address the problem impact in relevant contexts.		Students give evidence on how the ways to address the problem impact in relevant contexts.		Students clearly examine and weigh the impact of the ways to address the problem in all relevant contexts.

Comments

SCORING RULES

1. Review the general decision rules and scoring tips on the first page.
2. To score at the 2 level, student's considerations are superficial and potentially related to only one or two relevant areas.
3. To score at the 3 level, student considers impacts in meaningful ways in all major relevant contexts.
4. Consider assigning a subscore in each context, similarly as is done for individual performance indicators, but recognizing that some contexts are not necessarily as relevant as others to the scenario discussed.
5. Impacts should be related to solution approaches associated with relevant subproblems.

Source: EPSA Rubric May 2012 © 2012—ABET; Washington State University; University of Idaho; Norwich University; NSF DUE#: 1008755

ABET Skill 3i. Recognition of the Need for and Ability to Engage in Lifelong Learning

Rater Score for Skill _____

DEFINITION: Students consider what needs to be learned (what they know and do not know). Students verbalize a credible plan to retrieve and organize needed data. Students take action to respond to personal beliefs that might hinder attainment of a satisfactory solution.

	0-Missing	1-Emerging	2-Developing	3-Practicing	4-Maturing	5-Mastering
Sources/References	Students do not question sources or references.	Students begin to question sources/references cited in the scenario.		Students question sources/references cited in the scenario.		Students evaluate sources/references cited in the scenario.
Discern Fact/Opinion	Students do not distinguish between facts and opinions expressed in the scenario.	Students begin to distinguish between facts and opinions expressed in the scenario.		Students demonstrate some ability to distinguish between facts and opinions expressed in the scenario.		Students are successful in distinguishing facts from opinions expressed in the scenario.
Knowledge Status	Students do not distinguish between what they do and do not know.	Students begin to identify what they know as well as what they do not know, but have difficulty differentiating between the two.		Students identify what they know, as well as what they do not know.		Students identify why they still need to know and describe methods for obtaining that information.
Presumptions	Students do not recognize their own presumptions that may hinder their problem solving.	Students begin to recognize their own presumptions, but have difficulty recognizing how these presumptions may hinder their problem solving.		Students recognize their own presumptions that may hinder their problem solving.		Students take action to address their own presumptions that may hinder their problem solving.

Comments

SCORING RULES

1. Review the general decision rules and scoring tips on the first page.
2. Sources/references and discern fact/opinion relate to the scenario itself.
3. *Discern fact/opinion*: To score at a 2 level, students merely reference back to facts or opinions in the scenario, using indicators like: "it says."
4. *Knowledge status*: Simply asking questions is not necessarily a questioning of knowledge; it could indicate building on or clarifying other's ideas. Note type as well as number of questions raised as evidence of knowledge status.
5. Asking for validation/confirmation is a form of checking presumptions, using indicators like: "What do you think about xyz?", "Have you heard xyz?"
6. To score at the 5 level in any performance indicator category, the group should be significantly engaged and potentially transformed as a result of the discussion.

ABET Skill 3j. Knowledge of Contemporary Issues

Rater Score for Skill _____

DEFINITION: Students consider nontechnical issues such as societal, economic, and political concerns in their discussion, identification of the problem(s), and possible ways to address the problem(s). Students also display awareness of relevant technical issues/methods/tools surrounding the problem(s).

	0-Missing	1-Emerging	2-Developing	3-Practicing	4-Maturing	5-Mastering
Nontechnical Issues	Students do not consider any current societal, economic, and/or political issues.	Students give only a superficial consideration to current societal, economic, and/or political issues. Nontechnical issues may be treated in a condescending manner.		Students give some consideration to current societal, and or political issues.		Students give full consideration to current societal, economic, and/or political issues.
Technical Issues	Students do not consider modern methods, technologies, and/or tools.	Students give only passing consideration to modern methods, technologies, and/or tools.		Students give some consideration to modern methods, technologies, and/or tools.		Students give full consideration to modern methods, technologies, and/or tools.
Comments						

SCORING RULES
1. Review the general decision rules and scoring tips on the first page.
2. Keep track of the number and depth of different nontechnical issues raised/discussed.
3. Keep track of the number and depth of different technical issues raised/discussed.
4. To score at a 4 level in either performance indicator category, students consider relevant current topics in ways that inform their identification of the problem(s) and possible ways to address the problem(s).
5. Give equal weight for the investigation of secondary problems.

Source: EPSA Rubric May 2012 © 2012—ABET; Washington State University; University of Idaho; Norwich University; NSF DUE#: 1008755

Appendix B: Rubrics and Key Performance Indicators 405

Learning Outcome k: An Ability to Use the Techniques, Skills, and Modern Engineering Tools Necessary for Engineering Practice

Performance Indicator	0-Missing	1-Emerging	2-Developing 3-Practicing	4-Maturing 5-Mastering
1. Application of computer software to provide solutions to engineering problems.	Does not use a computer-based method and system software to provide solutions to engineering problems.		Uses a computer-based method and system software effectively in assignments/projects to provide solutions to engineering problems with some errors.	Uses a computer-based method and system software effectively in assignments/projects to provide an appropriate solution.
2. Application of engineering skills and techniques to data analysis and interpretation.	No systematic plan of data gathering; experimental data collection is disorganized, even random, and incomplete.		Shows a systematic plan for data gathering. Experimental data collection is organized and complete but with poor data interpretation.	Shows a systematic plan for data gathering; experimental data collection is organized and complete with good data interpretation.
3. An ability to use modern engineering tools to perform laboratory experiments.	Does not follow an experimental procedure and employ appropriate tools.		Experimental procedures most often followed with the use of modern tools, but occasional oversight leads to loss of experimental efficiency and/or loss of data.	Executes adequate experimental procedures with appropriate modern tools to generate informative results and solutions.

KEY PERFORMANCE INDICATORS TO ASSESS STUDENT LEARNING OUTCOMES

Student Outcome	Performance Indicators
An ability to apply knowledge of mathematics, science, and engineering.	1. Application of mathematical and scientific principles to engineering problems. 2. Engineering interpretation of mathematics and scientific theories and terms. 3. An ability to perform mathematical calculation.
An ability to design and conduct experiments, as well as to analyze and interpret data.	1. Conduct experiments in a logical way based on related theory by following safety rules. 2. An ability to analyze and interpret collected data using theory. 3. An ability to operate instrumentation and testing equipment.
An ability to design a system component or process to meet desired needs within realistic constraints such as economic, environmental, social, political, ethical, health and safety, manufacturability, and sustainability.	1. Set up a proper design strategy and develop solutions. 2. Use computer tools and engineering resources to find solutions. 3. Support the design procedure with documentation and references.
An ability to function on multidisciplinary teams.	1. Recognizes participants' roles in a team setting and fulfills appropriate roles to assure team success. 2. Integrates input from all team members and makes decisions in relation to objective criteria. 3. Improves communication among teammates, asks for feedback, and uses suggestions.
An ability to identify, formulate, and solve engineering problems.	1. Application of engineering principles and scientific concepts to solve engineering problems. 2. An ability to formulate a strategy for solving engineering problems. 3. An ability to identify engineering problems.
An understanding of professional and ethical responsibility.	1. Stakeholder perspective. 2. Problem identification. 3. Ethical consideration.
An ability to communicate effectively.	1. Group interaction among team members. 2. Problem identification in presentation and written reports.
The broad education is necessary to understand the impact of engineering solutions in a global, economic, environmental, and societal context.	1. Understanding the impact of engineering solutions in a global and societal context. 2. Understanding the impact of engineering solutions in an economic and environmental context.
Recognition of the need for and an ability to engage in lifelong learning.	1. Source/references. 2. Discern fact/opinion. 3. Knowledge status. 4. Presumptions.
Knowledge of contemporary issues.	1. Nontechnical issues. 2. Technical issues.
An ability to use techniques, skills, and modern engineering tools necessary for engineering practice.	1. Application of computer software to providing solutions to engineering problems. 2. Application of engineering skills and techniques to data analysis and interpretation. 3. An ability to use modern engineering tools to perform laboratory experiments.

Appendix C: ABET Outcomes for Each Chapter

Chapter	ABET Student Learning Outcomes
Chapter 1: Engineering and Technology Professions	h, f
Chapter 2: Engineering Education	a
Chapter 3: How to Learn	i
Chapter 4: Computer-Aided Design	g, j
Chapter 5: Statics	a, e
Chapter 6: Materials Engineering	e, h
Chapter 7: Design and Analysis	c, e
Chapter 8: Electric Circuits and Components	e, k
Chapter 9: Engineering Economics	c, e, k
Chapter 10: Probability and Statistics	a, e
Chapter 11: Computer Programming	k
Chapter 12: Product Design and Development	c, k
Chapter 13: Manufacturing Processes	k
Chapter 14: Engineering and Society	h, j
Chapter 15: Engineering Ethics	f
Chapter 16: Communication and Teamwork	d

Appendix D: Equation and Graph

D.1 How to Create Equations in Microsoft Word

1. Open a Microsoft Word document.
2. Now look at the menu bar on the top of the document. This menu bar has certain fields such as File, Home, Insert, Design, View, and others.

INSERT   DESIGN

3. Select Insert and click it to view the insert options. Go to the right corner in the Insert field menu where you can see Symbols.
4. A New menu will appear that will display Equation and Symbol at the top-right corner of the screen.

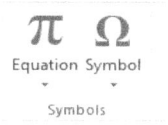

π   Ω
Equation Symbol
Symbols

5. Select Equation, and then the following menu bar will appear.

Type equation here.

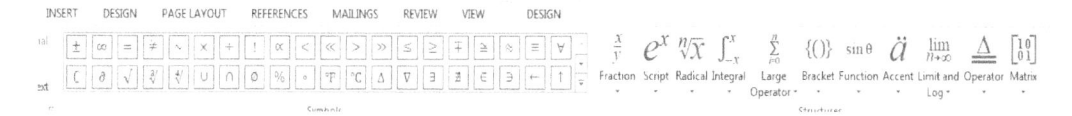

6. Create the Equations you want.

Example D.1

To calculate the moment of inertia of a beam with width b and height h, the following equation must be used: $I = 1/12\ bh^3$.

1. Select Insert and choose Equation.
2. Click, start to type.

Type equation here.

3. In "Design" of "Equation Tool," select Fraction.

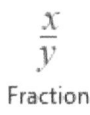

$$\frac{x}{y}$$
Fraction

4. Choose stacked fraction, then type 1 and 12, so that you have $\frac{1}{12}$
5. Type b, select Script, and then choose superscript.

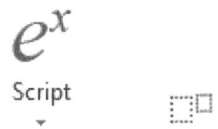

$$e^x$$
Script

6. Type h and 3. Let superscript become h^3.
7. The equation becomes $I = \frac{1}{12}bh^3$.

Example D.2

To calculate the moment required to raise an axially loaded nut (M) with W as axial load on nut, ϕ as helix angle thread, μ_s as the static coefficient of friction, and d_m as the thread diameter, the equation is

$$M = \left(\frac{\tan\phi + \mu_s}{1 - \mu_s \tan\phi}\right) W \frac{d_m}{2}$$

1. Select Insert from the top menu and choose Equation in the right corner.
2. Click inside the box and start to type the equation.

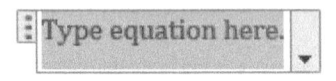

Type equation here.

3. From the "Design" of "Equation Tool," select Fraction.

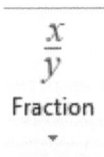

$$\frac{x}{y}$$
Fraction

4. Choose stacked fraction, then select and choose trigonometric functions.

Appendix D: Equation and Graph 411

sin θ
Function

Trigonometric Functions

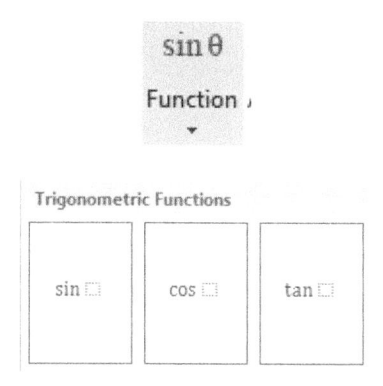

sin   cos   tan

5. Select the symbol, select script, then choose subscript and enter by selecting from the symbols. Type "s" in the subscript.

∅

e^x
Script

μ

6. Now in the denominator type "1-" and s and tanϕ and put the total fraction in parenthesis "()". Now type "W," select Fraction, choose and enter d_m /2.

μ

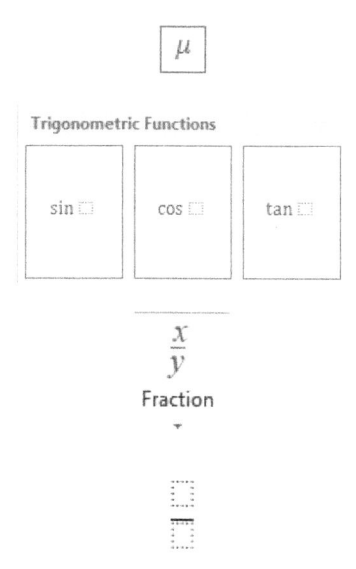

Trigonometric Functions

sin   cos   tan

$\dfrac{x}{y}$
Fraction

7. The final equation becomes

$$M = \left(\frac{\tan\phi + \mu_s}{1 - \mu_s \tan\phi}\right) W \frac{d_m}{2}$$

Example D.3

To calculate the maximum shear stress of a round wire spring with F as force, D as the helix diameter, and d as the wire diameter, the equation is $\tau = \frac{8FD}{\pi d^3}$

1. Select Insert and choose Equation.
2. Click, and start to type.

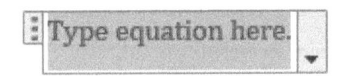

   [Type equation here.]

3. In "Design" of "Equation Tool," select the symbol.

   τ

4. Select Fraction and type 8, F, and D in the numerator.

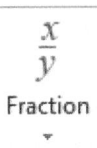

   $\frac{x}{y}$
   Fraction

5. Now select symbol from "Equation Tool." Select and choose superscript.

   π

   e^x
   Script

6. Type d and 3 in the denominator.
7. The equation is $\tau = \frac{8FD}{\pi d^3}$.

Appendix D: Equation and Graph

D.2 How to Create Charts and Graphs Using Microsoft Excel

1. Enter all data in worksheet cells.

2. Select the block of cells containing the data to be plotted, and then click "Insert" tab.

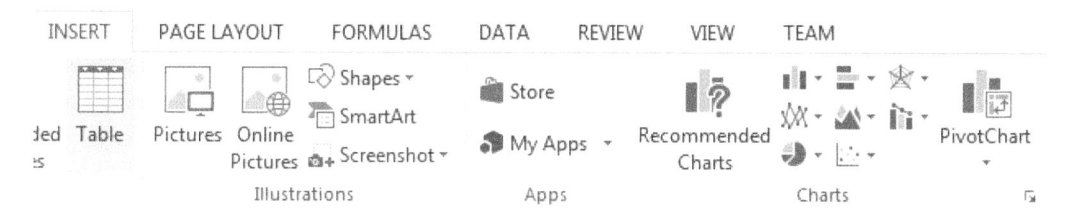

3. A New menu will appear that will display all charts at the center of the screen.
4. Choose the charts and graphs you want.
5. Or to create a blank chart first, right click the chart, and choose "Select Data."

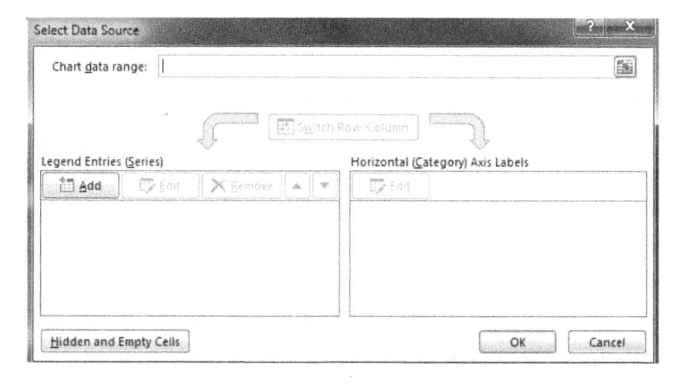

6. In the "Chart data range" field, enter the block of cells where you want to insert the data to be plotted.

7. Click ADD tab. A new window will appear that lets you choose x and y values.

D.3 How to Select Charts and Graphs of Best Fit

1. Once the chart appears on the worksheet, right click on a data point (make sure that all data points are highlighted) and choose "Add trendline."
2. A New menu named Format Trendline appears on the right side of the Excel sheet.

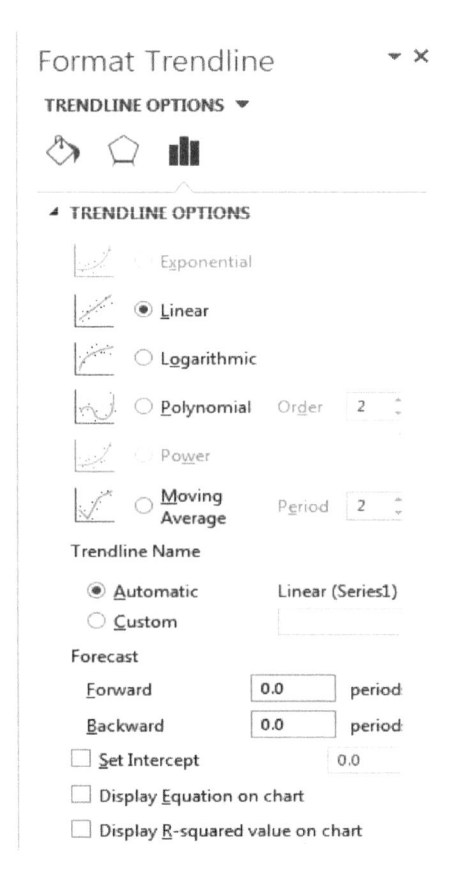

Appendix D: Equation and Graph   415

3. The trendline menu mentioned above has certain formatting options. Choose "Display Equation on chart" to get equation of plotted curve on the chart. And check best fit by selecting "Display R-squared value on chart" to view the R-squared value.
4. If the R-squared value is close to 1.0, then it is a good Fit. If the R-squared value is too low, change the type of trendline by right clicking TRENDLINE OPTIONS. Several options will be available (Exponential, Logarithmic, Polynomial, Power, and Moving Average).
5. Check again to see if the R-squared value is close to 1.0.

Example D.4

1. Get the equation and R-squared value of the spring test. The following table is given.

Mass (g)	Load (N)	Scale Reading (mm)	Extension (mm)
0	0	100	0
40	0.39	100	0
80	0.78	100	0
120	1.18	108	8
160	1.57	120	20
200	1.96	131	31
240	2.35	140	40
280	2.75	153	53
320	3.14	162	62
360	3.53	173	73
400	3.92	185	85
440	4.32	195	95
480	4.71	209	109

- Choose the columns you need (please notice the x and y values), and then plot the chart.

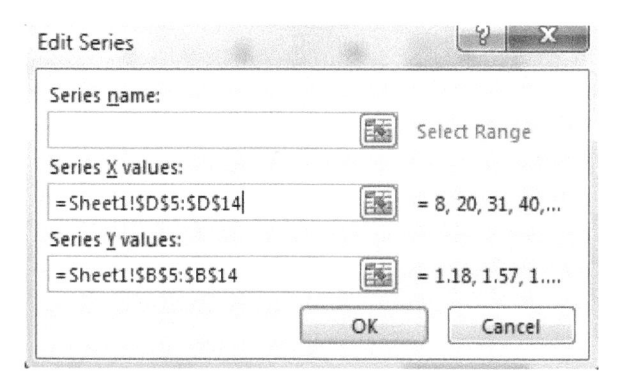

- Add trendline, equation, and R-squared value.

$$y = 0.0357x + 0.8869$$
$$R^2 = 0.9989$$

(Load vs. Extension chart)

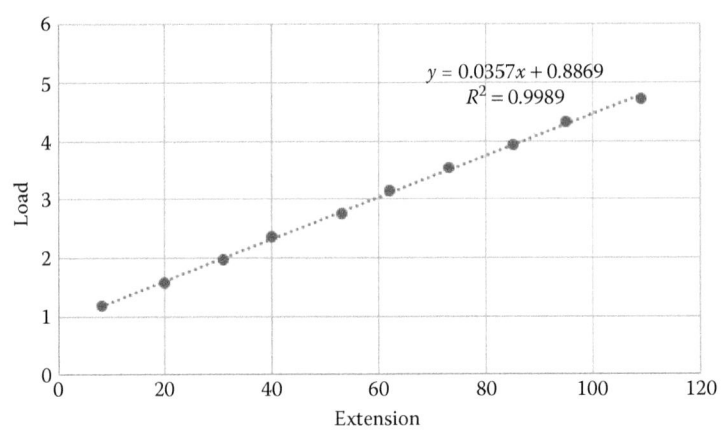

Example D.5

1. Get the equation and R-squared value of a table representing the time for a high-performance car with polynomial trendline. The following table is given.

Top speed (mph)	Time (sec)
30	1.9
40	2.8
50	3.8
60	5.2
70	6.5
80	8.3
90	10.4
100	12.7
110	15.6
120	19
130	23.2
140	31.2
150	45.1

2. Choose the columns you need (please notice the x and y values), and then plot the chart.

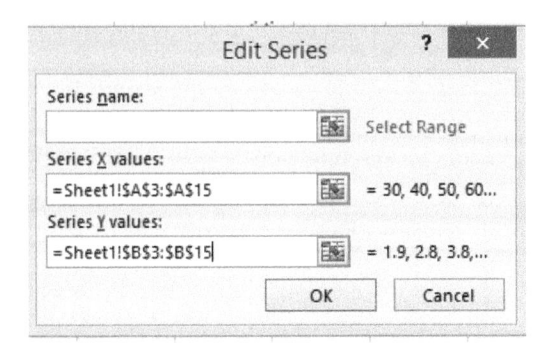

Edit Series

Series name:
    Select Range

Series X values:
=Sheet1!$A$3:$A$15   = 30, 40, 50, 60...

Series Y values:
=Sheet1!$B$3:$B$15   = 1.9, 2.8, 3.8,...

OK   Cancel

Appendix D: Equation and Graph

3. Add trendline, equation, and R-squared value.

$y = 0.0035x^2 - 0.3324x + 10.916$
$R^2 = 0.9657$

Top speed (mph)

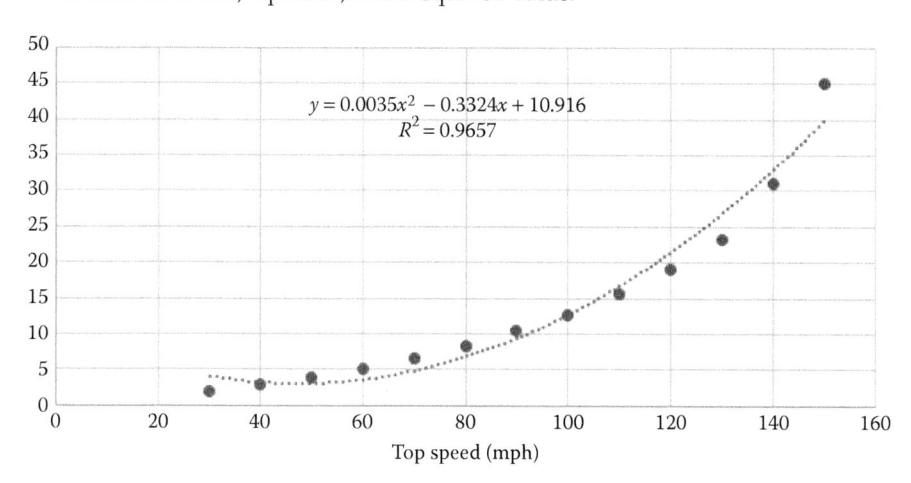

Example D.6

1. Get the equation and R-squared value from the table representing a power function.

(moles/cu ft)	(moles/sec)
100	2.85
80	2
60	1.25
40	0.67
20	0.22
10	0.072
5	0.024
1	0.0018

2. Choose the columns you need (please notice the *x* and *y* values), and then plot the chart.

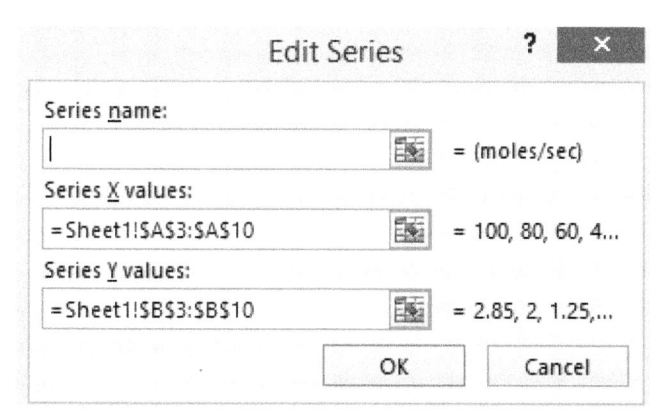

Edit Series

Series name:

= (moles/sec)

Series X values:
=Sheet1!$A$3:$A$10   = 100, 80, 60, 4...

Series Y values:
=Sheet1!$B$3:$B$10   = 2.85, 2, 1.25,...

OK   Cancel

3. Add power trendline, equation, and R-squared value.

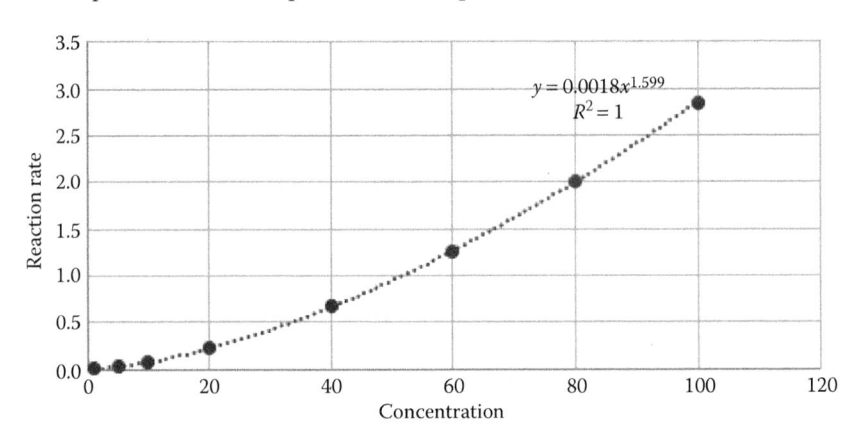

$y = 0.0018x^{1.599}$
$R^2 = 1$

D.4 How to Change Axis Title, Axis Levels

1. Click the chart, and then click to view the CHART ELEMENTS.

2. Customize required chart fields by clicking the appropriate box like Axis Title.

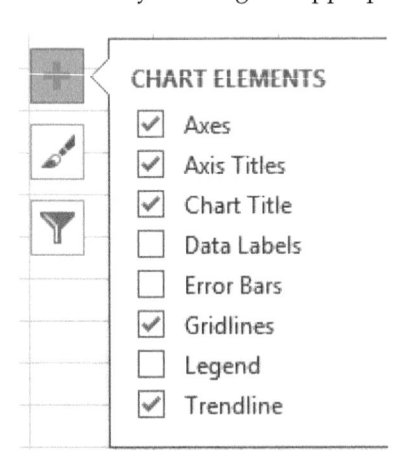

   CHART ELEMENTS
   ☑ Axes
   ☑ Axis Titles
   ☑ Chart Title
   ☐ Data Labels
   ☐ Error Bars
   ☑ Gridlines
   ☐ Legend
   ☑ Trendline

3. Then change the titles you want.

Appendix D: Equation and Graph 419

D.5 How to Change the Data Series Symbols (Circle, Square, Diamond, etc.)

1. Right click on a data point (all should be highlighted) and choose Format Data Series.
2. Choose "Fill" and "Line" and click "MARKER."

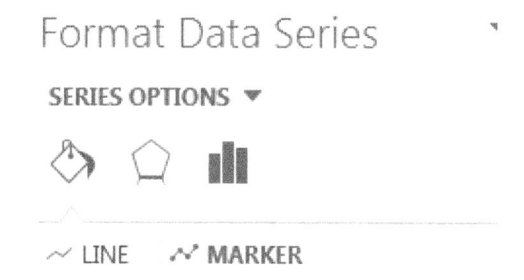

3. Click "MARKER OPTIONS," choose Built-in, and then change the type and size of series symbols.
4. Or change color in Fill and Border options.

D.6 How to Calculate a Value Using a Formula and Then Plot It

1. Enter known data in columns, and calculate a value in a blank column.
2. Enter "=" in blank column.
3. Enter the equation using data in columns in blank column or

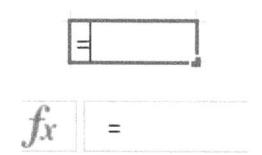

4. Once the equation is complete, press "Enter."
5. When you get one result, choose this result, and point to the right corner of the cell.
6. When the arrow becomes "+," drop down the cells, and answers will come out automatically.

For example, to calculate stress (stress = force/area).

- Enter force in the first column. Area keeps constant at 5.

	A	B	C
1	Force	Area	Stress
2	0	5	
3	4	5	
4	6	5	
5	8	5	
6	10	5	
7	12	5	
8	14	5	
9	16	5	
10	18	5	
11	20	5	

- Then enter "=" in the stress column.

Force	Area	Stress
0	5	=A2/B2
4	5	
6	5	
8	5	
10	5	
12	5	
14	5	
16	5	
18	5	
20	5	

- Press "Enter."
- Drop down "C2" cell, and then get all results.

Appendix E: Z-Tables

Z-Table: Negative Values

Body of the table gives area under the Z-curve to the left of z.

Example: $P[Z < -2.63] = .0043$

z	.00	.01	.02	.03	.04	.05	.06	.07	.08	.09
−3.80	.0001	.0001	.0001	.0001	.0001	.0001	.0001	.0001	.0001	.0001
−3.70	.0001	.0001	.0001	.0001	.0001	.0001	.0001	.0001	.0001	.0001
−3.60	.0002	.0002	.0001	.0001	.0001	.0001	.0001	.0001	.0001	.0001
−3.50	.0002	.0002	.0002	.0002	.0002	.0002	.0002	.0002	.0002	.0002
−3.40	.0003	.0003	.0003	.0003	.0003	.0003	.0003	.0003	.0003	.0002
−3.30	.0005	.0005	.0005	.0004	.0004	.0004	.0004	.0004	.0004	.0003
−3.20	.0007	.0007	.0006	.0006	.0006	.0006	.0006	.0005	.0005	.0005
−3.10	.0010	.0009	.0009	.0009	.0008	.0008	.0008	.0008	.0007	.0007
−3.00	.0013	.0013	.0013	.0012	.0012	.0011	.0011	.0011	.0010	.0010
−2.90	.0019	.0018	.0018	.0017	.0016	.0016	.0015	.0015	.0014	.0014
−2.80	.0026	.0025	.0024	.0023	.0023	.0022	.0021	.0021	.0020	.0019
−2.70	.0035	.0034	.0033	.0032	.0031	.0030	.0029	.0028	.0027	.0026
−2.60	.0047	.0045	.0044	.0043	.0041	.0040	.0039	.0038	.0037	.0036
−2.50	.0062	.0060	.0059	.0057	.0055	.0054	.0052	.0051	.0049	.0048
−2.40	.0082	.0080	.0078	.0075	.0073	.0071	.0069	.0068	.0066	.0064
−2.30	.0107	.0104	.0102	.0099	.0096	.0094	.0091	.0089	.0087	.0084
−2.20	.0139	.0136	.0132	.0129	.0125	.0122	.0119	.0116	.0113	.0110
−2.10	.0179	.0174	.0170	.0166	.0162	.0158	.0154	.0150	.0146	.0143
−2.00	.0228	.0222	.0217	.0212	.0207	.0202	.0197	.0192	.0188	.0183
−1.90	.0287	.0281	.0274	.0268	.0262	.0256	.0250	.0244	.0239	.0233
−1.80	.0359	.0351	.0344	.0336	.0329	.0322	.0314	.0307	.0301	.0294
−1.70	.0446	.0436	.0427	.0418	.0409	.0401	.0392	.0384	.0375	.0367
−1.60	.0548	.0537	.0526	.0516	.0505	.0495	.0485	.0475	.0465	.0455
−1.50	.0668	.0655	.0643	.0630	.0618	.0606	.0594	.0582	.0571	.0559
−1.40	.0808	.0793	.0778	.0764	.0749	.0735	.0721	.0708	.0694	.0681
−1.30	.0968	.0951	.0934	.0918	.0901	.0885	.0869	.0853	.0838	.0823
−1.20	.1151	.1131	.1112	.1093	.1075	.1056	.1038	.1020	.1003	.0985
−1.10	.1357	.1335	.1314	.1292	.1271	.1251	.1230	.1210	.1190	.1170
−1.00	.1587	.1562	.1539	.1515	.1492	.1469	.1446	.1423	.1401	.1379
−0.90	.1841	.1814	.1788	.1762	.1736	.1711	.1685	.1660	.1635	.1611
−0.80	.2119	.2090	.2061	.2033	.2005	.1977	.1949	.1922	.1894	.1867
−0.70	.2420	.2389	.2358	.2327	.2296	.2266	.2236	.2206	.2177	.2148
−0.60	.2743	.2709	.2676	.2643	.2611	.2578	.2546	.2514	.2483	.2451
−0.50	.3085	.3050	.3015	.2981	.2946	.2912	.2877	.2843	.2810	.2776

(*Continued*)

Z-Table: Negative Values

Body of the table gives area under the Z-curve to the left of z.

Example: $P[Z < -2.63] = .0043$

z	.00	.01	.02	.03	.04	.05	.06	.07	.08	.09
−0.40	.3446	.3409	.3372	.3336	.3300	.3264	.3228	.3192	.3156	.3121
−0.30	.3821	.3783	.3745	.3707	.3669	.3632	.3594	.3557	.3520	.3483
−0.20	.4207	.4168	.4129	.4090	.4052	.4013	.3974	.3936	.3897	.3859
−0.10	.4602	.4562	.4522	.4483	.4443	.4404	.4364	.4325	.4286	.4247
0.00	.5000	.4960	.4920	.4880	.4840	.4801	.4761	.4721	.4681	.4641

Z-Table: Positive Values

Body of the table gives area under the Z-curve to the left of z.

Example: $P[Z < 1.16] = .8770$

z	.00	.01	.02	.03	.04	.05	.06	.07	.08	.09
0.00	.5000	.5040	.5080	.5120	.5160	.5199	.5239	.5279	.5319	.5359
0.10	.5398	.5438	.5478	.5517	.5557	.5596	.5636	.5675	.5714	.5753
0.20	.5793	.5832	.5871	.5910	.5948	.5987	.6026	.6064	.6103	.6141
0.30	.6179	.6217	.6255	.6293	.6331	.6368	.6406	.6443	.6480	.6517
0.40	.6554	.6591	.6628	.6664	.6700	.6736	.6772	.6808	.6844	.6879
0.50	.6915	.6950	.6985	.7019	.7054	.7088	.7123	.7157	.7190	.7224
0.60	.7257	.7291	.7324	.7357	.7389	.7422	.7454	.7486	.7517	.7549
0.70	.7580	.7611	.7642	.7673	.7704	.7734	.7764	.7794	.7823	.7852
0.80	.7881	.7910	.7939	.7967	.7995	.8023	.8051	.8078	.8106	.8133
0.90	.8159	.8186	.8212	.8238	.8264	.8289	.8315	.8340	.8365	.8389
1.00	.8413	.8438	.8461	.8485	.8508	.8531	.8554	.8577	.8599	.8621
1.10	.8643	.8665	.8686	.8708	.8729	.8749	.8770	.8790	.8810	.8830
1.20	.8849	.8869	.8888	.8907	.8925	.8944	.8962	.8980	.8997	.9015
1.30	.9032	.9049	.9066	.9082	.9099	.9115	.9131	.9147	.9162	.9177
1.40	.9192	.9207	.9222	.9236	.9251	.9265	.9279	.9292	.9306	.9319
1.50	.9332	.9345	.9357	.9370	.9382	.9394	.9406	.9418	.9429	.9441
1.60	.9452	.9463	.9474	.9484	.9495	.9505	.9515	.9525	.9535	.9545
1.70	.9554	.9564	.9573	.9582	.9591	.9599	.9608	.9616	.9625	.9633
1.80	.9641	.9649	.9656	.9664	.9671	.9678	.9686	.9693	.9699	.9706
1.90	.9713	.9719	.9726	.9732	.9738	.9744	.9750	.9756	.9761	.9767
2.00	.9772	.9778	.9783	.9788	.9793	.9798	.9803	.9808	.9812	.9817
2.10	.9821	.9826	.9830	.9834	.9838	.9842	.9846	.9850	.9854	.9857
2.20	.9861	.9864	.9868	.9871	.9875	.9878	.9881	.9884	.9887	.9890
2.30	.9893	.9896	.9898	.9901	.9904	.9906	.9909	.9911	.9913	.9916
2.40	.9918	.9920	.9922	.9925	.9927	.9929	.9931	.9932	.9934	.9936
2.50	.9938	.9940	.9941	.9943	.9945	.9946	.9948	.9949	.9951	.9952
2.60	.9953	.9955	.9956	.9957	.9959	.9960	.9961	.9962	.9963	.9964
2.70	.9965	.9966	.9967	.9968	.9969	.9970	.9971	.9972	.9973	.9974
2.80	.9974	.9975	.9976	.9977	.9977	.9978	.9979	.9979	.9980	.9981
2.90	.9981	.9982	.9982	.9983	.9984	.9984	.9985	.9985	.9986	.9986
3.00	.9987	.9987	.9987	.9988	.9988	.9989	.9989	.9989	.9990	.9990

(Continued)

Z-Table: Positive Values

Body of the table gives area under the Z-curve to the left of z.

Example: $P[Z < 1.16] = .8770$

z	.00	.01	.02	.03	.04	.05	.06	.07	.08	.09
3.10	.9990	.9991	.9991	.9991	.9992	.9992	.9992	.9992	.9993	.9993
3.20	.9993	.9993	.9994	.9994	.9994	.9994	.9994	.9995	.9995	.9995
3.30	.9995	.9995	.9995	.9996	.9996	.9996	.9996	.9996	.9996	.9997
3.40	.9997	.9997	.9997	.9997	.9997	.9997	.9997	.9997	.9997	.9998
3.50	.9998	.9998	.9998	.9998	.9998	.9998	.9998	.9998	.9998	.9998
3.60	.9998	.9998	.9999	.9999	.9999	.9999	.9999	.9999	.9999	.9999
3.70	.9999	.9999	.9999	.9999	.9999	.9999	.9999	.9999	.9999	.9999
3.80	.9999	.9999	.9999	.9999	.9999	.9999	.9999	.9999	.9999	.9999

Index

Note: Locator followed by 'f' and 't' denotes figure and table in the text

3D printing tool, 313
555 timer, 223, 224f

A

Absolute dynamic viscosity (μ), 172
Academic activities, weekly priority of, 49t
Academic program, planning, 35–39
    balancing work and life, 38
    finding catalog year, 36
    for mechanical engineering program, 39t
    meeting with advisor, 37–38
    organizing materials, 37
    prerequisite requirements, 37
    taking courses from different areas, 38–39
    understanding program requirements, 36–37
Accreditation Board of Engineering and Technology (ABET), 26–29
    CAD with, 92–93
    commissions review, 27
    criteria for, 27–29
        continuous improvement, 29
        curriculum, 29
        program educational objectives, 28
        student outcomes, 28
        students, 27
    earning degree from, 27
    high-impact educational practices, 33
    outcomes, 407
Activities performed by engineers
    aerospace/aeronautical engineers, 10
    civil engineers, 11–12
    electrical engineers, 13
    electronic engineers, 14
    environmental engineers, 14
    hardware engineers, 12
    industrial engineers, 15
    mechanical engineers, 15
Actual parameter, program, 291
Additive law of probability, 254
Aerospace and aeronautical engineering, 10–11
Aircraft, categories of, 11
Air/nonmagnetic core inductors, 208

Algorithm, development of, 276–280
    flowcharts, 277–278, 277f
    guidelines for creating, 276–277
    pseudocode, 278–280
Alloys, 124
Alternating current (AC), 337
Aluminum alloys, 137
Aluminum capacitors, 203
Aluminum ore, 128
Amazon.com, 349
American Institute of Aeronautics and Astronautics (AIAA), 20
American Society for Engineering Education, 20
American Society of Civil Engineers, 20
American Society of Mechanical Engineers (ASME), 20
Amin, Ishtiaque, 5–6
Amorphous polymers, 146, 146f, 149–150
Analytical method, engineering economics, 233
Annealing process, 136
Annual worth (AW), 231
    analysis, 246
Anode, 213
Application software developers, 13
Archimedes' principle, 173
Arc welding, 335–336
Arithmetic gradient factors, 244
Arithmetic operators, 285–286, 285t
Assembler, program, 275
Assembly drawings, 76, 77f
Assembly language, 275
Assessment and continuous improvement, program, 31–33
Assignment operator, 285–286
Association of American Colleges and Universities, 33
Astable multivibrator, 225
Atactic polypropylene, 151
Aural (auditory-musical) learning style, 47
Automatic welding, 337
Automotive engine block, 133–134
Avalanche breakdown, 214, 218
Average flow velocity, 176
Axial loading, 166
Axial stress, 163–164

B

Baccalaureate-level engineering technology (BET) programs, 4
Ball mill, 141
Barometers, 172, 172f
Batching, ceramic materials, 141
Bauxite, 128
Bending stress in beam, 165, 165f
Bernoulli's equation, 175
Bias voltage, 215
Bill of materials (BOMs), 59, 76
    for circuit, 191, 192t
Bipolar junction transistor (BJT), 186–187
    operating regions, 188
    terminals, 187
Blast furnace iron, 127
Block of code, 279
Bloom's taxonomy, 50–51
    new, 50f
    with verbs and behaviors, 51f
Blow molding process, 150
Body-centered cubic (BCC) crystal structure, 125
    atom, 126
    of iron, 136
    unit cell, 125–126, 125f
Body-centered tetragonal (BCT) unit cell, 136–137
Body, FET, 188
Bracket tutorial, 65–68, 66f, 67f, 68f, 69f, 70f
    front view setup, 66
    two-dimensional sketch, 65f
Branched piping system, flow through, 176f
Bridge rectifier, 214
Buckling in column, 166
Bug, 292
Buoyancy force, 173–174

C

CamelCase variables, 283
Camera Toolbar, SketchUp Pro, 61, 64f
Capacitance, 202
Capacitors, 186
    block diagram, 187f
    common types of, 203
    designing low-pass RC filter, 207–208
    discharge circuit, 204f
    energy stored in, 202, 205t, 206–207
    in parallel, 204, 204f
    RC time constant, 205–206
    in series, 203, 204f
    series and parallel combination, 204–205, 204f
    types of, 184f
Capital recovery factor, 241–242
Capstone courses and projects, 35
Carriage mechanism, 321
Case sensitive, variables, 283
Cash flow, 238
    arithmetic gradient factors, 244
    capital recovery factor, 241–242
    diagram, 238–239
    geometric gradient series factors, 245–246
    single amount factors, 239–240
    sinking fund factor, 242–244
    uniform series compound amount factor, 242–244
    uniform series PW factor, 241–242
Casting process, metals, 133–134
    risers, 134–135
Catalog year, 36–37
Cathode, 213
Center of gravity, 110, 114
    and area of common geometries, 111t–112t
    and length of common geometries, 113t
Central limit theorem, 265–266
Central processing units (CPUs), 137
Central tendency, measure of, 262
Centroid, 110, 114
    axis, 165
    of T-shaped section, 114f
Ceramic capacitor, 203
Ceramic materials, 138–144
    application of, 144
    categories, 139
    liquid slurry of, 142
    mills, 141
    obtaining, 139
    process affecting properties of, 143
    processing, 141–143
    properties of, 139–141
    technologies made possible by, 144
Chain-type polymers, 145
Chamfering tool, lathe, 322
Channel, 188
Charge distribution, 183
Charged particles, 183
Chuck, lathe, 320f
Chvorinov's rule, 134
Circuit
    components, 185–189
        BJT, 186–187
        BJT operating regions, 188
        capacitors, 186

Index

diodes, 186
FET, 188
IC, 189
JFET, 188–189
MOSFET, 189
n-type semiconductor, 188
p-type semiconductor, 188
resistors, 185–186
transistors, 186
connection symbols, 187*f*
current-limiting resistor, 187*f*
current traveling through, 187*f*
diodes used in, 186
implementing inductor in, 204*f*
lamps and resistor, 194*f*
symbols and schematic design, 189–193, 190*f*
circuit design, 191–192
ground connections, 192
power connections, 192–193
reference designators, 191, 191*t*
schematics, 190–191
understanding, 193–195
parallel circuits, 195
series circuits, 193, 195
Civil engineering, 11–12
Closed circuit, 193
Coating process, 340
Cold forging process, 318
Cold rolling, metals, 135
Cold-worked metal, 136
Collaborative assignments and projects, 34
Collector–base voltage (V_{CBO}), 215
Collector current (I_C), 216
Collector–emitter voltage (V_{CEO}), 215, 217
Color codes, resistor, 197
five-band resistor, 198
four-band resistor, 197–198
Common intellectual experiences, 34
Communication skills, engineers, 370–386
communicating by listening, 387
oral communication. *See* Oral communication skills
technical communication mediums, 386–387
written communication. *See* Written communication skills
Community-based learning, 35
Compiler, program, 275
Composite materials, 152–154
manufacturing affect properties of, 154
obtaining/manufacturing, 153
overview, 152
properties of, 153
technologies made possible by, 154

Compounding period (CP), 232
Computational fluid dynamics (CFD) analysis, 313, 313*f*
Computational method, engineering economics, 233
Computational techniques and tools, product, 311–313
3D printing tool, 313
CAx systems, 311
E/P BDM system, 311
FEA, 312–313, 312*f*
KBE, 312
RP, 313
Computer-aided design (CAD) system, 57, 304, 312
with ABET program outcomes, 92–93
advantages of, 57–59
orthographic drawings in, 74–78
software, 59–70
abbreviations, 62–64
adding dimensions to model, 68–70
dynamic viewing function, 62
installing Google SketchUp Pro, 59
modeling 3D object, 65–68
sketch and extrusion, 65
starting SketchUp Pro, 59–60
toolbar, 60–62
Computer-aided product planning (CAPP), 311
Computer-aided (CAx) systems, 311
Computer engineering, 12–13
Computer hardware engineers, 12
Computerized numerical controlled (CNC) machine, 330–331
codes, 333
Computer programming, 273
constant, 282
data types, 280–281
importance of, 273–274
input and output, 284–285
languages and applications, 274–276
assembly language, 275
compiling and executing program, 275–276
high-level languages, 274
machine code, 275
variables, 282
Computer software developers, 12–13
Concentric loading, 166
Concurrent engineering (CE) system, 306–308, 306*f*
Conditional expressions, 286–288
examples of, 287–288
logical operators, 287–288, 287*t*
relational operators, 286, 286*t*

Conditional knowledge, 45
Conduction, heat transfer, 169
Conservation of energy, 175
Conservation of mass, 174
Constant, program, 282
 naming conventions, 283–284
Constant speed feeder, 337
Construction engineers, 12
Continuity equation, 174
Convection, heat transfer, 170
Conventional gradient, cash flow, 244
Copper, 127
Corning® Gorilla® Glass, 144, 144f
Coulomb's law, 182
Couple, force, 106–107
Critical load on column, 166, 167f
Crystalline ceramic materials, 139
Crystalline thermoplastic polymers, 149, 151
Crystallinity
 ceramic materials, 139
 in polymer, 146, 151
Crystal structures, 125
Cubic unit cell, 125, 125f
Cultural awareness/cultural convergence, 348
Current, 185, 201
 traveling through circuit, 187f
Curriculum mapping, 29–31
 approaches, 30
 goal of, 30
 for mechanical engineering program, 31
 practice in, 30
Custom-designed components, circuit, 185
Cutoff frequency (f_c), 207–208
Cutoff region, transistor, 218

D

Data series symbols, changing, 419
Data types, program, 280–281
DC current gain (h_{FE}), 217
Debugging, program, 292
Decision making, engineering economics, 247–249
Declarative knowledge, 45
DEDICT (Demonstrate, Explain, Demonstrate slowly and repeatedly, Imitate, Coach, and Test) method, 53
Deductive statistics, 262
Degree of crystallinity, polymer, 146
Delegator teaching style, 48
Demonstrator/personal model teaching style, 48
Density, fluid, 170

Depletion zone, 213
Deposition process, 220
Descriptive statistics, 260–261
Design and analysis, engineering, 157–158
Design calibration, 305
Design for assembly (DFA) tool, 305
Design for manufacture (DFM) method, 305
Design for X (DfX), 309–310
Design process, product, 298–299
 conceptual design, 299
 final design, 304
 objectives, 307
 preliminary design and testing, 304
 problem definition, 301
 product definition, 299
 product development, 299
 product discovery, 298
 to production, 298f
 product support, 299
 project planning, 299
 sources for, 298
Design simplification, 305
DFMA (design for manufacturing and assembling), 310
Digital sensors, 284
Dimensions, definition of, 71f
Diodes, 186, 212–215
 block diagram of, 212f
 circuit symbol of, 187f
 current to flow in one direction, 184f
 depletion zone, 213
 doping of silicon crystal, 213
 forward biasing, 212f
 full-wave rectification, 212f, 214
 half-wave rectification, 212f, 214
 regions of operation, 184f
 types of, 184f
 Zener, 214–215
Direct current (DC), 337
Disciplines of engineering, 9–17
 aerospace and aeronautical engineering, 10–11
 civil engineering, 11–12
 computer engineering, 12–13
 distribution of, 9f
 electrical and electronic engineering, 13–14
 employment trend, 8t
 environmental engineering, 14
 industrial and operations engineering, 15
 mechanical engineering, 15–16
Dislocations, metals, 136
Dispersion, measures of, 263–265
Diversity/global learning, 35, 389

Index

Doping process, IC, 221
DO…UNTIL loops, 291
Drag, force, 177
Drain, FET, 188
Drilling machine, 324–328, 324f
   construction, 324–325
   operations performed on, 326–328
      boring, 326, 327f
      core drilling, 328
      counterboring, 327, 327f
      countersinking, 327, 327f
      drilling, 326, 326f
      lapping, 328
      reaming, 326, 326f
      spot-facing, 328
      tapping, 328
   types, 325–326, 325f
Driving devices, 332
Drones, 349–350, 349f
Drying processes, 142
Ductility, metal, 131
Duty cycle, 225

E

Eccentric loading, 166
Edit Toolbar, SketchUp Pro, 60, 64f
Effective interest rate, 232–238
Electrical and electronic engineering, 13–14
Electrical conductivity, 132–133
Electrical motors, 332
Electrical testing, IC, 221
Electric current, 124, 185
Electric field, 183
   lines, 184f
   vectors, 185
Electricity, 182–185
   charge distribution, 183
   charged particles, 183
   conduction and insulation, 182
   Coulomb's law, 182
   current, 185
   electric charge, 182
   electric field, 183
   potential difference, 183, 185
Electrode, 183
Electrolytic capacitor, 203
Electromagnetic induction, 208
Electron cloud, 124–125
Elements, common, 124
Email and electronic communications, 379–380
Emitter–base voltage (V_{EBO}), 216

Engineering, 25–26
   activities, 5
   disciplines, 9–17
      aerospace and aeronautical engineering, 10–11
      civil engineering, 11–12
      computer engineering, 12–13
      distribution of, 9f
      electrical and electronic engineering, 13–14
      employment trend, 8t
      environmental engineering, 14
      industrial and operations engineering, 15
      mechanical engineering, 15–16
   employment outlook, 6
      job openings in STEM clusters (2012–2022), 7t
   and engineering technology, 3–6
      engineers, 4–5
      scientists, 4
      technologists, 5–6
   high-impact educational practices, 33
   political influence, 348
   programs, 3
   skills and competencies, 6–9
   social norms
      influence, 345–346
      pursuit of social understanding, 353
      in social networking, 346–347
      and technology, 346
Engineering design process, 29
Engineering economics
   analysis, 233
   decision making, 247–249
   factors and formulas, 233t
   nominal/effective interest rate, 232–238
   ROR analysis, 247
   solving problems, 231–237
   spreadsheet functions for, 234t
   study, 231–232
   symbols/terms, 249t
   time and interest on money, 238–246
Engineering education, 26
   ethics in, 358–359
   preventing plagiarism, 359
   teaching tips, 359
Engineering/product-based data management (E/P BDM) system, 311
Engineering Professional Skills Assessment (EPSA), 396–398
Engineering stress, 131
Engineering team, 3
Engineering technicians, 4

Engineering technologists, 4–6, 4f
Engineering technology (ET), 3
Engineer(s), 4–5
  activities performed by
    aerospace/aeronautical engineers, 10
    civil engineers, 11–12
    computer hardware engineers, 12
    electrical engineers, 13
    electronic engineers, 14
    environmental engineers, 14
    industrial engineers, 15
    mechanical engineers, 15
    software developers, 12–13
  communication skills. *See* Communication skills, engineers
  description, 25
  interest to become, 26
  licensed professional, 366–367
  NSPE code of ethics for, 360–365
  responsibilities for product development, 302
  soft skills, 369–370
  structural, 12
Engine lathe and tools, 319–320, 319f
Environmental engineering, 14
Equilibrium of forces, 103–105, 104f, 109f
Errors, types of, 292
Etching process, 221
Ethics
  in engineering education, 358–359
  in engineering profession, 360–367
    NSPE code of ethics for engineers, 360–365
    principles, 366–367
    questions before employee takes action, 367
    resolution of ethical dilemma in employment, 365–366
  student motivation, 358
Euler's formula, 166
Executable, program, 275
Expert learners, 42
Extended Bernoulli equation, 175
External force, 99, 103
Extrusion molding process, 150
Extrusion process, 65, 135

F

Fabrication of silicon wafer, 220
Facebook.com, 346–347
Facebook memorialization policy, 347–348
Face-centered cubic (FCC) crystal structure, 125
  lattice, 126
  unit cell, 125–126, 125f
Facilitator teaching style, 48
Factor of safety, 160–161
Failure mode and effects analysis (FMEA), 310, 310f
Federal Aviation Agency (FAA), 349
Ferrite core inductors, 208
Fiber-based composites, 153
Field-based experiential learning, 35
Field effect transistor (FET), 184f, 188
Field equation, 175
Film capacitors, 203
Finite element analysis (FEA) method, 312–313, 312f
First law of thermodynamics, 167
First-year seminars and experiences, 34
Five-band resistor, color code of, 198
Fixed resistors, 195–196
Flat drill, 326
Flowcharts, 277–278, 277f
Flow conservation, generalized, 174f
Flow distribution, 176–177
Fluid mechanics, 170–177
  barometers, 172, 172f
  conservation of energy, 175
  conservation of mass, 174
  drag, 177
  flow distribution, 176–177
  forces on submerged plane surfaces, 173–174, 173f
  lift, 177
  properties of fluids, 170–172
Force, 99
  act on rigid bodies, 103
  components of two-/three-dimensional, 101, 102f
  equilibrium of, 103–105, 104f, 109f
  multiple
    cantilever beam with, 106f
    simply supported beam with, 107f
  rectangular components, 102
  resultant of, 103f
  on submerged plane surfaces, 173–174, 173f
Forced convection, 170
Forging process, 135
FOR loops, 290
Formal authority teaching style, 47
Formal parameter, program, 291
Forward active region, transistor, 217
Forward biasing, electrons, 213
Four-band resistor, color code of, 197–198

Index

Fourier's law of conduction, 169
Four-point bending test, 140, 140f
Free-body diagram, 104f, 105f, 107–109, 109f
Free electrons, 124
Froth flotation separation, 127
Fukushima Daiichi disaster, 350–352
    hydrogen explosion at, 352f
    lessons from, 353
    nuclear plant location, 351f
    reactor at power plant, 352f
Full-wave rectification, 212f, 214
Functions, program, 291–292
Future worth (FW), 237
    analysis, 246

G

Gantt chart, 300f
Gas metal arc welding (GMAW), 337–338
Gas shielding, 336
Gate, FET, 188
General education courses, 38–39
Geometric gradient series factors, 245–246
Geotechnical engineers, 12
GET, pseudocode keyword, 284
Google SketchUp Pro
    download screen of, 60f
    drawings in, 70–74
    graphical user interface, 63f
    installation, 59, 61f
    preparing model for LayOut, 78–84
    starting, 59–60
    template
        for drawing, 62f
        with units, 63f
    toolbar, 60–62, 64f
Gradient, cash flow, 244
Graduates of associate degree (AS) programs, 4
Grains, metal, 136
Grand challenges for twenty-first century, 7, 9
Graphs
    best fit, 414–418
    calculating and plotting value using formula, 419–420
    changing Axis Title and Axis Levels, 418
    create using Microsoft Excel, 413–414
Ground connections, 192
Ground reference, circuit, 190

H

Half-wave rectifier, 212f, 214
Hardness, materials, 136

Head stock, 320
Heat sink, fan-cooled, 138, 138f
Heat transfer, 168–170
    conduction, 169
    convection, 170
    radiation, 170
    through wall, 169f
Helical drilling, 341
Hematite ore, 127
Hexagonal close packed (HCP) crystal structures, 125
    unit cells, 126
High-density polyethylene (HDPE), 150
High-level languages, 274
    comments and documentation, 280
Hole, electron, 213
Hollow shaft with torsional load, 164, 164f
Hooke's law, 130, 159
Hoop stress, 162–163
Horizontal coherence, curriculum, 30
Horizontal milling machine, 329f
Hot forging process, 318
Hydraulic motors, 332
Hydrogen explosion at Fukushima Daiichi, 352f

I

Identifier, program, 282
IF-THEN selection statement, 288–289
Impression-die forging process, 318
Inductance, 208
Inductive statistics, 262
Inductors, 208–211
    designing RL filter, 211
    energy stored in, 204f, 210, 211t
    in parallel connection, 204f
    RL time constants, 211
    series and parallel connection, 204f, 209–210
    in series connection, 204f
    types of, 184f
Industrial and operations engineering, 15
Industrial product design, 301
Inferential statistics, 260–261
Injection molding process, 150
Input, computer programming, 284–285
Institute of Electrical and Electronics Engineers, 20
Institute of Industrial Engineers, 20
Integrated circuits (IC), 185, 189, 219–228
    555 timer, 223
    absolute maximum ratings, 223
    assembly, 221
    astable multivibrator, 225

Integrated circuits (IC) (*Continued*)
  characteristic graphs, 222–223
  datasheets, 221–222
  deposition, 220
  doping, 221
  electrical testing, 221
  etching, 221
  fabrication of silicon wafer, 220
  functional description, 222
  inverting amplifier, 227–228
  manufacturing process, 220
  masking, 220–221
  monostable (one-shot) multivibrator, 223, 225
  noninverting amplifier, 226–227
  op-amp, 225, 226*f*
  passivation, 221
  supply current, 222
  supply voltage, 222
  unity gain buffer, 226*f*, 227
  wafer production, 220
Interest factor tables, 234–237, 235*t*, 236*t*
Internal force, 99, 103
Internships, 35
Interpreter, program, 275–276
Inverting amplifier, 227–228
Ionic bonds in ceramic materials, 138–139, 141–142
Ion stuffing, 144
Iron
  BCC/FCC of, 136
  from blast furnace, 127
  core inductors, 208
  ore, 126–127
  refining, 127
Isotactic polypropylene, 151

J

Joint probability, 257–258
Junction FET (JFET), 188–189
  n-channel, 184*f*
  p-channel, 184*f*

K

Keywords, 280
Kinematic viscosity (ν), 172, 175
Knowledge-based engineering (KBE) system, 312
Knowledge, types of, 45
Kowalski, Elisia Garcia, 16–17

L

Laminar flow, 175
  maximum velocity for, 176
Languages and applications, program, 274–276
  assembly language, 275
  compiling and executing program, 275–276
  high-level languages, 274
  machine code, 275
  syntax, 274
Large commercial aircraft, 11
Laser cutting process, 338–339
Laser drilling/piercing, 341
Laser hardening process, 340
Laser welding process, 338
Lathe machine
  chuck for, 320*f*
  engine, 319*f*
  operations, 322–324
  specification, 321, 321*f*
  tools, 322, 322*f*
  working principle of, 319*f*
LayOut, SketchUp Pro, 78–84
  adding scenes, 79*f*, 80*f*
  opening model in, 79–84
  select template, 81*f*
  update orientation for view, 81*f*
L-bracket design, 88–92
  with dimensions, 88*f*, 93*f*
  tutorial, 89*f*, 90*f*, 91*f*, 92*f*
Learning communities, 34
Learning model, 52
Learning pyramid, 51–52, 52*f*
Learning styles, 46–47
Lecture model, 51–52
Licensed professional engineer, 366–367
Lift, force, 177
Light emitting diode (LED) datasheets, 200–201
Linear strain (ϵ), 159
Liquid metal, 128
  internal pool of, 134
The Literary Digest, 260
Load, circuit, 193
Logical errors, 292
Logical (mathematical) learning style, 47
Logical operators, 287–288, 287*t*
Longitudinal stress, 163
Loop structures, 290–291
  DO…UNTIL loops, 291
  FOR loops, 290
  WHILE…DO loops, 290–291
Low-density polyethylene (LDPE), 150

Index

M

Machine code, 275
Machine control unit (MCU), 332, 332f
Machines
  drilling. *See* Drilling machine
  machining problem, 334f
  milling. *See* Milling machine
  overview, 317
  process flow, 333–335, 334f
  and tools, 319–333, 322f
    carriage mechanism, 321
    engine lathe and tools, 319–320
    head stock, 320
    lathe machine specification, 321, 321f
    lathe operations, 322–324
    tail stock, 320–321
    tool post, 320, 320f
  welding and cutting, 335–341
    arc welding, 335–336
    laser in manufacturing, 338–341
    process fundamentals, 336
Magnetite ore, 127
Manual control unit, 332
Manual welding, 337
Martensite, 136
Masking process, 220–221
Mass density (ρ), fluid, 171
Material(s)
  properties of engineering, 158t
  specific roughness of, 176t
Materials science, 123
Maximum normal stress theory, 160–162
Mean, 262, 264f
Mechanical engineering, 15–16
  program
    curriculum mapping for, 31t
    educational objectives, 28
    plan of study for, 39t
Mechanical forming processes, 135
Median, 263, 264f
Memos, 377–379
Metacognition, students, 42–44
  to become expert learner, 43f
  knowledge, 45
  learning strategies/study skills, 45–46
  learning styles, 46–47
  motivation. *See* Motivation in learning
  time management, 48–50
Metal cutting/forming processes, 317–318
Metallic atoms, 124, 124f
Metallic elements, 123–124
Metal oxide semiconductor FET (MOSFET), 189

Metals, 123–138
  conductors, 182
  liquid, 128
  obtaining, 126–128
  process affecting properties of, 135–137
    annealing, 136
    precipitation hardening, 137
    quenching, 136–137
  processing of, 133–135
  properties of, 128–133
  resistivity, 133
  rolling, 135
  in solid piece of, 124
  stress–strain curve for, 131f
  technologies made possible by, 137–138
Microsoft Excel, creating charts and graphs using, 413–414
Microsoft Word, creating equations in, 409–412
Military aircraft, 11
Milling machine, 328–333
  construction of, 328
  horizontal, 329f
  milling cutters, 328–329
  operations performed on, 330–333, 330f
    CNC machine, 330–331
    codes, 333
    driving devices, 332
    elements of NC machine, 331
    machine tool, 332
    manual control unit, 332
    MCU, 332, 332f
    NC programs, 333
    software, 331
  vertical, 329f
MIME (material increase manufacturing) techniques, 313
Minimum attractive rate of return (MARR), 246
Mnemonics, 275
Mode, 263, 264f
Modulus of elasticity, steel, 159
Modulus of rupture (MOR), 140–141
Moment, force, 106–107
Moment of inertia, 114–118
  of common geometries, 115t–116t
  of T-shaped section, 117f
Monostable (one-shot) multivibrator, 223, 225
Motivation in learning, 43–44
  self-assessment, 44
  self-efficacy, 43–44
  self-reflection, 44
Multilayer ceramic capacitors (MLCC), 203
Multipoint cutting tool, 318
Murray, Neil G., Jr., 18–19

N

Nanometer, 370
Nanoscale, 370
Nanotechnology, 370–372
    consumer products manufactured using, 371f
    emotions, 372
    IBM demonstration in 1989, 371f
National Academy of Engineers (NAE), 7
National Society of Professional Engineers (NSPE), 20–21, 360
    code of ethics for engineers, 360–365
        fundamental canons, 360
        professional obligations, 362–365
        rules of practice, 360–362
        sustainable development, 364
    resources available through, 367
n-channel JFET, 184f, 188
n-channel MOSFET, 187f
Negatively charged particle, 182–183
Negative polarity, 191
Neutral axis, 165
Neutral, object, 182
Newton's law of convection, 170
Next generation communication device, 25
Nominal interest rate, 232–238
Noncrystalline ceramic materials, 139
Noninverting amplifier, 226–227
Nonverbal communication, 375
Normal curve, 266
Normal frequency distribution, 266
Normal stress, 158
Nottingham, Marsha, 21–22
Novice learners, 42
n–p–n transistor, 186, 215, 216f, 219
n-type semiconductor, 184f, 188
Numerical controlled (NC) machines, 317
    elements of, 331
    layout of, 331f
    programs, 333
    software of, 331
numToTest number, 285

O

Ohm's law, 196–197
Open circuit, 193
Open communication, 388
Operands, 285
Operating regions, transistor, 217
Operational amplifiers (op-amp), 225
    in noninverting configuration, 226f
Operator, 285
Oral communication skills, 372–377
    concluding remarks on, 377
    knowing audience, 372–373
    knowing content, 373
    nonverbal messages, 375–376
    presentation software, 374–375
    presenting with groups, 376–377
    rehearsing presentations, 376
Order of execution, operators, 286, 286t
Orthogonal views, 74
    creation, 77f, 78f, 82f, 83f, 84f, 85f, 86f, 87f
    dimensioning, 84–87
    drawing with dimensions and, 87f
    projections, 75f
Orthographic drawings in CAD, 74–78
Orthographic projection, 74, 76f
Output, computer programming, 284–285
Over-the-wall approach, 306

P

Parallel circuits, 195
Parison, 150
Particulate composites, 153
Parting tool, lathe, 322
Part program, 331
Part tutorial, 71–74, 72f, 73f, 74f
Passivation process, 221
Passive components, circuit, 185
p-channel JFET, 184f, 188–189
p-channel MOSFET, 187f
Percussion drilling, 341
Performance indicators, students, 31, 32t
Photolithography, 220
Physical (kinesthetic) learning style, 47
Pig iron, 127
Plastic materials, 145
p–n junctions, 186
p–n–p transistor, 186–187, 215, 216f, 219
Poisson's ratio (v), 159
Polycarbonate polymer, 151, 152f
Polyethylene polymer
    crystallizing, 146f
    ethylene gas for, 145f
Polymers, 124, 145–152
    amorphous, 146, 146f
    chain-type, 145
    obtaining, 148–149
    overview, 145–147
    process affecting properties of, 150–151
    processing into usable forms, 149–150
    properties of, 147–148

semicrystalline, 148
   technologies made possible by, 151–152
Polypropylene, 149–151
Population mean, 262
Positively charged, electricity, 182
Posttraumatic stress disorder (PTSD), 349, 351
Potassium ions, bulky, 144
Potentiometer, 186
Power
   connections, 192–193
   dissipation ratings, 199
   supply, 193
Power dissipation (P_D), 217
Precipitation hardening process, 137
Prepolymer, 150
Present worth (PW), 239
   analysis, 246
Pressure vessel, 162–164
   axial stress, 163–164
   hoop stress, 162–163
   stresses in, 163f
Principal Toolbar, SketchUp Pro, 60, 64f
PRINT, reserved word, 284
Probability, 252–259
   permutations and combinations, 258–259
   theorems, 252–258
Procedural knowledge, 45
Procedures, program, 291–292
Product design process, 303–307
   factors to consider in, 305–307
      CE, 306–307
      DFM, 305
      product life cycle, 305–306
      remanufacturing, 307
   goals of, 307
   steps in, 303–304
Product development process, 301–302
   computational techniques and tools, 311–313
      3D printing tool, 313
      CAx systems, 311
      E/P BDM system, 311
      FEA, 312–313, 312f
      KBE, 312
      RP, 313
   flowchart, 302f
   objectives, 307
   responsibilities for engineer, 302
   techniques and tools for design improvement, 308–310
      CE, 308
      DfX, 309–310
      FMEA, 310, 310f
      QFD, 308–309, 309f

Product life cycle (PLC), 303
   diagram, 305f
   stages of, 305–306
Professional engineer (PE), 17–18
Professional engineering
   license, 17–19
   organizations, 19–22
Program educational objectives, mechanical engineering, 28
Prototype delivery drone, 349f
Pseudocode, 278–280
   examples using comments in, 281t
   reserved pseudocode words, 281t
p-type semiconductor, 184f, 188, 213
Pulse width, 223

Q

Quality function deployment (QFD), 308–309, 309f
Quenching process, 136–137

R

Radiation, heat transfer, 170
Radius tool, lathe, 322
Rapid prototyping (RP), 313
Rate of return (ROR) analysis, 247
RC time constant, 205–206
Reactance, inductor, 211
Reactions exerted on rigid body, 108t
READ, pseudocode keyword, 284
Recrystallization, 136
Rectifying diodes, 214
Regenerative systems, CAPP, 311
Relational operators, 286, 286t
Remanufacturing, products, 307
Remelting process, 340
Renewable energy systems, 25
Reserved words, 280
   reserved pseudocode words, 281t
Resistivity of metals, 133
Resistors, 185–186, 195–201
   color codes, 194f, 197
   current limiting with, 194f, 200–201
   fixed, 195–196
   Ohm's law, 196–197
   in parallel, 194f, 200
   power dissipation ratings, 199
   in series, 194f, 199
   series and parallel combination, 194f, 200
   tolerance, 198–199
   types of, 184f
   voltage divider circuit with, 194f, 201

Reverse active region, transistor, 217
Revision block, drawing, 76
Reynolds number (Re), 175–176
Right-hand/left-hand turning tool, lathe, 322
Risers, casting, 134–135
RL time constant, 211
Robotic-arm surgery, 181
Rockwell testers, 136
Rolling, metals, 135
Rolling process, 318
Rubrics, 31
   and key performance indicators, 393–398
   for student learning outcome, 32*t*

S

Saturation region, transistor, 217
Scalars, 99–103
Science, Technology, Engineering, and Mathematics (STEM) cluster, 6, 7*t*
Scientists (engineers), 4
Scope of Work (SOW), 16
Second law of thermodynamics, 168
SELECT…CASE selection statement, 289–290
Selection statements, 288–290
Self-assessment methods, 44
Self-efficacy, 43–44
Self-inductance, 208–209
Semiautomatic welding, 337
Semicrystalline polymer, 148
Senior capstones, 35
Series circuits, 193, 195
Service learning, 35
Shear strain (γ), 159
Shear stress, 158, 164–165
Shielded metal arc welding, 337
Short circuit, 193
Silicon crystal
   doping of, 213
   electrons, 213
   pure, 212*f*
Silicon wafer, 220
Single amount factors, 239–240
Single point cutting tool, 318, 318*f*
Single-shot drilling, 341
Sinking fund factor, 242–244
Sintering process, ceramic, 142–143
SketchUp Pro. *See* Google SketchUp Pro
Skill-based learning, 53
Slag, 127, 336
Sliding-plate viscometer, 171*f*
Small commercial aircraft, 11

Social (interpersonal) learning style, 47
Social media, 347
Social norms, engineering
   influence, 345–346
   pursuit of social understanding, 353
   in social networking, 346–347
   and technology, 346
Society of Automotive Engineers (SAE), 21
Society of Women Engineers, 21
Soft skills, engineers, 369–370
Software
   CAD, 59–70
      abbreviations, 62–64
      adding dimensions to model, 68–70
      dynamic viewing function, 62
      installing Google SketchUp Pro, 59
      modeling 3D object, 65–68
      sketch and extrusion, 65
      starting SketchUp Pro, 59–60
      toolbar, 60–62
   of NC system, 331
   presentation, 374–375
Software engineers/software developers, 12
Solid model, 70
Solitary (intrapersonal) learning style, 47
Source, FET, 188
Space shuttle, 11
Specific gravity, 171
Spray drying process, 142
Stable state, IC, 223
Standard deviation, 264
Standard normal probability distribution, 266–267, 267*f*
State licensing boards, 17
Statements, program, 274
Statistics, 251, 259–260
   data, 262–267
      central limit theorem, 265–266
      measures of central tendency, 262–263
      measures of dispersion, 263–265
      normal frequency distribution, 266
      standard normal probability distribution, 266–267, 267*f*
   descriptive and inferential, 260–261
   in organizations, 260–262
   sample *vs.* population, 261, 262*t*
   statistical difficulties, 260
Steel, 137
   external tension on, 141
   modulus of elasticity, 159
Stick welding, 337
Strain, 129, 159

Index

Stress, 129
    in beams, 165–166, 165f
    material, 158
Stress–strain curve for metals, 131f
Structural engineers, 12
Student learning outcomes
    and performance indicators, 32t
    program, 28
    sample rubrics for, 32t
Students' perception and instructional approaches, 42f
Study strategy, 43
Subject-area coherence, curriculum, 30
Subtraction of vector, 100, 101f
Surface learners, 42
Switch, transistors as, 218
Syndiotactic polypropylene, 151
Syntax errors, 292
Syntax, program, 274
Systems software developers, 13

T

Tail stock, 320–321
Tantalum capacitor, 203
Teaching styles, engineering, 47–48
Team, 387
Teamwork essentials, 387–389
    clear purpose, 388
    embracing diversity, 389
    leadership in teams, 388
    open communication, 388
    teams and team members, 387–388
Technical reports, 380–381
Technology-ready techniques, 308
Tensile tester, 129, 130f
TEPCO, 351
Testing, program, 292
Thermal dissipation, 189
Thermal radiation, 170
Thermal resistance ($R_{\theta JA}$), 217
Thermodynamics, 167–168
    first law, 167
    second law, 168
    third law, 168
Thermoplastic polymer, 145–146
    crystalline, 149, 151
    molding techniques with, 150
Thermosetting polymer, 145, 150
Third law of thermodynamics, 168
Threading tool, lathe, 322
Time and interest on money, 238–246

Time constant (τ), circuit, 205
Time management for students, 48–50
Title block, drawing, 76, 86
Tool post, 320, 320f
Torque, force, 106–107
Transistors, 186, 215–219
    avalanche breakdown, 218
    cutoff, 218
    datasheet parameters, 215–217
    forward active, 217
    inverter circuit with MOSFET, 219
    n–p–n transistor as NOT gate, 218
    operating regions, 217
    p–n–p transistor as NOT gate, 218–219
    reverse active, 217
    saturation, 217
    as switch, 218
    types of, 184f
Transportation engineers, 12
Trepanning drilling, 341
Truth tables, logical operators, 287
Tsunami, Japan, 350
Turbulent flow, 175–176
Turning tool, lathe, 322

U

Undergraduate research, 34–35
Uniaxial loading and deformation, 160, 160f
Uniform series compound amount factor, 242–244
Uniform series PW factor, 241–242
Unimodal, 263
Unit cell structures, ceramic, 138
Unit conversion, 391–392
Unity gain buffer, 226f, 227
Unity gain inverter circuit, 228
University of Michigan–Flint, 36

V

Valence electrons, 124
van der Waals forces, 145, 148
Variables, program, 282
    initialization, 282–283
    naming conventions, 283–284
Variance, 265
Variant systems, CAPP, 311
Vectors, 99–103
    addition of
        using parallelogram law, 100f
        using polygon rule, 101f
        using triangles, 101f

Vectors (*Continued*)
  with magnitude and direction, 100*f*
  subtraction of, 101*f*
Verbal (linguistic) learning style, 47
Vertical coherence, curriculum, 30
Vertical milling machine, 329*f*
Viscosity, fluid, 171
Visual (spatial) learning style, 47
Voltage, 208
Voltage-controlled FET, 189
Voltage divider circuit with resistors, 194*f*, 201
Voltage follower, 227
Voltage-sensing feeder, 337–338

W

Wafer, silicon, 220
Welding and cutting, machines, 335–341
  arc welding, 335–336
  laser in manufacturing, 338–341
    coating, 340
    cutting, 338–339
    drilling/piercing, 341
    hardening, 340
    remelting, 340
    surface treatment/modification, 339–340
    welding, 338
  process fundamentals, 336
    GMAW, 337–338
    shielded metal arc welding, 337
    shielding and fluxing, 336

WHILE...DO loops, 290–291
White, Bruce R., 10
Wire drawing process, 135
Word processing tools, 383, 384*f*
Workplace diversity, 389
WRITE, reserved word, 284
Writing-intensive courses, 34
Written communication skills, 377–386
  citations and bibliographies, 381–382
  email and electronic communications, 379–380
  figures, images, and tables, 386
  letters and memos, 377–379
  reports, 380–381
  source reliability, 382–386
    headings and subheadings, 386
    optional and special report, 384–385

Y

Yield point, 130
Young's modulus, 130

Z

Zener diodes, 214–215
Z-tables
  negative values, 421–422
  positive values, 422–423